Anaerobic treatment of mine wastewater for the removal of selenate and its co-contaminants

Lea Chua Tan

Joint PhD degree in Environmental Technology

Docteur de l'Université Paris-Est
Spécialité : Science et Technique de l'Environnement

Dottore di Ricerca in Tecnologie Ambientali

Degree of Doctor in Environmental Technology

Tesi di Dottorato - Thèse - PhD thesis

Lea Chua TAN

Anaerobic Treatment of Mine Wastewater for the Removal of Selenate and its Co-contaminants

Defended on December 18th, 2017

In front of the PhD committee

Prof. Dr. Jonathan Lloyd	Reviewer
Prof. Dr. Artin Hatzikioseyian	Reviewer
Prof. Dr. Ir. Piet N.L. Lens	Promotor
Dr. Hab. Eric van Hullebusch	Co-Promotor
Dr. Hab. Giovani Esposito	Co-Promotor
Prof. Eddy Moors	Examiner

European Commission
ERASMUS
MUNDUS

Erasmus Joint doctorate programme in Environmental Technology for Contaminated Solids, Soils and Sediments (ETeCoS³)

Thesis Committee:

Thesis Promotor

Prof. Dr. Ir. Piet N.L. Lens,
Professor of Environmental Biotechnology
UNESCO-IHE, Institute of Water Education
Delft, The Netherlands

Thesis Co-Promotors

Dr. Yarlagadda V. Nancharaiah,
Scientific Officer G
Bhabha Atomic Research Centre, Department of Atomic Energy
Mumbai, India

Dr. Hab. Eric D. van Hullebusch,
Hab. Associate Professor in Biogeochemistry
Université Paris-Est
Marne-la-Vallée, France

Dr. Hab. Giovanni Esposito,
Associate Professor of Sanitary and Environmental Engineering
University of Cassino and Southern Lazio
Cassino, Italy

Other Members

Prof. Dr. Jonathan Lloyd,
Professor of Geomicrobiology
University of Manchester
Manchester, United Kingdom

Prof. Dr. Artin Hatzikioseyian,
National Technical University of Athens
School of Mining and Metallurgical Engineering
Zografou, Greece

This research was conducted under the auspices of the Erasmus Mundus Joint Doctorate Environmental Technologies for Contaminated Solids, Soils, and Sediments (ETeCoS^3) and the Graduate School for Socio-Economic and Natural Sciences of the Environment (SENSE).

Published by:
CRC Press/Balkema
Schipholweg 107C, 2316 XC, Leiden, the Netherlands
Pub.NL@taylorandfrancis.com
www.crcpress.com – www.taylorandfrancis.com
ISBN 978-1-138-32841-9

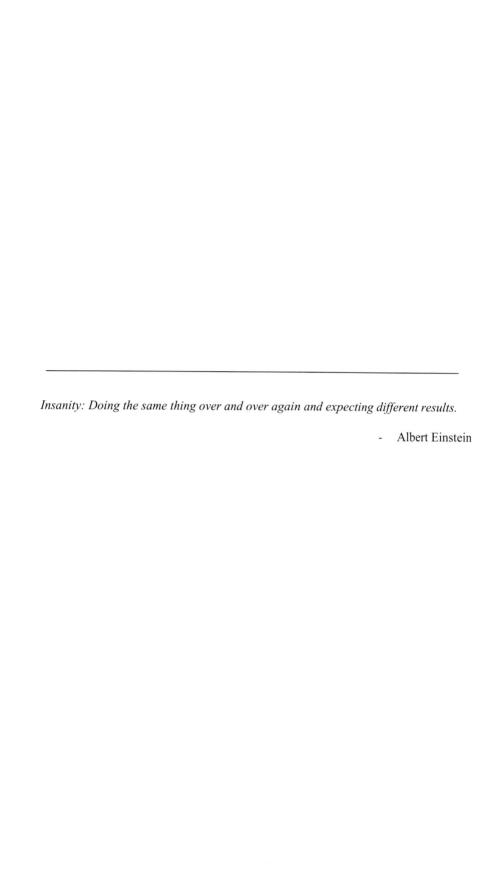

Insanity: Doing the same thing over and over again and expecting different results.

- Albert Einstein

Table of Contents   **Page No.**

Author information

Acknowledgments

Pursing a graduate study in itself is a challenge and though I have worked hard to accomplished this dissertation, I would not have completed my PhD research without all the guidance, help and support I have received from many people within the course of my three years' journey.

Firstly, I would like to thank Erasmus Mundus ETeCoS[3] program (FPA no. 2010-0009) for providing the financial support to carry out this dissertation. I would also like to thank UNESCO - IHE, Center for Biofilm Engineering - Montana State University, University of Naples Federico II, University of Paris-Est and University of Cassino for hosting me.

I would like to express my gratitude to my main promotor and supervisor Prof. Piet N.L. Lens (UNESCO-IHE, The Netherlands) for his constant supervision, encouragement and constructive comments that allowed me to grow as a researcher. You were a critical factor that helped me in completing my PhD. A special thanks to Prof. Eric van Hullebusch (University of Paris-Est, France), co-promoter of the thesis, for his support and encouragement. I would also like to thank Prof. Giovanni Esposito (University of Cassino and the Southern Lazio, Italy) for his constant support with the administrative affairs related to the PhD program as well as for kindly providing me with all the items I required in the laboratory during my stay in Naples. And many thanks to Dr. Eldon for always being there as well as his constant support.

I would like to give my whole heartedly thanks to Dr. Yarlagadda Venkata Nancharaiah (Bhabha Atomic Research Centre, Kalpakkam, India) for being a great mentor and support in this journey. You were the first person that I go to when I have research questions and you provided tremendous help in shaping not only my entire PhD but as well as improving my manuscripts. Both your scientific and moral support has been a great influence in improving my views on research as well as the quality of this dissertation. I look up to you and it was a great pleasure to have worked with you.

I am also extremely grateful to my mobility supervisors Prof. Robin Gerlach (Center for Biofilm Engineering, Montana State University) and Dr. Stefano Papirio (University of Naples Federico II). You both have made me feel very welcomed despite the short time I have spent in Montana and Italy. I cannot express with words how thankful I am to the both of you for diligently explaining and correcting any mistakes I made both in the laboratory work as well as in my writing. You both have lend ears to my rumbling be it work related or outside and have contributed much to support and uplift my spirits and confidence in my work and myself. It was definitely a pleasure getting to know you both and I have come to think that we have formed a friendship beyond a mentor-mentee relationship.

I am grateful to Dr. Erika Esposito-Ortiz for being my mentor in work and friend during my stay in US. Thanks to you, I was able to work well and meet great people there. Thank you to Dr. Shipeng Lu, Dr. Paola Cennamo, Dr. Chiachi Hwang and Neerja Zambare for their large contribution and help in conducting any microbial and imaging related analysis I needed for my work. Thank you to Vincenzo Luongo and Dr. Ludovico Pontoni for their laboratory assistance in Naples. My gratitude also goes to UNESCO-IHE lab staff: Fred, Frank, Berend, Ferdi, Peter and Lyzette. Thank you guys for all your help in the lab. It was a rough start but I have come to appreciate all that you guys do for the lab and considered you all as friends.

A very special thanks to Prof. Jonathan Lloyd (University of Manchester) and Prof. Artin Hatzikioseyian (National Technical University of Athens), jury members, for their critical comments and suggestions regarding my work in the dissertation which has given me thoughts on how to improve this work further.

This section would be incomplete without acknowledging my colleagues and friends: Gabrielle, Manivannan, Mink, Mohanned, Motassem, Mulele, Neerajan, Nirakar, Ramita, Rohan, Paolo, Saltana and Viviana for giving spiritual and intellectual support. Thank you to Angelica, Angelo, Bikash, Chris, Eugenia, Feishu, Jairo, Joy, Mohan, Neeraj, Ouafa, Samayita and Tejaswini, for always being open with me and engaging my silliness; I appreciate your friendship. I would like to give a special and warmhearted thanks to Iosef and Leon who have lend both an ear and hand to me. I can always count on you guys for everything and I am honored to call you my friend. Most importantly, I would like to give a tremendous appreciation to Shrutika who has become my partner in crime as well as my buddy for the entire PhD. The both of us have been together from the start even before we officially met until our defense. We have become so co-dependent with each other for everything from writing emails to working in the lab that I find it weird without you. I hope we can work together again in the future.

I would like to express my deepest gratitude to my family, especially my mom and sister, my sister from another mother Hazel and very close friends (Fatsy, Lina, Ito, and childhood friends), that despite the time and distance, they have given me their support and listen to my rants. I would have gone insane without them. And finally, thank you to Lord God almighty for making this entire journey possible.

Summary

This research aimed at addressing the effect of wastewater characteristics (i.e. co-contaminants, heavy metals and pH) on the biological reduction of selenate (SeO_4^{2-}) and evaluating process integration and configurations for selenium-laden wastewater treatment with co-contaminants.

The first part of the study focused on the effect of co-electron acceptors and low pH on the bioremediation of SeO_4^{2-}. Results from the experiments showed that the molar ratio of NO_3^- and SO_4^{2-} to SeO_4^{2-} has a controlling factor in either increasing or decreasing the selenium (Se) removal efficiency. Additionally, a study on biofilm-Se interactions revealed the presence of either NO_3^- or SO_4^{2-} influences the Se speciation, biogenic Se (Se^0) levels and biomass activity. Upflow anaerobic sludge blanket (UASB) reactor operation with a gradual decrease in the influent pH from 7.0 to 5.5 showed a stable removal performance of NO_3^-, SO_4^{2-} and SeO_4^{2-}, before a 20% decrease in removal of all these components was observed at pH 5.0. Furthermore, long-term operation of the UASB reactor at pH 5.0 showed the enrichment of *Geobacteraceae* and *Spirochaetaceae* families, which were not detected at pH > 5.0.

The second part of the study explored the effectiveness of different removal techniques for the treatment of SeO_4^{2-} with co-contaminants. Comparing the SeO_4^{2-} removal performance in the presence of SO_4^{2-} in a biotrickling filter (BTF) and UASB reactor revealed that SO_4^{2-} largely influenced the attached biofilm growth and increased SeO_4^{2-} removal by > 200%. On the other hand, SeO_4^{2-} removal was similar in the UASB reactor irrespective of the presence or absence of SO_4^{2-}. Biomass characterization revealed the formation of spherical Se^0 and poly-selenium sulfide in the biomass of both bioreactors. Addition of Ni in both bioreactors led to a decrease in SO_4^{2-} and SeO_4^{2-} removal performance by ~20-30%. Ni removal was > 80% in both bioreactors. Ni was removed via nickel sulfide precipitation. Evaluation of integrated process system for SeO_4^{2-} and SO_4^{2-} removal was conducted by coupling an ion exchange column (IX) and UASB bioreactor, using IX as either a pre-treatment (IX $\rightarrow$ UASB) or post-treatment (UASB $\rightarrow$ IX) unit for the bioreactor. The pre-treatment process scheme showed a better overall removal efficiency of 99% SO_4^{2-} and 94% total Se (Se_{tot}) reaching < 100 mg/L SO_4^{2-}, < 0.3 mg Se/L Se_{tot} and < 0.02 mg Se/L dissolved Se in the effluent over 42 days of continuous operation.

Keywords: Bioreactors; selenate; nitrate, sulfate; nickel; ion exchange; remediation

Sommaria

Questa ricerca è finalizzata allo studio dell'influenza di alcune caratteristiche delle acque reflue (come la presenza di co-contaminanti e di metalli pesanti ed il pH) sulla rimozione biologica dei selenati (SeO_4^{2-}) e all'integrazione della stessa con altri processi per il trattamento di acque contenenti selenio.

In una prima parte dello studio, l'attenzione è stata rivolta allo studio dell'effetto della presenza di altri accettori di elettroni e di un pH acido sulla rimozione biologica dei selenati. I risultati degli esperimenti hanno dimostrato che il rapporto molare tra NO_3^-, SO_4^{2-} e SeO_4^{2-} è un parametro di fondamentale importanza, per determinare sia l'efficienza dell'abbattimento dei selenati che la speciazione del selenio e l'attività della biomassa, anche in applicazioni con biofilm. La graduale diminuzione del pH da 7,0 a 5,5 in un reattore UASB non ha avuto alcuna influenza sulle rese della rimozione dei selenati, mentre un pH pari a 5,0 ha comportato una diminuzione dell'efficienza del processo biologico di circa pari il 20%. Inoltre, l'esercizio prolungato del reattore a pH 5,0 ha provocato l'arricchimento delle famiglie batteriche *Geobacteriaceae* e *Spirochataceae*, a differenza di quanto avvenuto a pH superiori a 5,5.

Nella seconda parte dello studio, si è esplorato la possibilità di utilizzare differenti configurazioni impiantistiche per la rimozione dei selenati in presenza di altri contaminanti. È stata confrontata l'efficienza ottenuta in un letto percolatore con quella raggiunta in un reattore UASB. Nel caso del letto percolatore, la presenza di solfati, insieme ai selenati, nella soluzione influente ha comportato un attecchimento e una crescita del biofilm nettamente più veloci e un miglioramento delle rese di abbattimento dei selenati superiore al 200%. D'altro canto, nel reattore UASB la rimozione dei SeO_4^{2-} è risultata la medesima sia in presenza che in assenza dei SO_4^{2-}. Un'analisi sulla biomassa formatasi nei reattori ha permesso l'individuazione di micro-particelle sferiche di selenio elementare (S^0) e di polisolfuro di selenio nel biofilm e nei granuli. L'aggiunta di nickel (Ni), infine, in entrambi i bioreattori ha comportato una riduzione della rimozione dei selenati e dei solfati di circa il 20-30%. Il nickel introdotto è stato abbattuto di circa l'80% soprattutto a causa della sua precipitazione come solfuro metallico.

In ultima istanza, la valutazione di un processo integrato per la rimozione simultanea dei selenati e dei solfati è stata condotta accoppiando una colonna di scambio ionico (IX) con un reattore UASB, utilizzando lo scambio ionico sia come pre-trattamento (IX → UASB) che come post-trattamento (UASB → IX). La prima configurazione ha comportato un'efficienza migliore e pari a circa il 99% di abbattimento dei SO_4^{2-} e il 94% di abbattimento del Se totale. Le concentrazioni finali sono risultate al di sotto di 100 mg/L per i SO_4^{2-}, < 0,3 mg Se/L per il selenio totale e < 0,02 mg Se/L di Se disciolto nell'effluente dopo 42 giorni di esercizio in continuo.

Parole chiave: Bioreattori; selenato; nitrato, solfato; nichel; scambio ionico; bonifica

Samenvatting

Dit onderzoek richtte zich op de eigenschappen van afvalwater (e.g. co-electron acceptoren, zware metalen, pH) en het effect hiervan op de biologische afname van selenaat (SeO_4^{2-}) en het evalueren van procesintegratie en configuraties voor de verwijdering behandeling van afvalwater vervuild met seleniumen co-electron aceeptoren.

Het eerste deel van het onderzoek richtte zich op het effect van co-elektron acceptoren en lage pH op het herstel van de SeO_4^{2-} verwijdering. Experimenten toonden aan dat de molaire ratio van NO_3^- en SO_4^{2-} met SeO_4^{2-} een controlerende factor in voor in ofwel de toename of afname van de verwijderingsefficiëntie van selenium (Se). Daarnaast toonde onderzoek naar biofilm-Se interacties aan, dat de aanwezigheid van ofwel NO_3^- of SO_4^{2-} de speciatie van Se beïnvloedt, evenals de concentratie van biogenische Se (Se^0) en biomassa activiteit. De werking van de opstroom anaërobe slibbed (UASB) reactor, met een geleidelijke afname van de pH instroom van 7.0 naar 5.5, toonde een stabiele afname van NO_3^-, SO_4^{2-} en SeO_4^{2-} aan, nog voordat er een afname van 20% werd waargenomen in de verwijdering van alle aanwezige componenten bij een pH van 5.0. Bovendien toonde de werking van de UASB reactor op lange termijn, met een pH van 5.0, een verrijking van *Geobacteraceae* en *Spirochaetaceae* families, die nog niet gedetecteerd zij bij pH > 5.0.

Het tweede deel van het onderzoek richtte zich op de efficiëntie van verschillende verwijderingstechnieken voor de behandeling van SeO_4^{2-} met co-infectiestoffen. Wanneer de verwijdering van SeO_4^{2-} vergeleken werd bij een aanwezigheid van SO_4^{2-} in een biotrickling filter (BTF) en UASB reactor, werd het duidelijk dat SO_4^{2-} voornamelijk beïnvloed wordt door de biofilm groei en de toegenomen SeO_4^{2-} verwijdering bij (> 200%). Anderzijds was de SeO_4^{2-} verwijdering gelijk aan die van de UASB reactor, ongeacht de aanwezigheid van SO_4^{2-}. De classificatie van biomass toonde de formatie van sferisch Se^0 en poly-selenium sulfide aan in de biomassa van beide bioreactoren. De toevoeging van Ni in beide bioreactoren zorgde voor een afname in de SO_4^{2-} en SeO_4^{2-} verwijdering bij ~20-30%. De verwijdering van Ni was > 80% in beide bioreactoren. Ni werd verwijderd via nikkel sulfide precipitatie. Evaluatie van het geïntegreerde processysteem voor SeO_4^{2-} en SO_4^{2-} verwijdering werd uitgevoerd door de koppeling van een ion wisseling (IX) met een UASB bioreactor, door IX als voor- (IX → UASB) of als nabehandelingsunit (UASB → IX) voor de bioreactor te scjakelen. Het processchema van de voor-behandeling toonde over het algemeen een betere verwijderingsefficiëntie aan van 99% SO_4^{2-} en 94% van het totale Se bereik < 100 mg/L SO_4^{2-}, < 0.3 mg Se/L totale Se en < 0.02 mg Se/L opgelost Se in de afvoer gedurende 42 dagen van doorlopend gebruik.

Trefwoorden: Bioreactoren; seleniet; nitraat, sulfaat; nikkel; ion exchange; herstel

Résumé

Cette recherche visait à aborder l'effet des caractéristiques des eaux usées (c'est-à-dire les Co-contaminants, métaux lourds et pH) sur la réduction biologique du séléniate (SeO_4^{2-}), et évaluer l'intégration des processus et des configurations pour le traitement des eaux usées chargé en sélénium et autres co-contaminants.

La première partie de l'étude portait sur l'effet des accepteurs de Co-électrons et le faible pH sur la bioremédiation du SeO_4^{2-}, les études expérimentales a montré que le rapport molaire NO_3^- et SO_4^{2-} sur SeO_4^{2-} est un facteur de contrôle qui augmente ou diminue l'efficacité de l'élimination du sélénium (Se). De plus, l'étude sur les interactions biofilm-Se a révélé que la présence de NO_3^- ou de SO_4^{2-} influence la spéciation Se, les niveaux de Se élémentaire d'origine biologique (Se^0) et l'activité de la biomasse. Le fonctionnement du réacteur UASB (lit de boues expansées anaérobie) avec une diminution progressive du pH influent de 7,0 à 5,5 a montré une performance d'élimination stable de NO_3^-, SO_4^{2-} et SeO_4^{2-}, avant une diminution de 20% de l'élimination de tous ces ions à pH 5,0. En plus, le fonctionnement à long terme du réacteur UASB à pH 5.0 montre un enrichissement des familles microbienne Geobacteraceae et Spirochaetaceae, qui n'ont pas été détectés à pH > 5,0.

La deuxième partie de l'étude a exploré l'efficacité de différentes techniques d'élimination pour le traitement de SeO_4^{2-} avec des Co-contaminants. La comparaison des performances d'élimination de SeO_4^{2-} en présence de SO_4^{2-} dans un filtre biologique (BTF) et un réacteur UASB a révélé que SO_4^{2-} a largement influencé la croissance du biofilm attaché et l'élimination de SeO_4^{2-} a augmentée de > 200%. D'autre part, l'élimination de SeO_4^{2-} était similaire dans le réacteur UASB indépendamment de la présence ou de l'absence de SO_4^{2-}. La caractérisation de la biomasse a révélé la formation de Se^0 sphérique et de sulfure de poly-sélénium dans la biomasse des deux bioréacteurs. L'addition de Ni dans les deux bioréacteurs a entraîné une diminution des performances de réduction de SO_4^{2-} et SeO_4^{2-} de ~ 20-30%. L'élimination du Ni était > 80% dans les deux bioréacteurs. Ni a été éliminé par précipitation sous la forme de sulfure de nickel. L'évaluation du flux de processus pour l'élimination de SeO_4^{2-} et SO_4^{2-} a été effectuée en couplant la colonne d'échange d'ions (IX) et le bioréacteur UASB en utilisant soit un prétraitement (IX → UASB), soit un post-traitement (UASB → IX) pour le bioréacteur. Le schéma du processus de prétraitement a montré une meilleure efficacité d'élimination globale de 99% de SO_4^{2-} et 94% de S totale atteignant < 100 mg/L de SO_4^{2-}, < 0,3 mg/L de Se total et < 0,02 mg/L de Se dissous dans l'effluent pendant 42 jours de fonctionnement continu.

Mots-clés: bioréacteurs; séléniate; nitrate; sulfate; nickel; échange d'ion; remédiation

CHAPTER 1

General Introduction

1.1 Background

Selenium (Se) is an essential trace element for all living cells. It is required for synthesizing as many as 25 selenoproteins with selenocysteine (21st amino acid) at the active site. Se plays a vital role in many metabolic functions in the human body such as immune response and reproduction and prevents cell damage through its antioxidant activity (Rayman 2012). Despite its biological importance, Se has been identified as a toxic element depending on the dose and speciation. Selenium has been referred to as a double-edged sword for its essentiality and toxicity with only a small difference in concentration (Brozmanová et al. 2010).

Sources of Se contamination in the environment can come from both anthropogenic activities, e.g. mining or agriculture, and natural geochemical cycles, e.g. weathering of rocks (Rosen and Liu 2009). Approximately 37-41% of Se emission to the environment can be attributed to anthropogenic activities (Wen and Carignan 2007). In the past, Se pollution has been largely neglected due to higher profiled and concentrated pollutants such as heavy metals and pesticides. In recent years, the advances in water quality and pollution monitoring have shown that Se is an important environmental concern (Lemly 2004).

Se contamination of water environments is largely both an industrial and environmental issue. It is mainly associated with mining of metals (e.g. copper), minerals (e.g. sulfides, phosphate), strategic materials (e.g. uranium) and fossil fuel (e.g. coal) (Anton and van Erveer 1997). However, compared to the other contaminants and heavy metals in the mining wastes, Se exists at low concentrations ranging from a few µg/L to a few mg/L (Mehdi et al. 2013). Despite the low concentration of Se, the release of wastewater with such low levels of Se without proper treatment can severely impact ecosystem well-being.

Se has the ability to bioaccumulate quickly in the soil and water environment and subsequently, can enter the food chain via plants and aquatic animals. Animals are most susceptible to Se poisoning with about 2-15 µg Se/g body weight potentially causing negative effects to the livestock and aquatic life (Lemly 2004). There are well-documented cases of Se pollution linked to reproductive failures, teratogenic effects, elimination of fishes and monetary loss to fishermen dependent on lakes. Due to bioaccumulation, higher concentrations of Se are found in the organisms than in the Se-contaminated surrounding water. To avoid Se pollution, the US Environmental Protection Agency (EPA) has set a regulatory discharge limit of 5 µg Se/L for effluent discharged to aquatic bodies. It was therefore recognized that the treatment and removal of Se from contaminated waters is essential for the protection of both animals and human health. As a consequence, a boom of Se related research has increased over the past decade, largely targeting Se removal from wastewaters like mine drainage, flue gas desulphurization (FGD) exhaust water, and agriculture drainage (Staicu et al. 2017; Tan et al. 2016; Lenz and Lens 2009).

1.2 Problem description

Se-laden wastewater not only contains Se oxyanions as electron acceptors, but also other oxyanions (e.g. nitrate (NO_3^-) and sulfate (SO_4^{2-})) as well. For example, mining wastewater is typically characterized by elevated concentrations of SO_4^{2-} (0.3-3 g/L), NO_3^- (0.2-1 g/L), various heavy metals (0.1-100 mg/L) and Se concentrations ranging from 50 µg/L to 10 mg/L depending on the original source of the mineral deposits (Mal et al. 2016; Tan et al. 2016; Sheoran and Sheoran 2006). The presence of other electron acceptors at relatively higher concentrations could compete with, or even inhibit SeO_4^{2-} reduction. Theoretical Gibbs energy (ΔG) calculations showed that the reduction hierarchy of oxyanions will follow $NO_3^- > SeO_4^{2-} > SO_4^{2-}$ (**Figure 2.5**). However, under actual experimental conditions, it is unclear how this would affect SeO_4^{2-} removal since nitrate and sulfate are present at approximately 100 and 1000 times higher concentrations compared to SeO_4^{2-}, respectively.

Over the last few decades, various technologies and approaches have been developed and patented for treating Se-containing industrial wastewater. These technologies involve physical, chemical and biological processes. Physical (e.g. membrane filtration and adsorption using ferrihydrite) and chemical (e.g. metal precipitation and electro-coagulation) methods are effective for removing Se, but these methods are costly in terms of energy and chemical consumption (Staicu et al. 2017). In addition, these processes employ complex process flow and may also generate Se oxyanion-bearing chemical sludge.

Biological processes for Se removal offer a cheap and greener alternative as compared to the conventional physical and chemical treatment methods (Gusek et al. 2008). Microorganisms are able to reduce selenate (SeO_4^{2-}) to insoluble biogenic elemental Se (Se^0), thereby removing Se from wastewaters. Moreover, biogenic Se^0 exhibits unique optical and spectral properties that can be recovered and reused as a valuable material for various industrial applications, i.e. medical, electrical, and manufacturing processes (Jain et al. 2014). Despite the advances in biotechnology for Se reduction, there are still many challenges in this process, particularly the application of biotechnology to treat real wastewaters.

To the best of our knowledge, though there are studies on SeO_4^{2-} reduction along with other oxyanions, this is typically paired with either NO_3^- or SO_4^{2-} alone and only a few studies have reported the reduction of all three oxyanions. The effect of co-electron acceptors on the SeO_4^{2-} removal and Se by-products must be established in order to have a better understanding on how to improve biological treatment. Additionally, despite the efforts made to develop and optimize various bioreactors for Se removal, there is still a lack of information with regard to treating SeO_4^{2-} in real wastewaters with a complex composition, such as the presence of heavy metals or low pH condition. Se-containing wastewaters, i.e. acid mine drainage (AMD) or flue-gas desulfurization (FGD), are typically acidic in nature (pH 4-5) (Sánchez-Andrea et al. 2014; Soda et al. 2011). This could cause some difficulty for microbial activity and the reduction of SeO_4^{2-} as most Se-reducing bacteria have been reported to work more efficiently under near neutral and alkaline conditions (Winkel et al. 2012). Finally, the current biological treatment processes have issues with meeting the stringent discharge limit set by regulatory bodies to avoid long term environmental impacts. Therefore, an understanding of the biofilm-Se interactions and development of newer methods or process flow configuration are required.

1.3   Research objectives

This research aims to provide a better understanding of the effect of co-contaminants on SeO_4^{2-} removal and exploring the different operating conditions and reactor configurations for treating a model mining wastewater. The specific objectives of this research were:

1) To study the effect of NO_3^- and SO_4^{2-} on SeO_4^{2-} bioreduction by anaerobic granular sludge in both batch and continuous operation using different bioreactors.
    a. To establish the reduction profiles of SeO_4^{2-}, SO_4^{2-} and NO_3^- under different molar ratio in batch tests.
    b. To investigate the effect of different co-electron acceptors on the biomass formation/growth and microbial community.
2) To investigate the SeO_4^{2-} removal performance in a simulated Se-laden wastewater at acidic influent pH and in the presence of a heavy metal.
3) To understand the biofilm-Se interactions under different growth conditions.
4) To investigate the performance of different bioreactors and an integrated process flow unit for Se-laden wastewater treatment:

a. Compare the removal performance of the upflow anaerobic sludge blanket (UASB) and biotrickling filter (BTF) configurations utilizing, respectively, attached and suspended biofilm.

b. Evaluate process flow treatment by combining physical (ion exchange column) and biological (UASB bioreactor) treatment processes for effective treatment of Se-laden wastewaters.

1.4   Structure of the thesis

This dissertation is comprised of nine chapters. An overview of the structure of this dissertation is given in **Fig. 1.1.** The following paragraphs outline the content of the individual chapters.

Chapter 1 presents a general overview of this dissertation, including background, problem description, research objectives and thesis structure.

Chapter 2 provides a literature review on the importance of Se and available technologies for the remediation of Se-laden wastewaters. The challenges and limitations encountered while applying Se remediation technology in practice are highlighted.

Chapter 3 investigates the effect of NO_3^- and SO_4^{2-} on SeO_4^{2-} bioreduction in batch tests. The potential for simultaneous removal of NO_3^-, SO_4^{2-} and SeO_4^{2-} at 20°C without biomass pre-adaptation was explored in a continuous laboratory-scale UASB bioreactor. **Chapter 4** gives an insight in the effect of different oxyanions (NO_3^-, SO_4^{2-} and SeO_4^{2-}) on the biofilm-Se interaction, removal performance, biomass growth and activity in a continuous short-term operation of a multi-panel drip flow reactor (DFR). **Chapter 5** evaluates the feasibility of treating slightly acidic ($5.0 < pH < 7.0$) influent with NO_3^-, SO_4^{2-} and SeO_4^{2-} using a laboratory-scale UASB reactor inoculated with anaerobic granular sludge. Optimal operation at low pH and 20°C was established. Detailed changes in microbial community shift during changing pH operation were also investigated.

Chapters 6, 7 and 8 are focused on the comparison and evaluation of treatment processes for Se-laden wastewaters with co-contaminants (SO_4^{2-} and Ni^{2+}). **Chapter 6** compares microbial aggregates (i.e. attached biofilm and suspended biofilm using granular sludge) grown in different reactor configurations (i.e. BTF and UASB bioreactor) for removing SeO_4^{2-} with SO_4^{2-} and Ni^{2+}. Biomass characterization for the formation of minerals/nanoparticles and changes in soluble extra-polymeric substances (EPS) under different operating conditions were described.

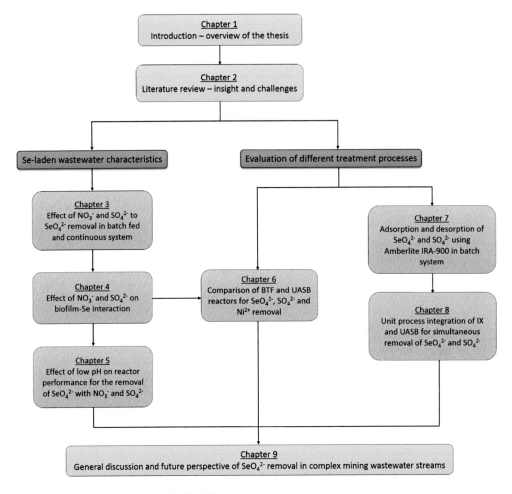

Fig. 1.1 Overview of the chapters in this PhD thesis

Chapter 7 describes the optimal parameters for effective adsorption of SeO_4^{2-} and SO_4^{2-} in binary anion solutions using ion exchange (IX) with Amberlite® IRA-900. A regeneration process was defined for optimal adsorption-desorption cycles to characterize resin reusability. Kinetic and isotherm model fitting was conducted to define the resin adsorption mechanisms in binary anion solutions. **Chapter 8** explores the combination of physical (adsorption using IX) and biological (UASB reactor) treatment in two different process flow configurations for the simultaneous removal of SeO_4^{2-} and SO_4^{2-} to achieve acceptable environmental discharge limits. Bio-regeneration of exhausted resin using anaerobic granular sludge was explored as an alternative for chemical regeneration.

Chapter 9 summarizes the research performed and draws general conclusions on knowledge gained from this dissertation. It also discusses practical applications and gives recommendations for future research.

1.5   References

Anton, S.P., van Erveer, W.D., 1997. Selenium toxicity to aquatic life: An argument for sediment-based water quality criteria. *Environ. Toxicol. Chem.* 16, 1255-1259.

Brozmanová, J., Mániková, D., Vlčková, V., Chovanec, M., 2010. Selenium: A double-edged sword for defense and offence in cancer. *Arch. Toxicol.* 84, 919-938.

Gusek, J., Conroy, K., Rutkowski, T., 2008. Past, present and future for treating selenium impacted water. In: Tailings and Mine Waste '08. CRC Press, pp. 281-290. Edited by The Organizing Committee of the 12th International Conference on Tailings and Mine Waste. 978-0-203-88230-6.

Jain, R., Gonzalez-Gil, G., Singh, V., van Hullebusch, E.D., Farges, F., Lens, P.N.L., 2014. Biogenic selenium nanoparticles: Production, characterization and challenges. In: Kumar, A., Govil, J.N. (Eds.), Nanobiotechnology. Studium Press LLC, USA, pp. 361-390.

Lemly, A.D., 2004. Aquatic selenium pollution is a global environmental safety issue. *Ecotoxicol. Environ. Saf.* 59, 44-56.

Lenz, M., Lens, P.N.L., 2009. The essential toxin: The changing perception of selenium in environmental sciences. *Sci. Total Environ.* 407, 3620-3633.

Mal, J., Nancharaiah, Y.V., van Hullebusch, E., Lens, P.N.L., 2016. Effect of heavy metals co-contaminants on selenite bioreduction by anaerobic granular sludge *Bioresour. Technol.* 206, 1-8.

Mehdi, Y., Hornick, J.L., Istasse, L., Dufrasne, I., 2013. Selenium in the environment, metabolism and involvement in body functions. *Molecules* 18, 3292-3311.

Sánchez-Andrea, I., Sanz, J.L., Bijmans, M.F.M., Stams, A.J.M., 2014. Sulfate reduction at low pH to remediate acid mine drainage. J. *Hazard. Mater.* 269, 98-109.

Sheoran, A.S., Sheoran, V., 2006. Heavy metal removal mechanism of acid mine drainage in wetlands: A critical review. *Miner. Eng.* 19, 105-116.

Soda, S., Kashiwa, M., Kagami, T., Kuroda, M., Yamashita, M., Ike, M., 2011. Laboratory-scale bioreactors for soluble selenium removal from selenium refinery wastewater using anaerobic sludge. *Desalination* 279, 433-438.

Staicu, L.C., van Hullebusch, E.D., Rittmann, B.E., Lens, P.N.L., 2017. Industrial selenium pollution: Sources and biological treatment technologies. In: van Hullebusch E. (Ed.), Bioremediation of Selenium Contaminated Wastewater. Springer International Publishing, pp.75-10.

Staicu, L.C., van Hullebusch, E.D., Lens, P.N.L., 2017. Industrial selenium pollution: Wastewaters and physical-chemical treatment technologies. In: van Hullebusch E. (Ed.), Bioremediation of Selenium Contaminated Wastewater. Springer International Publishing, pp. 103-130.

Rayman, M.P., 2012. Selenium and human health. *Lancet* 379, 1256-1268.

Rosen, B.P., Liu, Z., 2009. Transport pathways for arsenic and selenium: A minireview. *Environ. Int.* 35, 512-515.

Tan, L.C, Nancharaiah, Y.V., van Hullebusch, E., Lens, P.N.L., 2016. Selenium: Environmental significance, pollution, and biological treatment technologies. *Biotechnol. Adv.* 34, 886-907.

Wen, H., Carignan, J., 2007. Reviews on atmospheric selenium: Emissions, speciation and fate. *Atmos. Environ.* 41, 7151-7165.

Winkel, L.H.E., Johnson, C.A., Lenz, M., Grundl, T., Leupin, O.X., Amini, M., Charlet, L., 2012. Environmental selenium research: From microscopic processes to global understanding. *Environ. Sci. Technol.* 46, 571-579.

CHAPTER 2

Selenium: Environmental significance, pollution, and biological treatment technologies

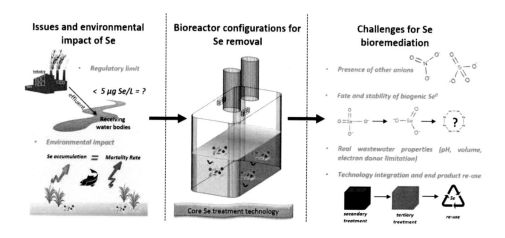

This chapter has been published in modified form:

Tan, L.C., Nancharaiah, Y.V., van Hullebusch, E.D., Lens, P.N.L. 2016. Selenium: Environmental significance, pollution, and biological treatment technologies. *Biotechnol. Adv.* 34, 886-907. doi:10.1016/j.biotechadv.2016.05.005.

Abstract

Selenium is an essential trace element needed for all living organisms. Despite its essentiality, selenium is a potential toxic element to natural ecosystems due to its bioaccumulation potential. Though selenium is found naturally in the earth's crust, especially in carbonate rocks, volcanic and sedimentary soils, about 40% of the selenium emissions to atmospheric and aquatic environments are caused by various industrial activities such as mining-related operations. In recent years, advances in water quality and pollution monitoring have shown that selenium is a contaminant of potential environmental concern. This has practical implications on industry to achieve the stringent selenium regulatory discharge limit of 5 µg Se/L for Se containing wastewaters set by the United States Environmental Protection Agency. Over the last few decades, various technologies have been developed for the treatment of Se-containing wastewaters. Biological selenium oxyanion reduction has emerged as the leading technology for removing selenium from wastewaters since it offers a cheaper alternative compared to physico-chemical treatments and is suitable for treating dilute and variable Se-laden wastewaters. Moreover, biological treatment has the advantage of forming biogenic elemental selenium nanospheres which exhibit unique optical and spectral properties for various industrial applications, i.e. medical, electrical, and manufacturing processes. However, despite the advances in biotechnology employing selenium reduction, there are still several challenges, particularly in achieving stringent discharge limits, the long-term stability of biogenic elemental selenium and predicting the fate of bioreduced selenium in the environment. This review highlights the significance of selenium in the environment, health, industry and biotechnological advances made in the treatment of selenium contaminated wastewaters. The challenges and future perspectives are overviewed considering recent biotechnological advances in management of these selenium-laden wastewaters.

Keyword: Bioremediation, selenium bioreduction, bioreactors, selenium pollution, selenium wastewaters, selenium environmental impact, selenium nanoparticles, selenium recovery

10

2.1   Introduction

Selenium (Se_{79}^{34}) is an essential element in living organisms. It is the 30th most abundant naturally occurring metalloid element on the Earth's crust. Selenium is present in the earth's crust at an estimated amount of 0.05-0.5 mg Se/kg (Lemly 2004), with total selenium concentrations in rocks amounting to 40% of the total earth's crust found mainly in sandstone, quartzite and limestone (Wang and Gao 2001). The selenium content in most soils is reported to be 0.4 mg Se/kg (world mean) with concentrations up to 1,200 mg Se/kg soil in seleniferous regions such as in the USA, Canada, Colombia, UK, China and Russia (Fordyce 2005). Selenium was first misidentified as tellurium, but was later discovered as a new element by the Swedish doctor Jöns Jakob Berzelius in 1817 during sulfuric acid production. It was named after the Greek moon goddess Selene (Kumar and Priyadarsini 2014) to complement its sister element tellurium (*Tellus*) which means Earth in Greek. Since then, selenium's importance has grown steadily in research and applications in fields ranging from environment to medicine and material science (**Fig. 2.1**).

Interest in selenium research has bloomed in the recent decades with new insights for molecular, genetic, biochemical and health areas applications. The sudden growth of interest in selenium stems from its double-edged sword characteristics of being both essential and toxic to living organisms, with the tipping scale of toxic and safe concentrations only being 3-5 fold (Brozmanová et al. 2010) up to 10 fold (40 to 100 µg/day; El-Ramady et al. 2014a) different. Selenium pollution has a wide environmental impact range in humans, ecosystems and even agricultural processes. Because environmental contamination by selenium typically occurs at low concentrations, the importance in remediating selenium-containing environments has been deemed less important. However, selenium is found in soil and water and can be mobilized to enter the food chain through the roots of plants or aquatic organisms, thus causing concern of possible long term environmental impacts (Brown and Arthur 2007; Staicu et al. 2015a).

Selenium treatment technologies have been reviewed in the last decade mainly in company reports, e.g., from MSE Technology Applications Inc. (2001), Sobolewski (2005), NSMP (2007), Golder Associates Inc. (2009), and CH2M HILL (2010). Many reviews for various areas in selenium reduction have also been published recently: Fernández-Martínez and Charlet (2009) on the chemistry of selenium cycling and bioavailability; Winkel et al. (2015) on selenium interaction in the soil-plant-atmosphere cycle; Nancharaiah and Lens (2015a,b) on selenium microbial reduction and biomineralization; Santos et al. (2015) focusing on selenium adsorbent materials; and Staicu et al. (2015a) on post-treatments for selenium recovery. The present review aims to emphasize the importance of selenium and give an overview of current biotechnologies available for pollution control as well as the challenges of biological treatment of selenium-laden wastewaters for future applications.

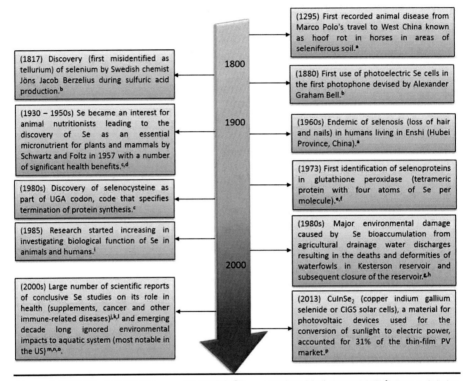

(1295) First recorded animal disease from Marco Polo's travel to West China known as hoof rot in horses in areas of seleniferous soil.[a]

(1817) Discovery (first misidentified as tellurium) of selenium by Swedish chemist Jöns Jacob Berzelius during sulfuric acid production.[b]

1800

(1880) First use of photoelectric Se cells in the first photophone devised by Alexander Graham Bell.[b]

(1930 – 1950s) Se became an interest for animal nutritionists leading to the discovery of Se as an essential micronutrient for plants and mammals by Schwartz and Foltz in 1957 with a number of significant health benefits.[c,d]

1900

(1960s) Endemic of selenosis (loss of hair and nails) in humans living in Enshi (Hubei Province, China).[a]

(1980s) Discovery of selenocysteine as part of UGA codon, code that specifies termination of protein synthesis.[e]

(1973) First identification of selenoproteins in glutathione peroxidase (tetrameric protein with four atoms of Se per molecule).[e,f]

(1985) Research started increasing in investigating biological function of Se in animals and humans.[i]

2000

(1980s) Major environmental damage caused by Se bioaccumulation from agricultural drainage water discharges resulting in the deaths and deformities of waterfowls in Kesterson reservoir and subsequent closure of the reservoir.[g,h]

(2000s) Large number of scientific reports of conclusive Se studies on its role in health (supplements, cancer and other immune-related diseases)[j,k,l] and emerging decade long ignored environmental impacts to aquatic system (most notable in the US)[m,n,o].

(2013) CuInSe$_2$ (copper indium gallium selenide or CIGS solar cells), a material for photovoltaic devices used for the conversion of sunlight to electric power, accounted for 31% of the thin-film PV market.[p]

[a]Yang et al. (1983); [b]Weeks (1932); [c]Allmang and Krol (2006); [d]Husen and Siddiqi (2014); [e]Tinggi (2003); [f]Pedrero and Madrid (2009); [g]Cantafio et al. (1996); [h]Ohlendorf (2002); [i]Wu and Huang (1991); [j]Duntas and Benvenga (2015); [k]Rayman (2012); [l]Kieliszek and Błażejak (2013); [o]Hamilton (2004); [l]Lemly (2002); [m]Lemly (2014); [p]Anderson (2015).

Fig. 2.1 Timeline of milestones in selenium research and applications.

2.2   Why is selenium important?

In the past, selenium was not considered as an important component in wastewater treatment compared to heavy metal contaminants or organic pollutants. It is neither part of the critical raw materials listed by the European Union (Hennebel et al. 2015; Nancharaiah et al. 2016) nor is it a worldwide issue due to the regionally uneven distribution of selenium in soil and atmosphere (Wu 2004; Wen and Carignan 2007). However, selenium is always tied to larger issues in many ways as described below.

2.2.1 The biogeochemical selenium cycle

Selenium is a unique trace element and has five existing stable isotopes (^{74}Se, ^{76}Se, ^{77}Se, ^{78}Se, and ^{80}Se) with ^{79}Se (fission product of ^{235}Uranium) receiving special attention due to the growing interest in nuclear waste disposal research (Fernández-Martínez and Charlet 2009). Selenium shares many similar physiological properties with arsenic and chemical behavior with sulfur. It has both chalcophilic (strong affinity with sulfur) and siderophilic (able to form a metallic bond with iron) properties (König et al. 2012). In its solid form, selenium is an allotropic element that can exist in different forms, i.e., crystalline, metallic and amorphous (El-Ramady et al. 2014b; Nancharaiah and Lens 2015a).

Since selenium shows strong metal-like properties, it can exist in different oxidation states (+6, +4, 0, -2) and inorganic forms (Nakamaru et al. 2005). **Table 2.1** lists the classes and individual selenium compounds found in environmental and biological systems. For more details on the properties of various selenium compounds, the reader is referred to the review article from the U.S. Department of Health and Human Services (2003). The inorganic selenium species are often found in surface waters, ground waters and wetlands in the form of selenide (Se^{2-}), selenite (SeO_3^{2-}), selenate (SeO_4^{2-}) and insoluble elemental selenium (Se^0) (Vesper et al. 2008). Organic selenium compounds prevail in air, soil and plants in the form of volatile methylselenides, trimethylselenonium ions or selenoamino acids (Pyrzynska 2002). Overall, depending on the concentration in soil, sediment or water, the selenium availability and mobility is a function of the geochemical reactions and selenium speciation (Persico and Brookins 1988). Various behaviors of selenium speciation depending on environmental and thermodynamic conditions are discussed in the review of Nakamaru and Altansuvd (2014). One such example given was that under conditions of high redox and non-flooded field soils, selenate is the predominant and bioavailable Se species for uptake by vegetation compared to selenite, due to the high adsorption preference of selenite onto soil particles.

Selenium is released into the environment from both natural and anthropogenic sources. It can be found in virtually all kinds of natural environments, particularly in geochemical locations such as volcanoes and various types of igneous and sedimentary rocks subjected to mining (Kessi et al. 1999). The biogeochemical cycle of selenium moves from rocks, sediments and soils to waters, where it enters plants, animals or humans via the food chain. The selenium cycle becomes complete through the activity and degradation of organisms, allowing for the incorporation of selenium into the sediments and finally to rock deposits. Thus, the selenium distribution in the environment is affected by a variety of physical, chemical and biological processes. **Fig. 2.2** illustrates the selenium cycle and depicts both natural and anthropological mobilization of selenium to water sources.

Table 2.1 List of most common individual aqueous or gaseous selenium compounds found in environmental and biological systems [modified from Fernández-Martínez and Charlet (2009) and Kieliszek and Błażejak (2013)].

Selenium Species	Description
Inorganic Species	
Elemental Selenium (Se^0)	Naturally non-soluble with at least 11 different allotropes. Formed as a precipitate after chemical and microbial reduction processes.
Selenide (Se^{2-} or H_2Se), Se -II (Se^- or HSe), Se -I	Stable in strongly reducing conditions as metal selenides and other organic compounds. Volatile compounds formed by microbial processes.
Selenate (SeO_4^{2-}), Se +VI	Exists as a tetrahedral oxyanion form that shares similarities with sulfate and is mainly predominant in soils, sediments and waters.
Selenite (SeO_3^{2-}), Se +IV	Exists as a pyramidal oxyanion that is more reactive, present in mildly oxidizing acidic environments and major mobile species governed by sorption or desorption processes on solid surfaces.
Organic Species	
Dimethylselenide (Me_2Se) Dimethyldiselenide (Me_2Se_2)	Volatile compounds that can be formed from biomethylation processes.
Trimethylselenonium cation (Me_3Se^+)	Metabolite from urine.
Dimethylselenosulfide (MeSSeMe)	Product of microbial methylation process.
Selenocyanate ($SeCN^-$)	Can be found in wastewater from petroleum refineries.
Amino Acids and Low-Molecular Mass Species	
Selenomethionine (SeMet)	Main selenium form found in plants.
Selenocysteine (SeCys)	21st amino acid component of selenoproteins. Main selenium form in organic tissues.
Selenoglutathione	Can be formed during *in vitro* experiments mimicking biochemical processes.
Selenoproteins	Selenium is a component of various forms of enzymes, peptides, and sugars that can be found in a biological system (i.e., GPX, TR, etc.).

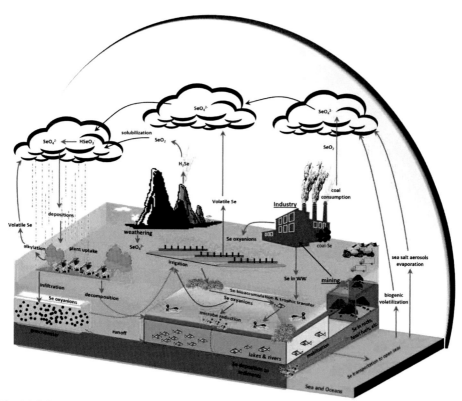

Fig. 2.2 Schematic diagram of the global selenium cycle and transportation in the environment focusing on water exposure and contamination. Selenium mobilization and speciation depends heavily on the interactions with and condition of the environment. Immobilization of selenium can incur through chemical and microbial reduction, uptake by plants and animals, adsorption through soils and sediments and co-precipitation from the atmosphere. Selenium deposited in sediments can stay in the system and bioaccumulate for long periods of time (Lemly 1999; Winkel et al. 2012; Nancharaiah and Lens 2015a).

In nature, selenium is closely associated with sulfur containing minerals, pyrites and fossil fuel sources (Mehdi et al. 2013). These account for 50-65% of the selenium emissions to the atmosphere (Sharma et al. 2015) by natural processes such as weathering and soil leaching (Kabata-Pendias, 2011). Human activities, however, are one of the major causes for selenium mobilization and accumulation in the environment (Winkel et al. 2012) with about 37-40% of the total selenium emissions to the atmosphere attributed to anthropogenic activities (Wen and Carignan 2007). Selenium is mainly a by-product in metal (i.e., copper) refinery and processing plants (i.e., sulfuric acid production) (Nancharaiah and Lens 2015b). Other anthropogenic activities include industrial activities of coal combustion, mining activities, glass, electronic and petroleum industry, utilization of rock phosphates as fertilizer and sewage sludge application to agricultural land and agricultural run-offs (Lemly 1997; Pyrzynska 2002; Kabata-Pendias 2011).

2.2.2   Global market and uses of selenium

Selenium's annual global production is estimated to be around 2,500-2,800 tons with Japan, Canada, Belgium and Germany being the main producers (Mehdi et al. 2013). Selenium in various forms has a wide range of applications in the industry. The selenium world consumption was estimated by USGS (2015) to be 40% for metallurgy, 25% in glass manufacturing, 10% for agricultural, 10% in chemicals and pigments manufacturing, 10% for electronics and 5% for other applications. More than 80% of the commercially available selenium is recovered from anode slime, a by-product generated in the electrolytic production and refining of copper (Young et al. 2010b; Anderson 2015) and from other metal mining activities such as lead, nickel, platinum-group metals and zinc mining (Anderson 2015).

There is an increasing supply risk of critical raw materials, as reviewed in Hennebel et al. (2015), being faced around the globe. This led to the listing of 14 critical elements considered as high risk of depletion in the future within the EU (Hennebel et al. 2015; Nancharaiah et al. 2016). Though selenium is not a part of the list, it is directly affected by the production of the parent material of certain metal deposits, with the majority found within copper ores (USGS 2015). As such, production of minor elements such as selenium heavily depends on the parent material resulting in a complex global market of selenium supply and demand. As an example, selenium showed a drastic price change in just over a year from 2003-2004 where its price increased by 440% and reached a staggering price of $160.94 per kilogram by 2011 as compared with $13.23 per kilogram from 1981-2010 (Hennebel et al. 2015). Commercial-grade selenium prices have, however, decreased from $119.05 per kilogram in 2012 (Anderson 2015) to $79.37 per kilogram in 2013. This clearly shows that though selenium is not yet a critical material, selenium sources can rapidly change, thus putting more pressure on finding alternative sources for this element.

Selenium has an intrinsic opto-electrical property that makes this element of commercial interest in nanotechnology. It is used in various industrial processes such as semi-conductors and photoelectric cells due to its electrochemical properties (Macaskie et al. 2010). Metal-selenium compounds have the ability to convert light into electricity or vice versa, leading to its potential implementation in applications, such as photovoltaic cells for solar energy, sensors for industrial and military applications, and as shields for protection from high energy light beams (Kapoor et al. 1995). In glass manufacturing, selenium is used to decolorize the green tint of glass from iron impurities, while cadmium sulfoselenide pigments are used in plastics, ceramics, and glass to produce a ruby-red color. Selenate has replaced the toxic lead (Pb) as a pigment along with bismuth vanadate in the glass and ceramic industries (Kabata-Pendias 2011). It is also used as a catalyst for oxidation enhancement in plating solutions and as a metallurgical additive for machinability (ability of a material to be machined) improvement of copper, lead, and steel alloys as well as thin-film photovoltaic copper indium gallium diselende (CIGS) solar cells (USGS 2015).

Selenium is furthermore a raw material used for the vulcanization of rubbers to increase abrasion resistance and a toner agent for photographic prints. For agricultural uses, sodium selenite is used as an additive for fertilizer, insecticides and foliar sprays. Moreover, selenium can also form covalent bonds with many other elements and through this, can detoxify toxic metals such as mercury and arsenic (Rosen and Liu 2009). Selenium supplements can alleviate mercury poisoning by increasing mercury excretion and can also decrease biomarkers related to oxidative stress in humans (Sears 2013). This is because selenium is a mercury chelator and able to form extremely stable compounds or organic complexes with mercury. Another study by Siscar et al. (2013) on the liver of deep-sea fish further observed that a higher Se concentration is required relative to the mercury concentration in order to lower the mercury toxicity.

Aside from technological applications, selenium is a dietary supplement for both humans and livestock. Selenium is supplied in multivitamin supplements typically present in < 5 mg Se per gram of supplement (Schrauzer 2001) or/and in food as reviewed by Navarro-Alarcon and Cabrera-Vique (2008). It can be found as a common component in various cosmetics and personal care products such as shampoo (Kabata-Pendias 2011). Additionally, selenium, in combination with other trace elements, can also have a stimulatory effect on the activated sludge wastewater treatment process (Burgess et al. 1999), such as improved process stability and microbial growth (Mao et al. 2015). Selenium dosing has been shown to be a useful approach for improving the performance of anaerobic digesters. Addition of selenium, along with other trace elements, at 0.062 mg Se/L showed enhancement of the methane yield by 40% and prevented accumulation of volatile fatty acids (Wall et al. 2014). Recent studies have also shown that application of organo-selenium on material surfaces has the ability to limit biofilm formation on reverse osmosis membranes with about 8%-15% less flux loss using a combination of the amine group of selenocystamine and the carboxyl group of selenocyanatoacetic acid (Vercellino et al. 2013a, 2013b) and as well as to prevent bacterial growth on bandages with over 8 log bacterial inhibition using selenium 2-(acetoacetoxy)ethylmethacrylate (Tran et al. 2012, 2014).

2.2.3 Human health and bioaccumulation of selenium

Selenium is of fundamental importance to human health and has a long history of interest in the field of both clinical medicine and environment (Rayman 2012). It has been recognized as an essential nutrient because it plays a key role in several major metabolic pathways such as thyroid hormone metabolism, antioxidant defense systems, and immune functions (Brown and Arthur 2007; Zimmerman et al. 2015). It is essential in humans and animals because of the role of selenium in antioxidant glutathione peroxidase, which allows for the protection of the cell membrane from damage caused by peroxidation of lipids (Tinggi 2003).

Despite the important physiological function of selenium in living organisms, there is a fine line between the concentration yielding a positive or a toxic effect, from 40 to 400 µg/day (El-Ramady et al. 2014a). Selenium deficiency and toxicity are largely dependent on the location and condition of the environment, i.e. selenium content of the soil. As such, only specific geographical regions have extreme Se concentrations (i.e. deficient/excess of Se content in the soil) and associated selenium related problems (El-Ramady et al. 2014b).

Worldwide, there are about 500-1100 million selenium-deficient people which account for about 15% of the estimated 7 billion people (White et al. 2012). Geographical regions known for selenium deficiency are volcanic regions of former Yugoslavia, Poland, China and Russia. Studies have reported that selenium has a protective effect against some forms of cancer (Brozmanová et al. 2010), decreased cardiovascular disease mortality, regulates the inflammatory mediators in asthma (Brown and Arthur 2007), maintains bone homeostasis and protects against bone loss (Zhang et al. 2014). Geographical regions with Se deficiency in soil have been associated with heart disease and bone disorders (Lemly 1999), which can be corrected through intake of dietary selenium. One of the clear-cut examples of selenium deficiency health issues were seen in China during the early 1900s. Epidemic cases of the Keshan (heart disease) and Kashin-Beck (endemic osteoarthropathy) diseases (Fordyce 2005) occurred around the geological region of northwest to southwest of China characterized by a low selenium content (0.125 mg Se/kg-soil).

On the other hand, problems associated with excess selenium occur in seleniferous regions due to high bioavailability and uptake of selenium by plants and animals (El-Ramady et al.

18

2014a). Selenium-related diseases for animals go all the way back to the time of Marco Polo during his travels in China where a hoof disease in horses was found and later identified as selenium toxicosis (Yang et al. 1983; Fernández-Martínez and Charlet 2009). Both the lowest (0.004-0.058 mg Se/kg) and highest (1.5-59.4 mg Se/kg) concentrations and the flux of selenium in the environment are reported in China, where endemic selenium deficiency and selenosis (hair and nail loss with addition of digestive and nervous system disruption) has occurred for the past few decades in both animals and humans (Wang and Gao 2001).

The recommended amount of selenium intake varies and largely depends on various factors such as gender, age, human condition and in particular geographical location (Kieliszek and Błażejak 2013). Essentially, international agencies have set the recommended selenium dietary values from 30 to 55 µg/day as the safe selenium intake levels (Winkel et al. 2012). Selenium intake heavily depends on the crops/food produced where sources rich in selenium can be found in eggs, seafood, wheat products, meat (e.g., chicken liver), vegetables (e.g., garlic bulbs, onion, broccoli) and yeast (Fairweather-Tait et al. 2010). In the world, the inhabitants of Finland consume the lowest amount of selenium at 30 µg/day, while the people of Venezuela consume the highest amount at 326 µg/day (Kieliszek and Błażejak 2013). Because of the low selenium intake in Finland, selenium has been made mandatory in fertilizer applications to increase the micronutrient provision for humans since the 1980s (White et al. 2012). A similar situation is found in New Zealand and Australia, with the latter focusing more on enhancement of selenium nutrition to livestock (Tinggi 2003). However, selenium fortification is a delicate task and must be performed in strict and precise conditions since it has a very narrow safety margin between toxicity and deficiency and its ability to bioaccumulate through the food chain should also be carefully considered (Ferrarese et al. 2012).

Though the implicit importance of selenium to human health is recognized universally, long term health implications in relation to selenium intake and chronic diseases have yet to be clearly linked together and as such there is still no clear cut recommendation regarding the use of selenium as a supplement or its direct effect to prevent some diseases such as cancer. Similarly, despite the long history of linking selenium and human health, a global standard for selenium dietary requirement is yet to be formulated (Brown and Arthur 2007).

2.2.4   Need for selenium management and pollution control

Selenium-laden water is a problem for most ecological wildlife (Luoma and Presser 2009), particularly for the aquatic-dependent, egg-laying vertebrates where most sensitive toxicity is found as embryo mortality of waterfowls and larva deformities of fishes (Chapman et al. 2010). Over the past few decades, interest has been steered towards minimizing selenium toxicity for aquatic life resulting in a number of recommendations regarding appropriate criteria for protecting aquatic life (Canton and van Derveer 1997).

Various allowable guidelines for safe selenium uptake for drinking water and selenium discharge limits of industrial effluents to freshwater have been adopted, depending on the country's policy (**Table 2.2**). Various nations have adopted an upper limit of 50 μg/L for drinking water and 5 μg/L for selenium effluent discharge (Gusek et al. 2008b; Santos et al. 2015). The wide differences in regulations are due to the different parameters (speciation and source), varying environmental conditions and hydraulics (habitats and stream flow rates) observed worldwide and conflicting toxicity results between different living systems (Swift 2002; Muscatello and Janz 2009; Vinceti et al. 2013).

Table 2.2 Selenium regulatory limits for drinking water and effluent discharge.

Limit	Agency/Country	References
Drinking water		
10 μg Se/L	Europe	Santos et al. (2015)
40 μg Se/L	World Health Organization	WHO (2011)
1 μg Se/L (as SeO_3^{2-})	Russian Federation	Bassil et al. (2014)
50 μg Se/L	USEPA	USEPA (2014)
No regulation	Europe	Luoma and Presser (2009)
Effluent water (freshwater discharge)		
5 μg Se/L (chronic aquatic life toxicity)	USEPA	USEPA (2014)
20 μg Se/L (acute aquatic life toxicity)	USEPA	
2 μg Se/L	Canada British Columbia	Lussier et al. (2003)
100 μg Se/L	Japan	Fujita et al. (2002)
1 μg Se/L	Korea	Lee et al. (2015)

One famous example of selenium poisoning was found in California's Kesterson Wildlife reservoir where selenium poisoning occurred, causing adverse health damage on inhabiting wildlife, which finally resulted in the closure of the reservoir (Ellis et al. 2003). Physical deformities, mutations, reproduction failures and even death were observed after exposure to toxic levels of selenium concentrations, particularly for aquatic life. As a result of the Kesterson selenium problem, the USA Department of Interior conducted a number of studies on irrigated lands throughout the western USA. The survey showed elevated selenium concentrations for other USA sites as well, such as the Arkansas River, St. Charles River and Belews Lake. The main source of high selenium concentrations was traced back to natural weathering of Cretaceous marine shales and shale-derived soils, while selenium concentrations in Belews Lake and the Kesterson National Wildlife Refuge were due to anthropogenic activities (Anton and van Erveer 1997). Further examples were reported by Young et al. (2010) for several ecotoxicological studies conducted mainly in North America, where unmonitored selenium contamination occurred in different surrounding freshwater regions as a direct result of nearby industrial activities (**Table 2.3**).

The most significant negative property of selenium is its propensity to bioaccumulate despite the low concentration levels associated with selenium-laden wastewaters as compared with other toxicants (Stewart et al. 2010) as illustrated in **Fig. 2.3**. There are various factors in considering selenium's enrichment along the food chain such as species involved in the chain, habitat interaction (sediment or water column) and feeding zone (Schneider et al. 2015). For example, within an aquatic food chain, inorganic selenium bioaccumulates up to 100-400 times, while organic selenium bioaccumulates in excess of 350,000 times. This means that even at a low waterborne concentration of 0.1 μg Se/L, selenium residues can easily be elevated to toxic levels in the food-chain organisms and can be harmful for fish diets at around 5-15 μg Se/g (Lemly 2004).

In a study conducted on selenium accumulation in the aquatic ecosystem surrounding the uranium mine in northern Saskatchewan, Canada (one of the world's top producing uranium mines), the selenium concentration along the food chain was enriched up to approximately 1.5 to 6 fold between planktonic, invertebrates and fish (Muscatello et al. 2008). The author concluded that though results did not show high Se concentrations in aquatic biota (Se concentration of 0.4 μg/L and 0.54 μg/g-dry weight in water and soil, respectively) or severe effects to the aquatic ecosystem at that time, the massive potential bioaccumulation has the possibility to exert a toxic impact in the future. Therefore, continual selenium discharge monitoring to the receiving waters is required to avoid potential Se toxicity in the future. Schneider et al. (2015) investigated the transfer of selenium from sediments and water into the food chain using stable isotopes. The study showed a mean magnification factor of 1.39 per trophic level (sediment/water to macroalgae to carnivorous fish). Additionally, the authors concluded that with a high sediment-water partition coefficient (K_d), calculated to be 4180, partitioning of selenium entering into the system was occurring. Schneider et al. (2015) determined that the initial partitioning of selenium into bioavailable particulate organic matter was a main driver of selenium bioaccumulation into the food web.

Table 2.3 Case studies of selenium contamination release to the water environment due to anthropogenic activities (compiled from Chapman et al. 2010).

Area	Source	Selenium Concentrations	Effect
Belews Lake, North Carolina, USA	Coal power plant, ash settling pond	100-200 µg/L	Widespread reproductive failure of fishery.
Hyco Lake, North Carolina, USA	Power plant, ash settling pond	7-14 µg/L	Decline in fish population was noticed.
Martin Creek Reservoir, Texas, USA	Coal power plant, ash settling pond	< 2,700 µg/L	Abnormalities in fish tissues were observed.
Savannah River, South Carolina, USA	D-area power plant	100-110 µg/L	Adverse effects on a variety of fauna in the area, i.e., changes in behavior, metabolism.
Appalachian Mountains, USA	Coal mining	~42 µg/L	Streams affected by waste material showed lower numbers of total species and benthic species compared to uncontaminated streams.
Kesterson Reservoir, San Joaquin Valley, California, USA	Agriculture irrigation drainage	140-1400 µg/L	Local fish extinction (up to 8 warm-water species) and deformities in waterfowls.
Blackfoot River Watershed, Idaho, USA	Phosphate mining	> 1,000 µg/L	Livestock death, waterfowl embryo deformities and elevated selenium concentrations in fish.
Lake Macquarie, New South Wales, Australia	Coal power plant, fly ash	50-300 mg/kg dry weight	Bioaccumulation in aquatic species that exceeded allowable selenium intake for human consumption.
Elk River Valley, Southeast British Columbia, Canada	Coal mining, waste rock leachate	> 300 µg/L	Indication of reproductive failure in aquatic wildlife, i.e., unviable embryos or unfertilized eggs.

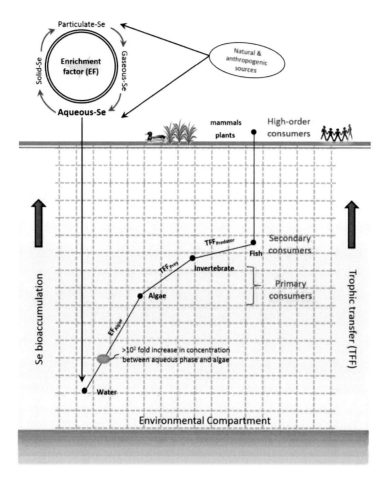

Fig. 2.3 Schematic diagram of aquatic selenium contamination dynamics in the water environment with selenium enrichment and trophic transfer, particularly for the aquatic food chain [modified from Stewart et al. (2010) and Young et al. (2010b)]. The enrichment factor (EF) represents the elevated concentration of selenium between the water and primary consumers living in it, which is typically a plant. The trophic transfer factor (TTF) represents the increase in selenium concentration along the food chain.

2.3 Selenium-laden wastewaters

2.3.1 Wastewater sources

Selenium is classified as a contaminant of potential concern in North America, Australia and New Zealand. Typical concentrations of selenium oxyanions encountered in wastewaters are dependent on anthropogenic activities (**Table 2.4**). Selenium contaminated waters are produced by various industrial activities such as mining (coal, hard rock, uranium and phosphate), refineries (metal and oil), power generation (coal-fired power plants), and agriculture (irrigation waters and selenium fortification) (Chapman et al. 2010).

In industrial wastewaters, selenium is available in the form of selenate (SeO_4^{2-}) and selenite (SeO_3^{2-}). Both selenium oxyanions are toxic to living systems, although SeO_3^- is potentially more toxic than SeO_4^{2-}, and should be removed from the wastewater. Reduction of selenium oxyanions leads to the formation of the less toxic and insoluble Se^0 form. Selenite is more likely to be adsorbed onto the solid inorganic surfaces (i.e., ferric hydroxide) or organic matter, while unbound SeO_4^{2-} is more likely to be flushed down the water stream (Vesper et al. 2008). Selenium speciation can change depending on the pH and redox potential of the wastewater as a function of the thermodynamic properties of the selenium species (Erosa et al. 2009). In most natural systems, SeO_4^{2-} exists in oxidizing conditions, while SeO_3^{2-} exists thermodynamically in the pH range from 3 to 8 and at a lower redox potential (Latorre et al. 2013).

2.3.2 Overview of selenium removal technologies

Several physical, chemical and biological technologies have been developed for the treatment of complex selenium containing waters. An overview of selenium removal technologies is presented in **Table 2.5**. The details of the different methods evaluated for remediation of selenium contaminated waters are further discussed below. This review focuses mainly on the biological selenium treatment technologies and as such physical and chemical methods are only briefly discussed. For more detailed information on physical and chemical selenium removal methods, the readers are referred to the articles of Golder Associates Inc. (2009), CH2M HILL (2010, 2013), and Santos et al. (2015). The latter specifically reviewed absorbents available for selenium treatment.

Physical methods

High-rate methods

Selenium removal was studied in different physical methods such as reverse osmosis (RO), nanofiltration (NF), and ion exchange (IX) (Golder Associates Inc. 2009; Lenz and Lens 2009; CH2M HILL 2010, 2013; Santos et al. 2015). So far, no known pilot or full-scale operational studies have focused on selenium removal using NF but there are some full-scale applications of RO in the industry. RO has been recognized by the USEPA as one of the best available technologies, along with chemical treatment using ferrihydrite, and has been successfully applied for treating mining wastewater from Barrick's Richmond Hill Mine (Sobolewski 2005). RO was used as a post-treatment after iron mediated SeO_4^{2-} or SeO_3^{2-} reduction and precipitation of selenium. The concentrations were decreased from 12-22 µg Se/L to 2 µg Se/L with a wastewater flow rate of 0.757 m³/min. Another RO application was established for a gold mine closure in California, USA (Golder Associates Inc. 2009). Se concentrations were decreased from 60 µg Se/L to below 5 µg Se/L at a flow rate larger than 75.7 m³/min.

Ion exchange (IX), on the other hand, is more cost effective for selenium removal as compared to RO. In an ion exchange system, a specific ionic contaminant is reversibly exchanged for a similarly charged ion that is attached to an immobilized solid surface material resin (synthetic or inorganic zeolites). Once full saturation of the resin has been achieved, the bed can be regenerated through solution washing. One famous example of a pilot-scale ion exchange reactor combined with electrochemical reduction (Selen-IX™) was developed by BioteQ and established in Elk Valley, British Colombia (The Free Press 2013). This technology has reported achieving a decrease in selenium concentration to as low as 1 µg/L. The removed selenium was stabilized in the form of an inorganic iron-selenium complex during the electrochemical reduction step. The Selen-IX™ units were made in a modular form for easy transport and have a treatment capacity of 2.8 m³/day (Mohammandi et al. 2014).

Physical treatment has the advantage of being well established particularly for drinking water and municipal wastewater treatment and therefore the mechanisms and operational parameters are well understood. Additionally, with the use of membrane systems, space requirements are minimized and regulatory selenium limits are achieved. However, membrane filtration has major disadvantages of high energy and capital cost for operation and maintenance. Moreover, generation of concentrated waste or brine adds the additional burden of post-treatment and disposal issues. In the case of ion exchange, the process is commonly employed in many industrial wastewaters but has seen little success for selenium removal due to competition with other anions, which are present at much higher concentrations as compared to selenium oxyanions, quickly saturating the resins and causing a low selenium removal efficiency (NSMP 2007). Pre-treatment of the wastewater feed for other anions (e.g., sulfate) is therefore necessary before applying the ion exchange system in order to avoid inhibition on the selenium removal.

Table 2.4 Concentration of selenium from natural sources and various industrial wastes generated.

Source	Se concentrations	References
Natural systems		
Groundwater	0.1-6000 µg/L	Pyrzynska (2002)
Soil	0.05-100 µg/g	Pyrzynska (2002); Yanai et al. (2015)
Surface water (CA, USA)	10-4200 µg/L	Twidwell et al. (1999)
Surface water (MT, USA)	1-560 µg/L	
Solid sources		
Earth's crust	0.05-0.09 µg/g	Wang and Gao (2001)
Coal mines	0.4-24 µg/g	Lemly (2004); Vesper et al. (2008)
Copper metal ores	20-82 µg/g	
Manganese production waste	50-1000 µg/g	Lemly (2004)
Phosphate mines	2-20 µg/g	
Mining industry		
Coal mine wastewaters	0.4-1500 µg/L	Twidwell et al. (1999); Lemly (2004); Santos et al. (2015)
Uranium mine discharge	1600 µg/L	Twidwell et al. (1999)
Gold mines wastewater	170-33000 µg/L	Twidwell et al. (1999); Astratinei et al. (2006)
Acid mine drainage	up to 570 µg/L	Lenz et al. (2008a)
Mining wastewater	3 µg/L up to > 12 mg/L	Santos et al. (2015)
Flue gas desulfurization		
FGD process water	1 µg/L-10 mg/L	Lemly (2004); Santos et al. (2015)
Fly ash leachate	40-610 µg/L	Lemly (2004)
Oil refinery plant		
Oil refinery wastewater	7.5-4900 µg/L	Twidwell et al. (1999); Santos et al. (2015)
Crude Oils	500-2200 µg/L	Lemly (2004)
Refined Oils	5-258 µg/L	
Mining industry		
Selenium wastewater	up to 620 mg/L	Fujita et al. (2002)
Copper wastewater	2000 µg/L	Lemly (2004)
Lead wastewater	1600-7000 µg/L	Twidwell et al. (1999); Santos et al. (2015)
Agriculture industry		
Irrigation wastewater	up to 1400 µg/L	Lemly (2004)
Agricultural drainage water	140-1400 µg/L	Santos et al. (2015)
Municipal wastewater		
Municipal landfill leachates	5-50 µg/L	Santos et al. (2015)

Evaporation ponds

Evaporation basins or ponds are commonly employed for the treatment of saline agricultural drainage to reduce the volume of selenium-contaminated wastewater using solar radiation as the main energy source (CH2M HILL 2010). Wastewater is typically pumped into lined ponds and allowed for water to evaporate until only patches of salt and concentrated Se brine solution is left in the water column, which is still a concern for potential risk of selenium toxicity to the exposed aquatic life and waterfowls (de Souza et al. 2001; Gao et al. 2007). Studies of evaporation ponds (Gao et al. 2000, 2007; de Souza et al. 2001; Ryu et al. 2011) showed that selenium accumulation and transformation mainly occurred via adsorption and immobilization onto surface sediments along the water column (decreasing with depth). Gao et al. (2000) revealed that elemental selenium (46%) and organic matter-associated selenium (34%) were the major selenium species formed in evaporation ponds.

A large scale connected evaporation pond was investigated by Ryu et al. (2011) in San Joaquin Valley (California, USA). The evaporation ponds were separated into 10 segments called cells, with cell 1 denoted as inlet and cells 9/10 as final cells. A significant decrease in total soluble selenium concentration, from an average value of 24.2 µg Se/L to 11.1 µg Se/L, was observed in cell 1 with an increase in selenium concentration in the sediments compared to the final cell 9, where the selenium concentration decreased from 1.94-2.50 mg/kg to 0.38-0.65 mg/kg. This showed active immobilization of selenium in the sediment of the Se loaded cells, but cells 9 and 10 also showed high total dissolved selenium concentrations (average 22.7 µg Se/L) due to evapo-concentration. Results showed that elemental Se and organic matter-associated selenium were mainly found in the sediments of the pond systems. Evaporation ponds have the advantage of low energy costs due to relying on solar radiation and no solid-liquid separation phase is required. However, it has the disadvantage of large space requirements and being dependent on climate conditions.

Chemical methods

Chemical technology mainly involves adsorption and precipitation of selenium species. Santos et al. (2015) recently published a comprehensive review detailing the development of selenium treatment using effective adsorbents (double layer hydroxides or binary metal oxides) and low-cost alternative materials such as unconventional biomaterial sorbents (e.g., algae and agricultural waste). One popular chemical technique for selenium removal is through the use of ferrihydrite as an absorbent. This adsorption method has been noted by the USEPA as the best demonstrated available technology (Golder Associates Inc. 2009), particularly for SeO_3^{2-} ions (Zelmanov and Semiat 2013). Ferrihydrite adsorption is a two-step chemical treatment process: (1) precipitation of ferrihydrite and (2) adsorption of selenium oxyanions on solid ferric oxyhydroxides particles (Ziemkiewicz et al. 2011).

Simultaneous precipitation and adsorption of selenium can occur onto the ferrihydrite surface or iron co-precipitate and was found to be effective in the removal of selenium at neutral to alkaline pH (CH2M HILL 2010). This method is noted for its capability in removing selenium over a wide spectrum of pH, but it is most effective if the predominant selenium oxyanion form is SeO_3^{2-}. Due to the strong affinity of SeO_3^{2-} ions for iron hydroxide surfaces, about 99% SeO_3^{2-} removal can be achieved from the solution at a pH between 3 to 8 (Persico and Brookins 1988), whereas SeO_4^{2-} tends to remain readily exchangeable forming weakly bound complexes (Lussier et al. 2003). Additionally, similar to the ion exchange process, SO_4^{2-} has shown to be a competitor for adsorption sites, thereby suppressing SeO_4^{2-} removal (Ziemkiewicz et al. 2011).

Other chemical treatments are cementation, electrocoagulation and photoreduction which are demonstrated for selenium removal in the laboratory, but yet to be tested in pilot and field scale. Major disadvantages of using chemical techniques include high operational costs and generation of high volumes of concentrated chemical sludge. Because chemical treatment methods work best with SeO_3^{2-}, due to its reactivity and formation of strong and stable bonds with surface ligands (Lussier et al. 2003), wastewater containing SeO_4^{2-} requires additional pre-treatment to reduce SeO_4^{2-} to SeO_3^{2-} before the chemical method can be utilized effectively (Hageman et al. 2013). A study conducted by Saeki and Matsumoto (1998) on selenite sorption onto goethite suggests that the mechanisms of ligand exchange reactions consist of either reaction of surface hydroxyl group oxygen atoms being released into the solution or reaction of selenite oxygen atoms being released from the goethite particles into the solution.

Biological treatment

Biological treatment has emerged in recent years as the leading technology for selenium removal from wastewaters. Biological treatment offers a cheaper alternative with the ability to adapt to variable wastewater characteristics and environmental conditions. Though there are limited full-scale biological treatment plants for selenium treatment currently in place, multiple small-scale experiments (laboratory and pilot-scale) have shown that biological treatment is effective in reaching regulatory limits for selenium discharge into the aquatic environment (Gusek et al. 2008; Golder Associates Inc. 2009; Rutkowski et al. 2013).

Biological treatment methods include reduction and volatilization of selenium oxyanions through different bioreactor configurations, phytoremediation or constructed wetlands. In biological systems, anaerobic and aerobic bacteria, algae, fungi and plants are able to catalyze the removal of selenium under environmental conditions (Golder Associates Inc. 2009; CH2M HILL 2013). Biological processes can use renewable organic carbon sources as electron donor, found either within the waste stream or added externally (Hageman et al. 2013).

In biological treatment of selenium wastewaters, anaerobic reduction is primarily used because the mechanism for SeO_4^{2-} reduction in anaerobic microorganisms is fairly more characterized (biochemistry of SeO_4^{2-} reductase) compared to reduction under aerobic conditions. Though there are studies on the reduction of selenium oxyanions by aerobic microorganisms (Kuroda et al. 2011; Kagami et al. 2013), the studies are performed using pure cultures at laboratory scale and this has not been demonstrated for large scale applications. Additionally, aside from the additional energy cost needed to aerate the bioreactor, dissolved oxygen can hamper the oxyanion reduction as oxygen is preferred as the terminal electron acceptor (Seghezzo et al. 1998; Chan et al. 2009). Furthermore, anaerobic reduction offers possibilities for recovery and re-use of bioreduced selenium from the aqueous phase and biomass.

The basis of the biological Se removal techniques is the conversion of inorganic soluble selenium oxyanions (SeO_4^{2-} and SeO_3^{2-}) to inorganic, insoluble and less toxic biogenic Se^0, with temperature, pH, and electron donor among the controlling factors (Lenz and Lens 2009). Insoluble Se^0 can be removed from the aqueous phase through a number of techniques, such as gravity settling, filtration, electrocoagulation (Staicu et al. 2014) and metal co-precipitation (Mal et al. 2016). Similar to chemical treatment methods, the main issue associated with bioremediation is the interference of other oxyanions such as nitrate and sulfate present in the wastewater as well as the re-oxidation and re-mobilization of bioreduced selenium (CH2M HILL 2010). The former can be overcome through proper cultivation of microbial communities in the bioreactor to allow for the simultaneous reduction of selenium oxyanions and other anions.

Table 2.5 Summary of treatment options available for selenium-laden wastewater: physical, chemical, and natural systems.

	Physical	Physical-Chemical	Physical-Chemical	Chemical-Biological	Bio-electrochemical	Natural System	Natural System
Technology	NF [a] / RO [b]	Selen-IX™ [c]	Ferrihydrite [d]	ZVI (*Enterobacter taylorae*) [e]	EBR coupled with PFAB [f]	Constructed wetlands (*Chlorella vulgaris*) [g]	Algal-bacterial (algae, NB & SeRB) [h]
Mechanism	Filtration	Adsorption	Adsorption + reduction	Biological reduction + adsorption	Electro-biochemical Reduction	Biological reduction + adsorption + volatilization	Biological reduction
Se range Inf.	< 1000 µg Se/L	~450 µg Se/L (85% SeO_4^{2-}, 15% SeO_3^{2-})	40–60 µg Se/L (SeO_3^{2-})	500 µg Se/L	< 1 mg Se/L	1.5 mg Se/L	200–400 µg/L
Se range Eff.	≥ 1 µg Se/L	< 5 µg Se/L	< 10 µg Se/L	< 4 µg Se/L	0.04 mg Se/L	< 7 µg Se/L	72–80 µg/L and 7–12 µg/L (after addition of $FeCl_3$)
Co-contaminant	Other multivalent ions	NO_3^-, SO_4^{2-}, alkalinity	Metals and other nutrients	None	NO_3^- and metals (Cd, Zn, Cu, etc.)	Micronutrients	NO_3^-
% Removal Eff.	> 95%	99%	85–90%	94.5–96.5%	85%, 65% (2 reactors)	96% (SeO_4^{2-}); 61% volatilize	
Operating Conditions	Ambient T, pH 4.4–8.5, 0.01-0.001 µm NF and <0.001 µm RO	Strong base anion exchange resin (SBA), pH 7.0, 5–10°C	pH 4.0–6.0, 14 mg/L Fe dosage (optimal)	ZVI and redox mediator (AQDS) were used to enhance bacterial reduction of SeO_4^{2-} using molasses-sucrose	30° C, low cell voltage (3 V potential), 8 h HRT (total 16 h), molasses (for anaerobic reactor)	Microcosm water column of 72 h HRT	High-rate aerobic-anoxic ponds for algae and anaerobic bacteria

Wastewater (WW) / scale-operation	Agricultural drainage WW / full-scale, USA	Mine-impacted WW / pilot-scale, 2.8 m³/day	Various WW / pilot-scales	Synthetic agricultural drainage WW / laboratory-scale	Gold mine WW / pilot-plant, 2 reactors	Synthetic farm run-off / laboratory-scale	Agricultural drainage water / full-scale, USA
Comments	TDS pre-treatment and post-treatment of concentrated brine solution needed.	Developed by BioteQ with 2 step process removal: (1) ion exchange and (2) spent regenerant is treated with Fe to remove Se as a solid product.	Most effective with SeO_3^{2-} but not with SeO_4^{2-}. Have high cost due to Fe addition and interference from other contaminants. Dubbed by USEPA as the best demonstrated available technology (BDAT).	Disturbed by SO_4^{2-}, PO_4^{3-} and CO_3^{2-} with long term efficiency questionable.	Landusky biotreatment system (LBS) in north-central Montana conventional reactors (anaerobic bioreactor) showed poor efficiency compared with EBR system due to the high conc. of NO_3^- that interfered with Se reduction. EBR also simultaneously reduced metal contaminants.	Se reduction occurs through the sediment by microbial mediated dissimilatory anaerobic reduction with algae Se uptake and subsequent volatilization into the atmosphere.	SeO_4^{2-} cannot be reduced to low levels unless dissolved O_2, NO_3^- are removed.

[a] Nanofiltration (Kharaka et al. 1996) ; [b] Reverse Osmosis (CH2M HILL 2010); [c] Ion Exchange (Mohammandi et al, 2014); [d] Ferrihydrite absorption (Twidwell et al. 2005); [e] Zero-valent Iron (Zhang et al. 2008a; Lenz and Lens 2009); [f] Electro-chemical reactor with plug flow anaerobic bioreactor (Opara et al. 2014); [g] Constructed wetland integrated with algal-treatment (Huang et al. 2013); [h] Algal-bacterial system using algae and anaerobic bacteria: denitrifying bacteria (NB) and selenium-reducing bacteria (SeRB) (Gerhardt et al. 1991).

Table 2.5 (continuation) Summary of treatment options available for selenium-laden wastewater: bioreactors.

		ABMet (microbe consortium) [a]	Envirogen FBR [b]	Methanogenic UASB reactor (granular sludge) [c]	Upflow Fungal Pelleted Reactor (*Phanerochaete chrysosporium*) [d]	Packed-bed medium-reactor (*Thauera selenatis*) [e]	Chemostat reactor (*Bacillus* sp. SF-1) [f]	Aerobic Jar Fermenter (*Pseudomonas stutzeri* NT-1) [g]	H₂-MBfR [h]
Technology		Biological	Biological	Biological	Biological	Biological	Biological	Biological	Biological
Mechanism		Biological reduction	Biological reduction	Biological reduction	Biological reduction	Biological reduction	Biological reduction	Biological + volatilization	Biological reduction
Se range	**Inf.**	57–1800 µg-T. Se/L	20–300 µg-Se/L	0.79 mg Se/L (SeO_4^{2-})	10 mg Se/L (SeO_3^{2-})	160–640 5 µg Se/L (mainly SeO_4^{2-})	41.8 mg Se/L (SeO_4^{2-})	111 mg Se/L (SeO_4^{2-})	260–1000 µg/L (SeO_4^{2-})
	Eff.	0.8–13.7 µg-T. Se/L	4–10 µg Se/L	≥ 28 µg Se/L		5–12 µg Se/L	≥ 50 µg Se/L	21.3 µg Se/L	12–50 µg Se/L
Co-contaminant		High salinity, SO_4^{2-}, and NO_3^-	NO_3^-	SO_4^{2-}	None	NO_3^-	None	None	SO_4^{2-}, NO_3^- and other ions
% Removal Eff.		~ 97%	~ 96%	99% (methanogenic condition) and 97% (SO_4^{2-} reducing condition)	70%	98%	99%	80% (soluble Se); 82% Se recovery in gas phase	~90%
Operating Conditions		pH 7.0, 15–40°C, 2–16 h EBCT using GAC, molasses (carbon source)	0.5–1.0 h HRT, 1- or 2-stage w/ sand filter, 8–10° C	30°C, pH 7.0, and HRT 6 h (both conditions) lactate (carbon source)	pH 4.5, 30°C and 24 h HRT, glucose (carbon source)	23°C, 3.25 h HRT and pH 6.2, acetate (carbon source)	pH 7.8–8.0, 95.2 h HRT, and 30° C.	pH 9.0, 48 h HRT, 38°C, 250 rpm	pH 7.0–9.0, H₂, hollow-fiber membrane

Wastewater (WW) / scale-operation	FGD WW / full-scale, 3 plants in Canada	Mine-impacted WW & Refinery WW / pilot-scales	Synthetic WW / laboratory-scale	Synthetic WW / laboratory-scale	Agricultural drainage WW / pilot scale, 11 m^3	Synthetic WW/ laboratory-scale	Synthetic WW / laboratory-scale	Various WW/ laboratory-scale
Comments	Requires large footprint and pre-treatment in order to remove suspended solids (SS).	More flexible for SS conc. but, pretreatment is still needed and in-bed cleaning frequency would be increase.	Se reduction works better at a high SeO_4^{2-} :SO_4^{2-} ratio (> 8.0×10^{-4}) but Se reduction is done through alkylation and nanoparticle formation requiring careful process control.	Able to handle fluctuating loads of SeO_3^{2-} and shows potential to synthesize Se^0 in continuous operation.	Works under unsterile condition but post treatment necessary for Se^0 and alternative carbon source for acetate must be tested to lower cost.	Bioaugmentation would be needed and SeO_3^{2-} accumulation was observed at low HRT. Additionally, post-treatment needed to remove Se^0 ($FeCl_3$ or ultrafiltration).	Simultaneous reduction of soluble Se and recovery of pure Se^0 through volatile Se trapping. But careful aeration and agitation of the system is necessary in order to fully optimize the recovery of Se.	No residual donor carry-over in the effluent due to use of H_2 but has competition with NO_3^- and membrane biofouling.

[a] ABMet® fixed-film, pack-bed, plug-flow bioreactor (Sonstegard et al. 2007, 2010; CH2M HILL 2010); [b] Envirogen Fluidized Bed Reactor (CH2M HILL 2013); [c] Upflow anaerobic sludge bed reactor (Lenz et al. 2008a, 2008c); [d] Fungal pellet upflow reactor (Espinosa-Ortiz et al. 2015b); [e] Pack-bed 4 tank medium reactor (Cantafio et al. 1996); [f] Chemostat reactor (Fujita et al. 2002); [g] Aerobic reaction using jar fermenter for selenium volatilization (Kagami et al. 2013); [h] Hydrogen-based membrane biofilm reactor (Chung et al. 2006a, 2006b, 2007, 2010; van Ginkel et al. 2011).

Table 2.6 Selected examples of microorganisms capable of reducing selenate under anaerobic conditions.

Organism	Source	Reaction	Electron donor(s)	Reference
Sulfurospirillum barnesii strain SES-3	Freshwater sediments	Respire SeO_4^{2-} and NO_3^-, and reduces SeO_4^{2-} to Se^0.	Lactate $\rightarrow$ Acetate + CO_2	Oremland et al. (1994)
Enterobacter cloacae strain SLD1a-1	Selenium contaminated agricultural drainage water in San Joaquin Valley, USA	Respire SeO_4^{2-} and NO_3^- and reduces SeO_4^{2-} to Se^0 only in the presence of NO_3^-	Glucose	Losi and Frankenberger (1997); Yee et al. (2007)
Thauera selenatis	Seleniferous sediments in San Joaquin Valley, USA	Grows anaerobically using SeO_4^{2-}, NO_3^- and NO_2^-; SeO_4^{2-} is completely reduced to Se^0 only when NO_3^- is present	Acetate	Rech and Macy (1992)
Bacillus sp. strain SF-1	Selenium polluted sediment	Respires SeO_4^{2-} and NO_3^-, and reduces SeO_4^{2-} to Se^0 with transient accumulation of SeO_3^{2-}	Lactate	Fujita et al. (1997)
B. arsenicoselenatis strain E-1H	Mono Lake California, USA (alkaline and hypersaline soda Lake)	Grow anaerobically using SeO_4^{2-} and reducing it to SeO_3^{2-}	Lactate	Blum et al. (2001)
B. selenitireducens strain MLS-10		Grow anaerobically using SeO_3^{2-} and reducing it to Se^0	Lactate, glucose	
E. taylorae	Rice straw bioreactor channel system	Respires SeO_4^{2-} following the order: $SeO_4^{2-} \rightarrow SeO_3^{2-} \rightarrow Se^0 \rightarrow Se^{2-}$	Rice straw	Zahir et al. (2003)

2.4   Biotreatment technologies

2.4.1   Microbial reduction mechanism

Numerous microorganisms capable of tolerating and transforming selenium oxyanions have been isolated from pristine and contaminated environments for better understanding of the mechanism enabling bioremediation purposes. A variety of microbial genera capable of reducing selenium oxyanions have been observed over the past few years (Nancharaiah and Lens 2015a). These selenium reducing microorganisms use various selenium conversions (**Fig. 2.4**), such as reducing toxic selenium oxyanions into non-toxic and insoluble Se^0, volatilization through hydrogen selenide (H_2Se) or dimethylselenide (Me_2Se) as well as conversion to methylselenocysteine (El-Ramady et al. 2014a).

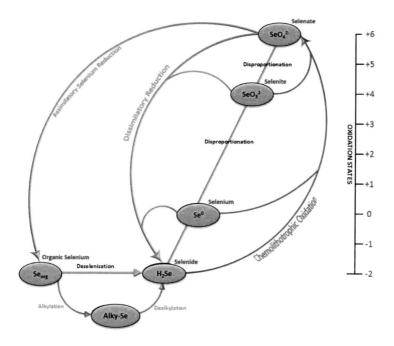

Fig. 2.4 Gradient representation of biological selenium transformations with selenium oxidation states. Pathways are shown through the arrows. Yellow lines indicate reduction while blue lines are for oxidation reactions. The cycle was modeled after sulfur transformation from Sánchez-Andrea et al. (2014) with microbial reduction information taken from Nancharaiah and Lens (2015a).

In most microorganisms (archaea and eubacteria), compounds containing selenium are metabolized along two pathways: (1) dissimilatory reduction of selenium oxyanions to Se^0 and possibly further to selenides or (2) their direct incorporation into amino acids (e.g., SeCys) and then to selenoproteins (Fisher and Hollibaugh 2008). In nature, selenium shows a complete biogeochemical cycle with dissimilatory reduction as a main reduction mechanism catalyzed by anaerobic microorganisms inhabiting anoxic sediments (Oremland et al. 2004). Direct incorporation into seleno-metabolites causes selenium to be substituted for sulfur. This antagonistic relationship can both alleviate selenium toxicity and intensify selenium deficiency in an organism (Kieliszek and Błażejak 2013).

Selenite is more readily reduced to Se^0 as compared to SeO_4^{2-} by a number of identified selenite reducing microorganisms (Losi and Frankenberger 1997), while relatively few selenate-reducing bacteria are known (Maiers et al. 1988; Oremland et al. 1994; Fujita et al. 1997). Selenite reduction has been found in both aerobic and anaerobic microorganisms (Viamajala et al. 2006). Selenite is highly reactive (Ziemkiewicz et al. 2011) and the conversion of SeO_3^{2-} to Se^0 is often mediated through detoxification mechanisms. This can be achieved using glutathione, an abundant thiol in bacteria (Turner et al. 2012), glutaredoxin (Nogueira and Rocha 2011), siderophores (Zannoni et al. 2008), nitrite reductase (Schröder et al. 1997), or fumarate reductase (Li et al. 2014).

In the past two decades, few selenate-reducing bacteria were successfully isolated from different environmental sources (**Table 2.6**). Formation of Se^0 as the major stable end product of selenium oxyanion reduction with methylated Se or selenide as minor by-products were predominately observed in mixed microbial cultures (Hageman et al. 2013). Identified selenate-reducing bacteria have been detected in various bacterial taxonomy such as *Pseudomonas*, *Desulfovibrio* sp., *Wolinella succinogenes*, *Bacillus* sp. and other unidentified obligate anaerobic cocci (Fujita et al. 1997; Nancharaiah and Lens 2015a). These microorganisms showed distinct physiological properties and reduction mechanisms, e.g., *Thauera selenatis* and SES-3, which can grow through the anaerobic respiration of SeO_4^{2-} (dissimilatory selenate reduction), while *Pseudomonas stutzeri*, a facultative anaerobe, reduces SeO_4^{2-} solely for detoxification (Oremland et al. 1994). Nancharaiah and Lens (2015) gave an in-depth review of the ecological role, mechanism and phylogenetic characterization of various selenium-reducing microorganisms and their role in biotechnological applications.

2.4.2 Inocula for bioreactors

Usage of defined cultures of selenium-reducing bacteria (Kashiwa et al. 2000; de Souza et al. 2001; Ghosh et al. 2008) and bioaugmentation with defined selenate-reducing bacteria (Lenz et al. 2009) have been studied for achieving reduction of selenium oxyanions in bioreactors. Compared to defined pure cultures, the use of mixed microbial cultures is more applicable for practical scale bioreactors treating large volumes of selenium-laden wastewater under non-sterile conditions. Mixed consortium of bacteria provide a better opportunity for syntrophic biodegradation and allow for simultaneous reduction of selenium oxyanions along with co-occurring electron acceptors such as NO_3^- and SO_4^{2-}.

Numerous microorganisms capable of reducing selenium oxyanions to Se^0 were found in biological wastewater treatment processes such as activated sludge, denitrifying sludge, as well as sulfate-reducing and methanogenic sludge (Soda et al. 2011). Astratinei et al. (2006) compared the capacity of different anaerobic granular sludges to remove SeO_4^{2-} from wastewater in batch experiments. Methanogenic anaerobic granular sludge was effective to remove high SeO_4^{2-} concentrations (up 10 mM) using different electron donors, i.e. ethanol, methanol, acetate, propionate, lactate, glucose, sucrose, and casein hydrolysate.

2.4.3   Bioreactors

Biological treatment has been evaluated for removing selenium from various waters, including mine runoff, agricultural drainage and flue gas desulfurization wastewaters from coal-fired power plants. Biological treatment systems are less expensive, do not use expensive or hazardous chemicals, avoid formation of chemical sludge and offer possibilities for Se recovery. Bioreactors allow for the use of various microorganisms/seed sludges for selenium (along with other co-oxyanions) removal without the use of chemical means. Bioreactors also offer flexibility in terms of manipulating operating parameters. As such, bioreactor systems have emerged as the new "best practice" for treatment of selenium wastewaters and are setting the standard for treatment of wastewaters generated in mining, power generation, agricultural drainage, and other industries (Pickett and Harwood 2014). **Table 2.5** presents the summarized information of the bioreactor configurations and operating conditions.

Biofilm reactors

Biofilm based bioreactors (i.e., static or moving bed filter, fluidized bed reactors, and rotating biological contactors) have been explored over the years for bioremediation of metal-laden wastewaters. Biofilm reactors are attached growth systems wherein microbial communities grow on the surface of a solid substratum (e.g., rock, granular activated carbon, or sand) that is dampened by the wastewater. Biofilm microorganisms produce extracellular polymeric substances (EPS) that allow for the biofilm bacteria to better attach themselves to the carrier material. Apart from their role in cell-substratum and cell-cell interactions, EPS play a pivotal role in entrapment and complexation of metal cations and biomineralization of Se^0 precipitates. Due to the growth of suspended microorganisms, these reactors rely on additional liquid-solid separation, e.g., plate separator, gravity clarification or membrane separation (CH2M HILL 2010). Treatment of selenium wastewaters has been tested mainly in three biofilm reactor configurations: ABMet®, fluidized biofilm reactor (FBR) and membrane biofilm reactors (MBfR).

ABMET® biological technology

The ABMet® (advanced biological metals removal) biological system is a full-scale fixed-film, packed-bed anoxic bioreactor treatment process that has been patented and pioneered by General Electric (GE), USA. The ABMet® technology is both a biological and filtration system where the bioreactor is filled with porous granular activated carbon. This technology uses a molasses based solution as electron donor for the reduction of selenium oxyanions by naturally occurring, non-pathogenic consortium of bacteria. The bacterial consortium are able to thrive in the harsh water chemistry of wet flue-gas desulphurization (WFGD) as they were isolated from this contaminated site (Sonstegard et al. 2010). This bacterial consortium requires only a nominal nutrient supplement, thus resulting in low operating and maintenance costs. The granular activated carbon (GAC) bed serves as a substratum for bacterial growth, where they form a biofilm and prevent it from being washed out from the reactors. In addition, the technology was able to remove co-contaminants such as NO_3^- and heavy metals. The plug-flow design of the system allows development of an oxidation reduction potential (ORP) gradient ranging from 0 to -200 mV along the vertical axis of the reactor tank facilitating reduction of oxyanions with varied redox potentials (Sonstegard et al. 2010).

The ABMet® technology is able to reduce and precipitate selenium within 2 to 16 h empty bed contact time (EBCT), where EBCT refers to the residence time of the water in the reactor assuming there was no biofilm or solids present in the bioreactor. The bioreduced selenium is removed from the bioreactor by a backwash cycle. So far, ABMet® systems have been installed for two power plants at Duke Energy and Progress Energy in North Carolina for treating waters from WFGD units. Sonstegard et al. (2008) reported that from the beginning of June 2008, total selenium effluent values were consistently < 10 µg Se/L for an average influent concentration of 1500 µg Se/L and attaining an average removal efficiency of 99.3%. Nitrate concentrations were decreased from 5-100 mg NO_3^--N/L to 0.1 mg NO_3^--N/L.

The ABMet® technology has the advantage of being commercially available from GE as well as the ability to utilize molasses, which is a cheap carbon source. Additionally, this technology was reported to be independent of feed water temperature showing consistent selenium removal throughout the year (Sonstegard et al. 2010). However, this technology has the disadvantage of large footprint requirements due to the low HRT, pre-treatment requirements to remove high suspended solids (SS) loads, biomass loss due to washout and competition for carbon source due to the presence of other ions.

Envirogen fluidized bed reactor (FBR) system

The Envirogen FBR system makes use of a solid carrier material where wastewater can be passed through at high velocities to suspend the media through re-circulation, such that it maintains a plug flow mixing regime with a suspended bed in the column (CH2M HILL 2013). The FBR is similar to the ABMet® technology with only having the flow configuration as fluidized bed as opposed to a down-flow system applied in the ABMet® technology. In an FBR, a large amount of biomass is generated while maintaining thin films, allowing for higher mass transfer rates and high volumetric efficiencies. The solid substratum includes sand or activated carbon, allowing for large specific surface areas for microorganisms to grow while creating anoxic/anaerobic conditions for SeO_4^{2-} and SeO_3^{2-} reduction. This system was shown to work at a minimum temperature of 10°C, and with a short HRT ranging between 0.5 to 1.0 h (based on expanded bed height). The system can be designed as either single-stage or two-stage. Both system design and HRT largely depend on the NO_3^- content of the wastewaters and the required ORP range of -150 to -200 mV. A high total selenium concentration was noticed in the effluent of the FBR effluent due to biofilm detachment. Additional liquid-solid separation is needed to remove the biomass and biogenic selenium from the effluent. A bench-scale FBR was used to treat mining water with 5 to 100 mg/L NO_3^--N and 20 to 300 µg Se/L. Nitrate and selenium concentrations were decreased to < 0.1 mg/L for NO_3^--N and 4-10 µg/L dissolved selenium (CH2M HILL 2013). This technology has the main advantage of having a small footprint and lower installation cost due to the complete mixing within the reactor, but has similar disadvantages as the ABMet® technology (carbon source competition and biomass loss).

Membrane biofilm reactor (MBfR)

Several studies (Chung et al. 2006a, 2006b, 2007, 2010; van Ginkel et al. 2011) have reported high SeO_4^{2-} removal rates by employing a H₂ based membrane biofilm reactor

(MBfR). The MBfR contains a bundle of bubble-less gas transfer membranes that allow for H_2 gas to travel to the interface of the hollow-fiber membrane and biofilm consisting of autohydrogenotrophic bacteria, e.g. *Cupriavidus metallidurans* (van Ginkel et al. 2011). Microorganisms harvest energy by coupling reduction of selenium oxyanions to H_2 oxidation with the following stoichiometric reaction: $SeO_4^{2-} + 3 H_2 + 2 H^+ \rightarrow Se^0 + 4 H_2O \quad \Delta G^0 -$ 71 KJ/e$^-$ (Chung et al. 2006a). Supplying H_2 directly to the substratum-biofilm interface allows effective removal and reduction of SeO_4^{2-} (Chung et al. 2006b). The membrane can be made from both organic and inorganic materials as sheet or hollow-fiber bundles and is the main cost component of MBfRs.

Hollow-fiber membranes are used since they can be operated at high gas pressures without bubbling and also offer a smaller outside diameter of 0.1 mm while providing a higher surface-to-volume ratio. Membranes made of hydrophobic materials are preferred to keep the pores dry permitting for gas molecules to diffuse quickly compared to liquid filled pores, thus improving the mass transfer and reducing the potential for biofouling.

In the study by Chung et al. (2006a), SeO_4^{2-} removal increased over the first three weeks with an influent SeO_4^{2-} concentration range of 260-1000 µg/L, resulting in an average of 12 µg/L total selenium in the effluent. Furthermore, both SeO_4^{2-} and SeO_3^{2-} were reduced to Se^0 between pH 7.0-9.0. Nitrate and SO_4^{2-} were also reduced within the reactor, although SO_4^{2-} reduction was affected by hydrogen gas pressure and SeO_4^{2-} concentration. This system has the major advantages of removing several oxidized toxic contaminants (i.e., chlorate, nitrate, chromium, arsenic and selenium) and has no residual COD carry-over since H_2 gas is used as the electron donor. However, the MBfR system has the disadvantage of high cost due to H_2 gas usage and storage issues, membrane biofouling and possibility of dysfunctioning under low pH conditions. Thus far, no pilot-scale studies have been conducted.

Sludge-based bioreactors

Upflow anaerobic sludge bed (UASB) reactor

Upflow anaerobic sludge blanket (UASB) reactors are the most implemented process for anaerobic treatment of industrial effluents worldwide (Nnaji 2014). The success of the UASB concept relies on the establishment of a dense sludge bed at the bottom of the reactor, in which the biological processes take place (Seghezzo et al. 1998). Under certain conditions, natural aggregation of the bacteria in flocs or granules occurs allowing for the formation of millimeter sized granules that have good settling properties (Nnaji 2014). Therefore, limited biomass wash-out occurs from the system under full-scale reactor conditions (Bhunia and Ghangrekar 2007). This allows for the reactor to have good performance even at high organic loading rates due to the retention of active anaerobic sludge, whether granular or flocculent. Additionally, good contact between the wastewater and biomass is maintained through the turbulence caused by the upflow influent flow as well as the biogas produced by the anaerobic granular sludge (Seghezzo et al. 1998). Precipitated selenium particles can be incorporated into sludge granules or suspended in the treated wastewater, from which Se^0 can be recovered in a post-treatment filtration or settling step (Staicu et al. 2015a).

Optimal design considerations for UASB reactors should take into account the water composition, loading rates, upflow velocity and pH of the incoming wastewater. Golder Associates Inc. (2009) reported that in the early 1900s pilot-scale UASB reactors (operated by Adams Avenue Agricultural Drainage Research Center) were operated to remove selenium from San Joaquin Valley agricultural drainage for over three years. Selenium and NO_3^- concentrations in the influent were 500 µg Se/L and 3 mg NO_3^--N/L. Methanol was fed as electron donor into the system at 250 mg/L for selenium oxyanion reduction and microbial growth. The selenium removal efficiency varied from 58% to 90%, while NO_3^--N was completely removed (Golder Associates Inc. 2009).

A series of mesophilic (30°C) lab-scale UASB reactors were investigated for the removal of SeO_4^{2-} (Lenz et al. 2008a, 2008b, 2008c, 2009). Lenz et al. (2008b) investigated the removal of SeO_4^{2-} (790 µg Se L^{-1}) from synthetic wastewater in both methanogenic and SO_4^{2-} reducing conditions using lactate as the electron donor. It was shown that SeO_4^{2-} was successfully reduced to 24 µg Se/L in the SO_4^{2-} reducing reactor, while a concentration of 8 µg Se/L was attained in the methanogenic reactor. It was further concluded that selenium removal efficiencies were limited by the SO_4^{2-} loading rate with a recommended ratio of SeO_4^{2-} to SO_4^{2-} exceeding 1.92×10^{-3} to minimize competition with SO_4^{2-} and maintain complete SeO_4^{2-} removal. Additionally, in the methanogenic bioreactor, both SeO_4^{2-} reduction and total dissolved selenium (detected in the effluent) were inhibited by sulfide (HS^-) at a concentration of 101 mg/L.

Interestingly, formation of methylated selenium compounds (DMSe 91 µg Se/L and DMDSe 11 µg Se/L) was noticed at lower temperature (20°C) amounting to a maximum of 15% of the initial selenium concentration (Lenz et al. 2008c). However, a drastic decrease in SeO_4^{2-} reduction efficiency was also observed at this temperature. It was further concluded that careful process control is required in UASB reactors since any change in operational conditions can induce an elevation of the effluent selenium concentration through alkylation or colloid formation.

UASB reactors can be bioaugmented with a SeO_4^{2-} reducing microbial strain like *Sulfurospirillum barnesii*. Reduction of SeO_4^{2-} was possible in such a bioaugmented UASB reactor in the presence of elevated NO_3^- (1500 times) and SO_4^{2-} (200 times) concentrations in comparison to SeO_4^{2-} (790 µg Se/L), which are the concentrations typically found in agriculture drainage waters (Lenz et al. 2009). Addition of the SeO_4^{2-} reducing strain to the UASB reactor allowed for simultaneous and improved removal of NO_3^- and SeO_4^{2-}. This also demonstrated that the self-immobilization of specialized bacterial strains can supersede wash-out and out-competition of newly introduced strains in continuously bioaugmented bioreactor systems.

Dessì et al. (2016) showed that biological reduction of SeO_4^{2-} under thermophilic conditions (55˚C) can be an option to have better retention of biogenic selenium and increased selenium removal efficiency, challenges that are commonly encountered during the treatment of selenium laden wastewater in UASB reactors. The study showed that UASB bioreactors are able to treat high temperature conditioned wastewater such as WFGD scrubbing waters without the need for cooling. Different operating conditions gave different selenium removal efficiencies: 94% and 70% under thermophilic conditions (55˚C), while 82% and 43% under mesophilic conditions (30˚C), in absence and presence of 5 mM nitrate, respectively. In addition, different biogenic Se^0 developed: nanospheres and nanorods in mesophilic and thermophilic conditions, respectively. Furthermore, thermophilic conditions showed better retention of the biogenic selenium and therefore achieving a higher total Se removal efficiency as compared to the mesophilic conditions.

UASB reactors have the major advantage of allowing for high wastewater loading rates while using a smaller footprint compared to other systems, no major loss of biomass or attached growth carrier material due to the granular sludge configuration, while having similar disadvantages as the other bioreactor configurations (COD competition).

Sequencing batch reactor (SBR)

A sequencing batch reactor (SBR) is a bioreactor where the biodegradation and solid separation occurs in one reactor. Most anaerobic-aerobic SBR systems are exploited in the treatment of textile wastewater due to their efficient color and COD removal performance (Chan et al. 2009). A SBR system is operated with batch steps of fill, react, settle, decant and idle in a cyclic operation with a complete aeration or alternating anaerobic-aerobic phase during the reaction time period (Chan et al. 2009). Enrichment of desired microbial populations can be accomplished by the alternating anaerobic-aerobic phases.

Kashiwa et al. (2000, 2001) investigated the ability of *Bacillus* sp. SF-1 to reduce high concentrations of SeO_4^{2-} (20 mM) and SeO_3^{2-} (2 mM) in a lab-scale sequencing batch reactor system using synthetic sewage supplemented with a variety of electron donors, i.e., acetate, citrate, lactate, pyruvate, fructose, glycerol, glucose, and sucrose. Up to about 2 mM (160 mg/L) of soluble selenium can be removed without inhibitory effects under anaerobic conditions at a removal rate of 40 $g/m^3 \cdot d$ in the SBR within a 24 h cycle. However, transient accumulation of SeO_3^{2-} was observed in the reactor which may lead to toxicity to the bacteria (Kashiwa et al. 2001). Chemical precipitation with $FeCl_3$ allowed removal of SeO_3^{2-} from the wastewater but led to the generation of excess sludge. Additionally, the co-presence of NO_3^- at 1 or 5 mM delayed the reduction of SeO_4^{2-} (1 mM), while the co-presence of SO_4^{2-} at 20 mM showed no effect.

Rege et al. (1999) tested the ability of a denitrifying bacterial consortium obtained from the Pullman (Washington, USA) wastewater treatment facility for the reduction of SeO_4^{2-} and SeO_3^{2-} in a 1.5 L volume sequencing batch reactor using acetate as the electron donor. Onset of SeO_3^{2-} or SeO_4^{2-} reduction was observed only after a 150 h lag phase during which all of the NO_3^- and NO_2^- were completely denitrified. The SeO_4^{2-} kinetic reduction rate (0.93 $L/g \cdot mol \cdot min$) was determined to be considerably lower than the SeO_3^{2-} kinetic reduction rate (4.11 $L/g \cdot mol \cdot min$).

SBR have several advantages such as good removal efficiency obtained even with minimum operator intervention and flexibility in incoming water flow. Most importantly, the SBR configuration allows integration of both aerobic and anaerobic conditions in the same reactor tank, establishing different treatment conditions without the need for another tank and, as a result, allows for a smaller footprint. However, the SBR has the main disadvantage of a complex operation set-up and difficult control of the anaerobic-aerobic microbial consortia (Chan et al. 2009).

Fungal-pellet bioreactor

Compared to bacterial reduction, studies on selenium oxyanion reduction by fungi are limited. Recently, reduction of selenium oxyanions (2-10 mg/L Se) by fungal pellets of *Phanerochaete chrysosporium* was studied by Espinosa-Ortiz et al. (2015a). This was taken further by placing the fungal pellets in an upflow bioreactor supplied with oxygen at a 24 h

HRT, pH 4.5, 30°C, under sterile condition and at a glucose loading rate of 5 g COD/L·day (Espinosa-Ortiz et al. 2015b). A 70% removal efficiency of the total soluble selenium was achieved from a SeO_3^{2-} concentration of 10 mg/L with the reactor performance unaffected when the influent selenite concentration was doubled. Selenite reduction coupled to intracellular accumulation of Se^0 nanoparticles in the fungal filaments was observed. The use of fungi has an advantage in terms of growing in harsher conditions, i.e., low moisture content, acidic conditions (pH < 4-5) and low nitrogen or phosphorus concentration (Espinosa-Ortiz et al. 2015b). But it has the disadvantages of only SeO_3^{2-} reduction and not SeO_4^{2-}, possible outcompetition of the fungi by bacteria when treating (unsterile) wastewater and added costs for aeration.

2.4.4   Other biotreatment configurations

Volatilization of selenium

Volatilization of selenium has attracted interest mainly because selenium is transformed into the gaseous phase which enables selenium recovery through gas trapping. However, the slow rates of microbial volatilization limit the application of this approach for the treatment of high volumes of wastewaters generated in industrial operations. The toxicity of the volatile selenium compounds is another concern when adopting these technologies.

Most research on aerobic reduction of selenium oxyanions has focused on the isolation of strains able to work under aerobic conditions. *Pseudomonas stutzeri* strain NT-I, isolated from a drainage wastewater of a selenium refinery plant (Kuroda et al. 2011), was able to efficiently reduce high concentrations of SeO_4^{2-} (5-11 mM) to Se^0 with limited SeO_3^{2-} accumulation under aerobic conditions. Kagami et al. (2013) further investigated this strain in a 5 L jar fermenter attaining simultaneous reduction of selenium oxyanions (80%) and recovering volatilized selenium (82%) through gas trapping of the volatilized selenium with 70% HNO_3 in 48 h at a rate of 14 µmol/L·h. Despite the potential for selenium treatment in volatilization, aerobic reduction and volatilization studies are limited and their applicability for wastewater treatment is unclear.

Algal-pond systems

An algal-bacterial selenium removal (ABSR) system was utilized to treat agricultural drainage water containing NO_3^- and selenium oxyanions (Gerhardt et al. 1991). The ABSR system allows for the use of microalgae as the carbon and energy source for bacteria which perform the selenium bioreduction. In the ABSR system, the influent is introduced into a high rate algal pond (algal growth pond) containing an average algal concentration, measured as volatile suspended solids, of 178 mg/L allowing for partial uptake of NO_3^--N by the microalgae. Thereafter, the algal-laden water (200-400 µg/L Se) is sent to an anoxic reduction unit with both denitrifiers and selenate-reducing bacteria using the algal biomass as the carbon feed source for the bacteria (Gerhardt et al. 1991). The last step of the system is the solid-liquid separation of the elemental selenium particles, biomass and excess algae prior to discharge. Nitrate removal was about 90%, attaining a NO_3^- concentration in the effluent to be less than 10 mg/L, while SeO_4^{2-} was completely reduced to SeO_3^{2-}. Addition of ferric chloride was introduced to the anoxic unit in order to further reduce SeO_3^{2-} to 7-12 µg/L. Methane was recovered at an average of 0.16 L CH_4 per g of volatile solid introduced from the fermentation of the unused algae.

Algal systems have the potential to be used as a pre-treatment unit for a constructed wetland. Huang et al. (2013) investigated this combination through the use of microcosm water columns. Their study showed about 96% selenium removal (initial SeO_4^{2-} concentration was 1580 µg Se/L) with 61% of that accounted to selenium volatilization from the microcosm water column using *Chlorella vulgaris* with a 72 h retention time. The study suggested that the combination of an algal system as pre-treatment for constructed wetlands can improve the overall selenium removal efficiency of the treatment system as well as abate the potential bioaccumulation risk of selenium in the wetlands.

The main advantages of incorporating algae systems in selenium treatment are the potential low-cost technology-wise, *in situ* application and potential harvesting of algae for re-use applications (nutrient recovery and food or fertilizer sludge supplement). Disadvantages are the additional nutrients required for algal growth, climate dependency and solid-liquid separation difficulties (separating algae from effluent water).

2.5   Challenges in selenium biotechnologies

2.5.1   Toxicity of selenium oxyanions to biomass

Toxicity studies of selenium to sludge used in bioreactors, particularly in long-term operation have, to the author's knowledge, not yet been investigated. One study by Lenz et al. (2008b) investigated the inhibitory effect of both selenite and selenate on hydrogenotrophic and acetoclastic methanogenesis in anaerobic treatment with simultaneous methane production. The study concluded that methane production can be hindered by the presence of high concentrations of selenium oxyanions, where selenite completely and irreversibly inhibited methanogenic activity upon single exposure at concentrations exceeding 0.01 mM while selenate inhibited methanogens after repeated exposure. Jain et al. (2015b) similarly recognized selenite's toxicity to activated sludge for the crashing of a lab-scale activated sludge reactor treating 1 mM selenite, 2 g/L COD glucose and > 4 mg/L dissolved oxygen. Further studies should be done on selenium toxicity to bioreactor sludge/biomass. This is important to consider in scaling-up bioreactors to large-scale processes and to gather relevant information on operational conditions, such as how long it would take for sludges to acclimatize to selenium, biomass exposure time to Se before reactor performance failure due to Se toxicity as well as how long to achieve biomass Se oversaturation due to Se bioaccumulation.

2.5.2 Removal and recovery of biogenic selenium (Se^0) nanoparticles

A major issue is that the biological reduction of selenium oxyanions results in the production of colloidal Se^0 nanoparticles that will be partially present in the bioreactor effluent (Buchs et al. 2013; Jain et al. 2015a; Staicu et al. 2015b). Studies in the San Francisco Bay area have shown that colloidal selenium can exhibit toxicity on filter-feeding mollusks and the food web (Schlekat et al. 2002; Purkerson et al. 2003; Presser and Luoma 2006). Moreover, colloidal Se^0 has the potential to reoxidize back to selenium oxyanions when discharged to oxic surface waters with higher redox potentials (Zhang et al. 2004). Additionally, the initial selenium content as well as selenium conversion rates determine the concentration of colloidal Se^0 produced in the anaerobic granular sludge in UASB bioreactors (Staicu et al. 2015a).

Mal et al. (2016a) investigated the toxicity of chemically and biologically formed Se^0 on zebrafish embryos. The lethal concentration of biogenic Se^0 where 50% of the zebra embryos were negatively impacted (LC_{50}) was 1.77 mg/L which was 3.2 and 10 fold less toxic compared to selenite (LC_{50} 0.55 mg/L) and chemically synthesized Se^0 (LC_{50} 0.16 mg/L). The study also confirmed that the EPS layer formed on the surface of the biogenic Se^0 was a contributing factor in lowering the toxic impact to zebrafish embryos by decreasing the bioavailability of the nanoparticles to the embryos. This study indicates that despite the issue of colloidal selenium discharge, biogenic Se^0 is still less toxic to the aquatic biota compared to selenium oxyanions (SeO_3^{2-}) and chemically synthesized selenium nanoparticles.

Staicu et al. (2015b) have assessed the solid-liquid separation of colloidal Se^0 by different methods such as filtration, centrifugation, coagulation-flocculation and electrocoagulation. The study concluded that colloidal selenium can be effectively settled through coagulation-flocculation by using aluminum sulfate as a coagulant. However, if the concentration of nanoparticles in the effluent is too high, a second treatment step is required to reach the discharge limit, thus increasing the operating costs. It was also reported that a rise of temperature from mesophilic to thermophilic can increase crystallinity (Lee et al. 2007; Zhang et al. 2012; Jain et al. 2015b) and size (Lee et al. 2007; Tam et al. 2010) of the selenium nanoparticles, allowing them to settle without the addition of a coagulant. However, increased energy costs are incurred in operating bioreactors at higher temperatures. Additionally, the biologically produced Se^0 particles are not pure and are either retained in the biomass or coated by a layer of EPS that provides them poor settling properties and thus colloidal stability (Hageman et al. 2013).

Losi and Frankenberger (1997) and Fujita et al. (2002) have suggested to use biological reduction to produce Se^0 with the additional post-treatment of sedimentation/filtration and centrifugation/ultrafiltration as a means to separate selenium particles, adding to the cost of the remediation process. Viamajala et al. (2006) on the other hand made use of immobilized cultures within a packed-bed bioreactor observing both SeO_3^{2-} reduction and immobilization of Se^0 within the immobilized cell matrix (inside the cell or formed extracellularly trapped within the biofilm). Another study conducted by Gonzalez-Gil et al. (2016) used anaerobic granular sludge and observed that biogenic selenium were deposited within the surface of granules and largely within the bacterial cells, thereby allowing for a better retention of the Se^0 within a continuous system. However, though elemental selenium immobilization within bacterial cells might be useful in retaining selenium particles within the bioreactor, there is still the question of how to remove and recover the selenium nanoparticles. Additionally, eventual release of the biogenic selenium can be expected by damage or rapturing of the bacterial cells, as was observed in the study by Gonzalez-Gil et al. (2016). As such, alternative processes or improvements on the bioreactor configuration are required in order to achieve complete removal and recovery of selenium particles for re-use.

Oremland et al. (2004) found that biogenic selenium nanospheres have a lower band-gap energy compared to chemically produced Se^0, making them of interest in the realm of nano-photonics. Since various bacteria can produce biogenic selenium nanoparticles at ambient operational conditions (Mal et al. 2016b), their synthesis can essentially eliminate the need for employing harsher means, such as high temperatures and pressures, and the use of hazardous chemicals (e.g., hydrazine) allowing for a "green" nanoparticle synthesis technique compared to current physico-chemical methods (Narayanan and Sakthivel 2010). Biogenic selenium can thus be a source for meeting future selenium scarcity, with the additional benefit of remediation (Jacob et al. 2016).

2.5.3 Wastewater composition: presence of NO_3^- and SO_4^{2-}

Nitrate and sulfate are common constituents in most industrial wastewaters such as mining effluents, in which selenium is also present (Johansson et al. 2015). Nitrate and SO_4^{2-} concentrations present in wastewater are 100 to 1000-fold higher compared to selenium oxyanions concentrations (**Table 2.7**). They can affect SeO_4^{2-} reduction and thus decrease the selenium removal efficiency of bioreactors (Lenz et al. 2009). The majority of studies have focused on SeO_4^{2-} reduction in the presence of either NO_3^- or SO_4^{2-}.

The theoretical hierarchy of oxyanion reduction based on thermodynamic calculations shows that NO_3^- will be reduced first followed by selenium oxyanions, while SO_4^{2-} will not compete with selenium oxyanion reduction (**Fig. 2.5**). But contradictory results on inhibition or stimulation of selenium oxyanion reduction by the presence of either NO_3^- and/or SO_4^{2-} have been reported. Lai et al. (2014) showed that the reduction of SeO_4^{2-} is dramatically inhibited by the presence of NO_3^- at a surface loading rate higher than 1.14 g N/m·day in a H_2-MBfR (10 mg/L fed in the reactor). Moreover, the presence of NO_3^- will reshape the SeO_4^{2-} reducing microbial community (Lai et al. 2014). On the other hand, Oremland et al. (1999) showed that the presence of NO_3^- enhanced the ability of *S. barnesii* to reduce SeO_4^{2-} by keeping the cells at a constant state of high metabolic activity. Conversely, the reduction of trace concentrations of SeO_4^{2-} by *Desulfuvibrio desulfuricans* subsp. *Aestuarii* was inhibited by SO_4^{2-} at concentrations of 1 mM and higher (Zehr and Oremland 1987). However, Lenz et al. (2009) observed that SO_4^{2-} (2 mM) was reduced simultaneously with SeO_4^{2-} (10 μM), but only after NO_3^- (15 mM) was fully denitrified in a UASB reactor inoculated with anaerobic granular sludge bioaugmented with *S. barnesii*.

Few studies have investigated SeO_4^{2-} reduction in the presence of both NO_3^- and SO_4^{2-}. The bioreduction outcomes of NO_3^- and SO_4^{2-} in terms of alkalinity and HS^- on SeO_4^{2-} reduction, co-reduction of different electron acceptors (NO_3^-, SO_4^{2-} and SeO_4^{2-}), ecological interactions between different types of microorganisms and effects of different operational parameters on the reduction capability of microorganisms are unknown. Optimizing reactor operational parameters for establishing and maintaining a microbial community with co-existing denitrifying, SO_4^{2-} reducing and SeO_4^{2-} reducing microorganisms has yet to be fully understood and operated in a full-scale bioreactor for treating selenium-rich wastewaters.

Table 2.7 Selenium-laden wastewater composition (co-contaminants) from different industrial sources.

	Mining-impacted WW	Se refinery plant WW	Flue-gas desulfurization WW	Agricultural WW
Se (mg/L)	0.002 – 12	up to 620	0.015 – 162	~0.35
pH	2.1 – 6.6	0.8 – 1	4.5 – 5.5	7.0 – 8.0
Alkalinity (mg/L)	730	n.i.	10 -250	300
Anions				
SO_4^{2-} (mg/L)	525 – 6,837	0 – 35.6 (mg-S/L)	3,000 – 20,000	607 – 10,100
NO_3^- (mg/L)	255	n.i.	1 – 400	3.2 – 234
Heavy metals				
As (mg/L)	0.512	1.4 – 2.4	0.0075 – 341	
Fe (mg/L)	514	19 – 288		0.04 – 12
Cu (mg/L)	223	n.i.	0.04 – 251	n.i.
Pb (mg/L)	n.i.	n.i.	0.01 – 527	n.i.
Hg (mg/L)	n.i.	n.i.	0.073 – 39	n.i.
Cd (mg/L)	1.38	n.i.	0.005 – 81.9	n.i.
Zn (mg/L)	n.i.	n.i.	0.01 – 5070	n.i.
References	Tabak et al. (2003); Lenz and Lens (2009); Smith et al. (2009); Mohammandi et al. (2014); Santos et al. (2015)	Fujita et al. (2002); Soda et al. (2011)	Lenz and Lens (2009); Smith et al. (2009); Meawad et al. (2010)	Green et al. (2003); Smith et al. (2009)

*n.i. – No Information

It is unclear to what extent selenium oxyanion reduction is caused by denitrifying bacteria and/or sulfate-reducing bacteria (SRB). While previous studies have shown that both SRB and denitrifying bacteria can reduce SeO_4^{2-}, there is an apparent need to carefully control the concentration of both NO_3^- (Lortie et al. 1992; Takada et al. 2008) and SO_4^{2-} (Lenz et al. 2008a) to avoid inhibitory effects on selenium oxyanion reduction.

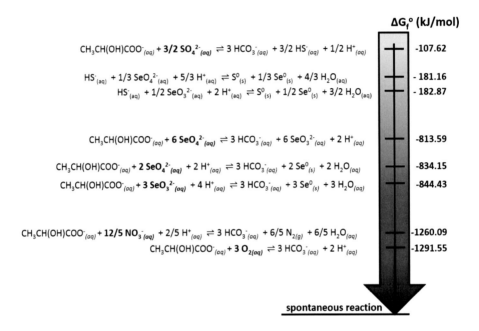

Fig. 2.5 Theoretical calculated Gibbs free energy for typical co-anions reduction (oxygen, nitrate, and sulfate) present in selenium-laden wastewater compared to selenate and selenite reduction using lactate as electron donor at pH 7.0 and oxyanion concentration of 1 M. Abiotic reaction of selenium oxyanions with sulfide is also included.

Thus far, no reports have been concluded on the proper ratio of NO_3^-, SO_4^{2-}, and SeO_4^{2-} for simultaneous reduction. **Chapter 3** investigated the effect of different concentrations of nitrate and sulfate on selenate reduction using anaerobic granular sludge in serum batch bottles. Selenium removal was slowed down with nitrate concentrations > 7 mM (70 molar ratio NO_3^- /SeO_4^{2-}) and sulfate concentrations > 10 mM (100 molar ratio SO_4^{2-}/SeO_4^{2-}) were present. On the other hand, at molar ratios of $\leq$ 70 NO_3^-/SeO_4^{2-} and 100 SO_4^{2-}/SeO_4^{2-}, a substantial increase in Se reduction efficiency ($\geq$ 30%) was observed compared to reduction with just selenate alone. It was further shown in a continuous UASB reactor operated at ambient temperature, selenate removal was not hindered by high concentrations of nitrate (4 mM) and sulfate (15 mM), achieving a removal efficiency of 77% for total selenium and 86% for total dissolved selenium with coupled to 97% nitrate and 29% sulfate removal efficiency. The study demonstrated that concentrations of nitrate and sulfate as an additional factor in improving or lowering Se removal efficiencies in bioreactors.

2.5.4   Operating parameters

There is no universal treatment configuration that can remediate selenium-laden wastewaters, particularly since most wastewater compositions are industry-specific. Despite bioreactors being well studied and, in some, optimized, there are still challenges in long-term full-scale operating bioreactors. Though there is a full-scale biological based selenium treatment system commissioned on February 2013 (for surface mining water in USA) that showed a compliant treatment of a 4.7 µg/L monthly average discharge selenium limit, there is still a lack of full-scale and long-term consistent biotechnologies demonstrating efficient selenium removal to less than 5 µg/L.

One of the major drawbacks in selenium bioremediation is the long time needed to achieve high removal efficiencies. In most bioreactors, a HRT ranging from a minimum of 6 h to more than 48 h was needed. The required HRT depends on the selenium concentration, selenium species and the presence of other co-contaminants that can interfere with selenium bioreduction. Systems, such as wetlands, need a long-term monitoring plan of the selenium concentration in order to fully assess the effectiveness of this technology in remediating selenium contaminated waters. Though this method is low cost, the cost of treating or processing the contaminated medium after remediation might offset the benefits of this system.

There is also the challenge of the additional cost of the electron donor since most industrial wastewaters have a low carbon content, requiring addition of an external source for energy, carbon and electrons such as lactate, acetate and glucose (Zhang and Frankenberger 2003). It is therefore important to select an inexpensive, but effective, organic electron donor for reducing the cost of selenium remediation. Losi and Frankenberger (1997) demonstrated that SeO_4^{2-} at 10 mg/L was completely reduced by *Enterobacter cloacae* SLD1a-1 in 8 h using glucose, while Zahir et al. (2003) and Zhang and Frankenberger (2003) used rice straw attaining 95% SeO_4^{2-} reduction, with Se^0 transformation to Se^{2-} of 5% and 75% in 9 and 14 days, respectively. Zhang and Frankenberger (2005) and Zhang et al. (2008) investigated the use of molasses (from sugar cane containing high levels of sucrose) as a cheap carbon source for treatment of agricultural containing SeO_4^{2-} drainage wastewater. The research showed that more than 92% of the SeO_4^{2-} was successfully reduced to Se^0 when passed through an agar-coated sand column flow-through bioreactor for 45 days using 0.05-0.01% molasses in the influent colonized by *E. taylorae*. The full-scale ABMet® system also uses molasses as the electron donor for the treatment of selenium wastewaters (**see section 2.4.3.1.1**). However, aside from molasses, there is still a lack of an alternative cheap electron donor for selenium reduction as well as knowledge on how to maximize electron donor utilization in bioremediation of selenium-laden wastewater.

Another parameter in bioremediation of selenium-laden wastewater is the pH. Redox conditions vs. pH have shown that the calculated Gibbs free energy per electron for the selenium reduction increases with higher pH and therefore, a near neutral or alkaline system is desired for microbial reduction. So far, most studies have adjusted the pH to neutral during laboratory testing or used buffer medium to avoid large pH changes. This would indicate that additional steps are required when translated to full-scale application, especially since most mining wastewaters are acidic at source (Sánchez-Andrea et al. 2014). Kenward et al. (2006) investigated the effect of pH (3.0-7.0), ionic strength (1-100 mM) and initial SeO_4^{2-} concentration (0.01-5 mM) on the sorption ability of *Shewanella putrefaciens* 200R in batch experiments. The bacterium showed the ability to reduced SeO_4^{2-} to Se^0 (selenium sorption onto the *S. putrefaciens* cell wall by outer-sphere complexation and cell surface reduction) without the addition of external electron donor at the optimal condition of pH 3.0 for all solid/solute ratios showing the feasibility in directly applying real wastewater into a bioreactor without additional pH adjustment. However, long-term reuse of this sorption system with repeated selenium removal usage must be investigated.

2.6 Future perspective in selenium biotechnologies

Selenium contamination is generally a site specific case and depends on the geological location of selenium containing materials and the industrial processes occurring at that location. There is a possibility that selenium issues will grow in the coming years due to the exploitation of lower grade coal, fossil fuels, and mineral ores by the mining industry and irrigation in semi-arid regions (Lenz and Lens 2009). Achieving the stringent regulatory limit of selenium levels is a big challenge. While biological reduction of selenium has been investigated well in both natural and laboratory settings, this process still requires further investigation in order to make an applicable and cost-effective technology that can tackle real applications.

Secondary wastewater treatment plants are the core selenium treatment technologies. These are based on biological selenium reduction, thus the amount of reduced selenium particles increases proportionally with the input selenium concentration. Tertiary liquid-solids separation technology is often required to achieve better efficiencies of the biological treatments and to allow reuse of the Se^0. As such, improvements can be made through the integration of various technologies in order to fully remove selenium in effluent waters as well as fully utilize and close the reuse and recovery of selenium. A schematic diagram for integration and combination of various technologies is proposed in **Fig. 2.6** along with other end process applications for selenium treated water and possible by-products.

Lines A and B show the option between *in situ* treatments, while lines 1-6 show possible post-treatment or further bioreactor effluent applications. Wastewater with low concentrations of selenium (< 30 µg/L) can be directly sent to a low-cost technology treatments (line A) such as phytoremediation or constructed wetland (Hansen et al. 1998; García-Hernández et al. 2000), while other complex selenium-laden wastewaters should be treated in a bioreactor (line B), which is considered as a core selenium treatment technology (CH2M HILL 2013). Volatile selenium produced can be trapped and mixed with acid or fermenter wastewater (line 1) allowing for acid digestion of volatile selenium simultaneously supplementing Se to the anaerobic digester, thus preventing the accumulation of volatile fatty acids (Wall et al. 2014).

Hageman et al. (2013) showed that manipulation of operating conditions can be done in order to promote SeO_3^{2-} formation in the bioreactor. Selenite (Hockin and Gadd, 2003; Geoffroy and Demopoulos, 2011) and Se^0 can react with HS^- and can be combined with effluent from SO_4^{2-} treated wastewater (line 2a) to form Se_xS_y molecules (Taavitsainen et al. 1998). Selenite can also be easily absorbed (line 2b) onto organic materials or absorbent materials (Sheoran and Sheoran, 2006; Santos et al. 2015). It can also be reduced by fungi with elemental selenium absorbed onto the fungal pellets (Espinosa-Ortiz et al. 2015a; 2015b).

Elemental selenium or possibly organo-selenium remaining in the effluent can be post-treated by NF or RO membrane systems (line 3) simultaneously producing selenium free water (Kharaka et al. 1996; Sobolewski 2005) and possibly preventing biofouling formation on the membrane due to selenium's antimicrobial ability (Kochkodan and Hilal 2015). Sulfide from the effluent can be fed into another reactor with metal-containing wastewater (line 4) allowing for metal precipitation (Fellowes et al. 2013). Colloidal selenium remaining in the effluent can be post-treated with electrocoagulation (line 5) in order to sediment Se^0 (Staicu et al. 2015a) or with coagulation/precipitation using commercial products (Cantafio et al. 1996). Selenium accumulated in the sludge (line 6) can be dewatered (Jain et al. 2015b) or extracted (Soda et al. 2011) and reused for agricultural purposes. The selenium end-product in lines 2b, 4, and 5 can be recovered and reused in various industrial applications (Nancharaiah and Lens 2015b).

2.7   Conclusions

In spite of the above-mentioned challenges, biological treatment has emerged as the "best treatment" practice for Se-wastewaters due to its advantages such as low cost, scalability, lack of chemical sludge formation and ability to remove Se in a recoverable form. Biological treatment is able to competently reduce the total effluent selenium while allowing for lower operation costs and easier system operation. One major advantage of microbial reduction of selenium is the production of biogenic selenium nanoparticles that have technical applications. Bioreduction of selenium oxyanions by various microorganisms that occur in natural and engineered settings under different operating conditions are relatively well investigated. However, further improvements of the treatment efficiency are still needed, particularly in the aspect of selenium removal and recovery and the simultaneous reduction of SeO_4^{2-} and other oxyanions (NO_3^-/SO_4^{2-}). Another challenge is that operation conditions should account for the real wastewater properties such as pH, temperature and volume. Lastly, though bioreduction has the clear advantage of biogenic Se^0 formation, there is still a lack of knowledge and actual application in place to fully utilize and recover this potential source from the bioreactors (sludge and effluent as colloidal particle). Overall, despite the extensive studies conducted for biotechnology for selenium removal, additional investigations are still needed in view of the complex nature of wastewater, low discharge limits, and effect of biogenic selenium on the long term functional stability of the microbial population.

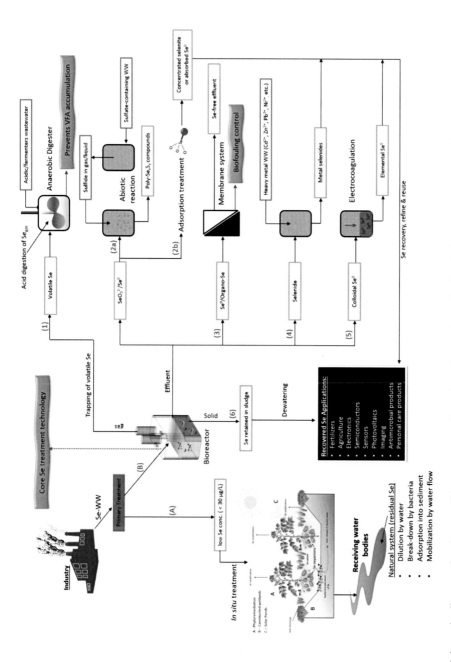

Fig. 2.6 Schematic diagram showing possible process configurations for treating selenium-laden wastewater as well as integration technology and reuse application after biotreatment and reduction.

2.8   References

Allmang, C., Krol, A., 2006. Selenoprotein synthesis: UGA does not end the story. *Biochimie* 88, 1561-1571.

Anderson, C.S., 2015. 2013 Minerals Yearbook: Selenium and tellurium. U.S. Geological Survey (advance release).

Astratinei, V., van Hullebusch, E., Lens, P.N.L., 2006. Bioconversion of selenate in methanogenic anaerobic granular sludge. *J. Environ. Qual.* 35, 1873-1883.

Bassil, J., Naveau, A., Bodin, J., Fontaine, C., Tullo, P.D., Razack, M., Kazpard, V., 2014. The nature of selenium species in the hydrogeological experimental site of Poitiers. *Proced. Earth Planet. Sci.* 10, 159-163.

Bhunia, P., Ghangrekar, M.M., 2007. Required minimum granule size in UASB reactor and characteristics variation with size. *Bioresour. Technol.* 98, 994-999.

Blum, J.S., Stolz, J.F., Oren, A., Oremland, R.S., 2001. *Selenihalanaerobacter shriftii gen. nov., sp. nov.*, a halophilic anaerobe from dead sea sediments that respires selenate. *Arch. Microbiol.* 175, 208-219.

Brown, K.M., Arthur, J.R., 2007. Selenium, selenoproteins and human health: A review. *Public Health Nutr.* 4, 593-599.

Brozmanová, J., Mániková, D., Vlčková, V., Chovanec, M., 2010. Selenium: A double-edged sword for defense and offence in cancer. *Arch. Toxicol.* 84, 919-938.

Buchs, B., Evangelou, M.W.H., Winkel, L.H.E., Lenz, M., 2013. Colloidal properties of nanoparticular biogenic selenium govern environmental fate and bioremediation effectiveness. *Environ. Sci. Technol.* 47, 2401-2407.

Burgess, J.E., Quarmby, J., Stephenson, T., 1999. Micronutrient supplements for optimisation of the treatment of industrial wastewater using activated sludge. *Water Res.* 33, 3707-3714.

Cantafio, A.W., Hagen, K.D., Lewis, G.E., Bledsoe, T.L., Nunan, K.M., Macy, J.M., 1996. Pilot scale selenium bioremediation of San Joaquin drainage water with *Thauera selenatis*. *Appl. Environ. Microbiol.* 62, 3298-3303.

Canton, S.P., van Derveer, W.D., 1997. Selenium toxicity to aquatic life: An argument for sediment-based water quality criteria. *Environ. Toxicol. Chem.* 16, 1255-1259.

CH2M HILL, 2010. Review of available technologies for the removal of selenium from water - Final report prepared for North American Metal Council.

CH2M HILL, 2013. Technical addendum: Review of available technologies for the removal of selenium from water. NAMC White Paper Report Addendum.

Chan, Y.J., Chong, M.F., Law, C.L., Hassell, D.G., 2009. A review on anaerobic-aerobic treatment of industrial and municipal wastewater. *Chem. Eng. J.* 155, 1-18.

Chapman, P.M., Adams, W.J., Brooks, M.L., Delos, C.G., Luoma, S.N., Maher, W.A., Ohlendorf, H.M., Presser, T.S., Shaw, D.P., 2010. Ecological assessment of selenium in the aquatic environment: Summary of a SETAC Pellston Workshop. Florida, Society of Environmental Toxicology and Chemistry (978-1-4398-2677-5).

Chung, J., Nerenberg, R., Rittmann, B.E., 2006a. Bioreduction of selenate using a hydrogen based membrane biofilm reactor. *Environ. Sci. Technol.* 40, 1664-1671.

Chung, J., Ryu, H., Abbaszadegan, M., Rittmann, B.E., 2006b. Community structure and function in a H_2-based membrane biofilm reactor capable of bioreduction of selenate and chromate. *Appl. Microbiol. Biotechnol.* 72, 1330-1339.

Chung, J., Rittmann, B.E.,Wright, W.F., Bowman, R.H., 2007. Simultaneous bio-reduction of nitrate, perchlorate, selenate, chromate, arsenate, and dibromochloropropane using a hydrogen-based membrane biofilm reactor. *Biodegradation* 18, 199-209.

Chung, J., Rittmann, B.E., Her, N., Lee, S.-H., Yoon, Y., 2010. Integration of H_2-based membrane biofilm reactor with RO and NF membranes for removal of chromate and selenate. *Water Air Soil Pollut.* 207, 29-37.

DeMoll-decker, H., Macy, J.M., 1993. The periplasmic nitrite reductase of *Thauera selenatis* may catalyze the reduction of selenite to elemental selenium. *Arch. f* 160, 241-247.

de Souza, M.P., Amini, A., Dojka, M.A., Pickering, I.J., Dawson, S.C., Pace, N.R., Terry, N., 2001. Identification and characterization of bacteria in a selenium-contaminated hypersaline evaporation pond. *Appl. Environ. Microbiol.* 67, 3785-3794.

Dessì, P., Jain, R., Singh, S., Seder-Colomina, M., van Hullebusch, E.D., Rene, E.R., Ahammad, S.Z., Carucci, A., Lens, P.N.L., 2016. Effect of temperature on selenium removal from wastewater by UASB reactors. *Water Res.* 94, 146-154.

Duntas, L.H., Benvenga, S., 2015. Selenium: An element for life. *Endocrine* 48, 756-775.

Ellis, A.S., Johnson, T.M., Herbel, M.J., Bullen, T.D., 2003. Stable isotope fractionation of selenium by natural microbial consortia. *Chem. Geol.* 195, 119-129.

El-Ramady, H., Abdalla, N., Alshaal, T., Domokos-Szabolcsy, É., Elhawat, N., Prokisch, J., Sztrik, A., Fári, M., El-Marsafawy, S., Shams, M.S., 2014a. Selenium in soils under climate change, implication for human health. *Environ. Chem. Lett.* 13, 1-19.

El-Ramady, H.R., Domokos-Szabolcsy, É., Abdalla, N.A., Alshaal, T.A., Shalaby, T.A., Sztrik, A., Prokisch, J., Fári, M., 2014b. Selenium and nano-seleniumin agroecosystems. *Environ. Chem. Lett.* 12, 495-510.

Erosa, M.S.D., Höll, W.H., Horst, J., 2009. Sorption of selenium species onto weakly basic anion exchangers: I. Equilibrium studies. *React. Funct. Polym.* 69, 576-585.

Espinosa-Ortiz, E.J., Gonzalez-Gil, G., Saikaly, P.E., van Hullebusch, E.D., Lens, P.N.L., 2015a. Effects of selenium oxyanions on the white-rot fungus *Phanerochaete chrysosporium. Appl. Microbiol. Biotechnol.* 99, 2405-2418.

Espinosa-Ortiz, E.J., Rene, E.R., van Hullebusch, E.D., Lens, P.N.L., 2015b. Removal of selenite from wastewater in a *Phanerochaete chrysosporium* pellet based fungal bioreactor. *Int. Biodeterior. Biodegrad.* 102, 361-369.

Fairweather-Tait, S.J., Collings, R., Hurst, R., 2010. Selenium bioavailability: Current knowledge and future research requirements. *Am. J. Clin. Nutr.* 91, 1484S-1491S.

Fellowes, J.W., Pattrick, R.A.D., Lloyd, J.R., Charnock, J.M., Coker, V.S., Mosselmans, J.F.W., Weng, T.-C., Pearce, C.I., 2013. Ex situ formation of metal selenide quantum dots using bacterially derived selenide precursors. *Nanotechnology* 24, 145603.

Fernández-Martínez, A., Charlet, L., 2009. Selenium environmental cycling and bioavailability: A structural chemist point of view. *Rev. Environ. Sci. Biotechnol.* 8, 81-110.

Ferrarese, M., Mahmoodi S.M., Quattrini, E., Schiavi, M., Ferrante, A., 2012. Biofortification of spinach plants applying selenium in the nutrient solution of floating system. *Veg. Crop Res. Bull.* 76, 127-136.

Fisher, J.C., Hollibaugh, J.T., 2008. Selenate-dependent anaerobic arsenite oxidation by a bacterium from Mono Lake, California. *Appl. Environ. Microbiol.* 74, 2588-2594.

Fordyce, F., 2005. Selenium deficiency and toxicity in the environment. British Geological Survey, Essentials Med Geol. NERC, pp. 373-415.

Fujita, M., Ike, M., Nishimoto, S., Takahashi, K., Kashiwa,M., 1997. Isolation and characterization of a novel selenate-reducing bacterium, *Bacillus sp.* SF-1. *J. Ferment. Bioeng.* 83, 517-522.

Fujita, M., Ike, M., Kashiwa, M., Hashimoto, R., Soda, S., 2002. Laboratory-scale continuous reactor for soluble selenium removal using selenate-reducing bacterium, *Bacillus sp.* SF-1. *Biotechnol. Bioeng.* 80, 755-761.

Gao, S., Tanji, K.K., Peters, D.W., Herbel, M.J., 2000. Water selenium speciation and sediment fractionation in a California flow-through wetland system. *J. Environ. Qual.* 29, 1275-1283.

Gao, S., Tanji, K.K., Dahlgren, R.A., Ryu, J., Herbel, M.J., Higashi, R.M., 2007. Chemical status of selenium in evaporation basins for disposal of agricultural drainage. *Chemosphere* 69, 585-594.

García-Hernández, J., Glenn, E.P., Artiola, J., Baumgartner, D.J., 2000. Bioaccumulation of selenium (Se) in the Cienega de Santa Clara wetland, Sonora, Mexico. *Ecotoxicol. Environ. Saf.* 46, 298-304.

Geoffroy, N., Demopoulos, G.P., 2011. The elimination of selenium (IV) from aqueous solution by precipitation with sodium sulfide. *J. Hazard. Mater.* 185, 148-154.

Gerhardt, M.B., Green, F.B., Newman, R.D., Lundquist, T.J., Tresan, R.B., Oswald,W.J., 1991. Removal of selenium using a novel algal-bacterial process. *Res. J. Water Pollut. Control Fed.* 63, 799-805.

Ghosh, A., Mohod, A.M., Paknikar, K.M., Jain, R.K., 2008. Isolation and characterization of selenite- and selenate-tolerant microorganisms from selenium-contaminated sites. *World J. Microbiol. Biotechnol.* 24, 1607-1611.

Golder Associates Inc., 2009. Literature review of treatment technologies to remove selenium from mining influenced water. Report to Teck Coal Limited, Calgary (08-1421-0034 Rev. 2, AB).

Gonzalez-Gil, G., Lens, P.N.L., Saikaly, P.E., 2016. Selenite reduction by anaerobic microbial aggregates: microbial community structure, and proteins associated to the produced selenium spheres. *Front. Microbiol.* 7, 571.

Green, F.B., Lundquist, T.J., Quinn, N.W., Zarate, M.A., Zubieta, I.X., Oswald,W.J., 2003. Selenium and nitrate removal from agricultural drainage using the ALWPS technology. *Water Sci. Technol.* 48, 299-305.

Gusek, J., Conroy, K., Rutkowski, T., 2008. Past, present and future for treating selenium impacted water. Tailings and Mine Waste '08. CRC Press, pp. 281-290 Edited by The Organizing Committee of the 12th International Conference on Tailings and Mine Waste (978-0-203-88230-6).

Hageman, S.P., van der Weijden, R.D., Weijma, J., Buisman, C.J., 2013. Microbiological selenate to selenite conversion for selenium removal. *Water Res.* 47, 2118-2128.

Hamilton, S.J., 2004. Review of selenium toxicity in the aquatic food chain. *Sci. Total Environ.* 326, 1-31.

Hansen, D., Duda, P.J., Zayed, A., Terry, N., 1998. Selenium removal by constructed wetlands: Role of biological volatilization. *Environ. Sci. Technol.* 32, 591-597.

Hennebel, T., Boon, N., Maes, S., Lenz, M., 2015. Biotechnologies for critical raw material recovery from primary and secondary sources: R&D priorities and future perspectives. *New Biotechnol.* 32, 121-127.

Hockin, S.L., Gadd, G.M., 2003. Linked redox precipitation of sulfur and selenium under anaerobic conditions by sulfate-reducing bacterial biofilms. *Appl. Environ. Microbiol.* 69, 7063-7072.

Huang, J.-C., Suárez, M.C., Yang, S.I., Lin, Z.-Q., Terry, N., 2013. Development of a constructed wetland water treatment system for selenium removal: Incorporation of an algal treatment component. *Environ. Sci. Technol.* 47, 10518-25.

Husen, A., Siddiqi, K.S., 2014. Plants and microbes assisted selenium nanoparticles: Characterization and application. *J. Nanobiotechnol.* 12, 28.

Jacob, J.M., Lens, P.N.L., Balakrishnan, R.M., 2016. Microbial synthesis of chalcogenide semiconductor nanoparticles: A review. *Microb. Biotechnol.* 9, 11-21.

Jain, R., Jordan, N., Schild, D., van Hullebusch, E.D.,Weiss, S., Franzen, C., Farges, F., Hübner, R., Lens, P.N.L., 2015a. Adsorption of zinc by biogenic elemental selenium nanoparticles. *Chem. Eng. J.* 260, 855-863.

Jain, R., Seder-Colomina, M., Jordan, N., Dessi, P., Cosmidis, J., van Hullebusch, E.D.,Weiss, S., Farges, F., Lens, P.N.L., 2015b. Entrapped elemental selenium nanoparticles affect physicochemical properties of selenium fed activated sludge. *J. Hazard. Mater.* 295, 193-200.

Johansson, C.L., Paul, N.A., de Nys, R., Roberts, D.A., 2015. The complexity of biosorption treatments for oxyanions in a multi-element mine effluent. *J. Environ. Manag.* 151, 386-392.

Kabata-Pendias, A., 2011. Trace elements in soils and plants. 4th ed. CRC Press: Taylor and Francis Group (LLC 978-1-4200-9370-4).

Kagami, T., Narita, T., Kuroda, M., Notaguchi, E., Yamashita, M., Sei, K., Soda, S., Ike, M., 2013. Effective selenium volatilization under aerobic conditions and recovery from the aqueous phase by *Pseudomonas stutzeri* NT-I. *Water Res.* 47, 1361-1368.

Kapoor, A., Tanjore, S., Viraraghavan, T., 1995. Removal of selenium from water and wastewater. *Int. J. Environ. Stud.* 49, 137-147.

Kashiwa, M., Nishimoto, S., Takahashi, K., Ike, M., Fujita, M., 2000. Factors affecting soluble selenium removal by a selenate-reducing bacterium *Bacillus sp.* SF-1. *J. Biosci. Bioeng.* 89, 528-533.

Kashiwa,M., Ike,M., Mihara, H., Esaki, N., Fujita, M., 2001. Removal of soluble selenium by a selenate-reducing bacterium *Bacillus sp.* SF-1. *Biofactors* 14, 261-265.

Kenward, P.A., Fowle, D.A., Yee, N., 2006. Microbial selenate sorption and reduction in nutrient limited systems. *Environ. Sci. Technol.* 40, 3782-3786.

Kessi, J., Ramuz, M., Wehrli, E., Spycher, M., Bachofen, R., 1999. Reduction of selenite and detoxification of elemental selenium by the phototrophic bacterium *Rhodospirillum rubrum. Appl. Environ. Microbiol.* 65, 4734-4740.

Kharaka, Y.K., Ambats, G., Presser, T.S., Davis, R.A., 1996. Removal of selenium from contaminated agricultural drainage water by nanofiltration membranes. *Appl. Geochem.* 11, 797-802.

Kieliszek, M., Błażejak, S., 2013. Selenium: Significance and outlook for supplementation. *Nutrition* 29, 713-718.

Kochkodan, V., Hilal, N., 2015. A comprehensive review on surface modified polymer membranes for biofouling mitigation. *Desalination* 356, 187-207.

König, S., Luguet, A., Lorand, J.-P.,Wombacher, F., Lissner, M., 2012. Selenium and tellurium systematics of the Earth's mantle from high precision analyses of ultra-depleted orogenic peridotites. *Geochim. Cosmochim. Acta* 86, 354-366.

Kumar, B.S., Priyadarsini, K.I., 2014. Selenium nutrition: How important is it? *Biomed. Prev. Nutr.* 4, 333-341.

Kuroda, M., Notaguchi, E., Sato, A., Yoshioka, M., Hasegawa, A., Kagami, T., Narita, T., Yamashita, M., Sei, K., Soda, S., Ike, M., 2011. Characterization of *Pseudomonas stutzeri* NT-I capable of removing soluble selenium from the aqueous phase under aerobic conditions. *J. Biosci. Bioeng.* 112, 259-264.

Lai, C.Y., Yang, X., Tang, Y., Rittmann, B.E., Zhao, H.P., 2014. Nitrate shaped the selenate-reducing microbial community in a hydrogen-based biofilm reactor. *Environ. Sci. Technol.* 48, 3395-3402.

Latorre, C.H., García, J.B., Martín, S.G., Peña, R.M., 2013. Solid phase extraction for the speciation and preconcentration of inorganic seleniumin water samples: A review. *Anal. Chim. Acta* 804, 37-49.

Lee, J.-H., Han, J., Choi, H., Hur, H.-G., 2007. Effects of temperature and dissolved oxygen on Se(IV) removal and Se(0) precipitation by *Shewanella sp.* HN-41. *Chemosphere* 68, 1898-1905.

Lee, W., Park, S.-H., Kim, J., Jung, J.-Y., 2015. Occurrence and removal of hazardous chemicals and toxic metals in 27 industrial wastewater plants in Korea. *Desalin. Water Treat.* 54, 1141-1149.

Lemly, A.D., 1997. Environmental implications of excessive selenium: A review. *Biomed. Environ. Sci.* 10, 415-435.

Lemly, A.D., 1999. Selenium transport and bioaccumulation in aquatic ecosystems: A proposal for water quality criteria based on hydrological units. *Ecotoxicol. Environ. Saf.* 42, 150-156.

Lemly, A.D., 2002. Symptoms and implications of selenium toxicity in fish: the Belews Lake case example. *Aquat. Toxicol.* 57, 39-49.

Lemly, A.D., 2004. Aquatic selenium pollution is a global environmental safety issue. *Ecotoxicol. Environ. Saf.* 59, 44-56.

Lemly, A.D., 2014. Teratogenic effects and monetary cost of selenium poisoning of fish in Lake Sutton, North Carolina. *Ecotoxicol. Environ. Saf.* 104, 160-167.

Lemly, A.D., Skorupa, J.P., 2012.Wildlife and the coal waste policy debate: Proposed rules for coal waste disposal ignore lessons from 45 years of wildlife poisoning. *Environ. Sci. Technol.* 46, 8595-8600.

Lenz, M., Lens, P.N.L., 2009. The essential toxin: The changing perception of selenium in environmental sciences. *Sci. Total Environ.* 407, 3620-3633.

Lenz, M., van Hullebusch, E.D., Hommes, G., Corvini, P.F., Lens, P.N.L., 2008a. Selenate removal in methanogenic and sulfate-reducing upflow anaerobic sludge bed reactors. *Water Res.* 42, 2184-2194.

Lenz, M., Janzen, N., Lens, P.N.L., 2008b. Selenium oxyanion inhibition of hydrogenotrophic and acetoclastic methanogenesis. *Chemosphere* 73, 383-388.

Lenz, M., Smit, M., Binder, P., van Aelst, A.C., Lens, P.N.L., 2008c. Biological alkylation and colloid formation of selenium in methanogenic UASB reactors. *J. Environ. Qual.* 37, 1691-1700.

Lenz, M., Enright, A.M., O'Flaherty, V., van Aelst, A.C., Lens, P.N.L., 2009. Bioaugmentation of UASB reactors with immobilized *Sulfurospirillum barnesii* for simultaneous selenate and nitrate removal. *Appl. Microbiol. Biotechnol.* 83, 377-388.

Li, D.-B., Cheng, Y.-Y.,Wu, C., Li,W.-W., Li, N., Yang, Z.-C., Tong, Z.-H., Yu, H.-Q., 2014. Selenite reduction by *Shewanella oneidensis* MR-1 is mediated by fumarate reductase in periplasm. *Sci. Rep.* 4, 3735.

Lortie, L., Gould,W.D., Rajan, S., McCready, R.G.L., Cheng, K.-J., 1992. Reduction of selenate and selenite to elemental selenium by a *Pseudomonas stutzeri* isolate. *Appl. Environ. Microbiol.* 58, 4042-4044.

Losi, M.E., Frankenberger Jr.,W.T., 1997. Reduction of seleniumoxyanions by *Enterobacter cloacae* SLD1a-1: Isolation and growth of the bacterium and its expulsion of selenium particles. *Appl. Environ. Microbiol.* 63, 3079-3084.

Luoma, S.N., Presser, T.S., 2009. Emerging opportunities in management of selenium contamination. *Environ. Sci. Technol. Viewp.* 43, 8483-8487.

Lussier, C., Veiga, V., Baldwin, S., 2003. The geochemistry of selenium associated with coal waste in the Elk River Valley, Canada. *Environ. Geol.* 44, 905-913.

Macaskie, L.E., Mikheenko, I.P., Yong, P., Deplanche, K., Murray, A.J., Paterson-Beedle, M., Coker, V.S., Pearce, C.I., Cutting, R., Patrick, R.A.D., Vaughan, D., van der Laan, G., Lloyd, J.R., 2010. Today's wastes, tomorrow's materials for environmental protection. *Hydrometallurgy* 104, 483-

487.

Maiers, D.T.,Wichlacz, P.L., Thompson, D.L., Bruhn, F., 1988. Selenate reduction by bacteria from selenium-rich environment. *Appl. Environ. Microbiol.* 54, 2591-2593.

Mal, J., Nancharaiah, Y.V., van Hullebusch, E.D., Lens, P.N.L., 2016a. Effect of heavy metal co-contaminants on selenite bioreduction by anaerobic granular sludge. *Bioresour. Technol.* 206, 1-8.

Mal, J., Nancharaiah, Y.V., van Hullebusch, E.D., Lens, P.N.L., 2016b. Metal chalcogenide quantum dots: Biotechnological synthesis and applications. *RSC Adv.* 6, 41477-41495.

Mal, J., Veneman, W.J., Nancharaiah, Y.V., van Hullebusch, E.D., Peijnenburg, W.J.G.M., Vijver, M.G., Lens, P.N.L., 2016. A comparison of fate and toxicity of selenite, biogenically and chemically synthesized selenium nanoparticles to the zebrafish (*Danio rerio*) embryogenesis. *Nanotoxicology* 11, 87-97.

Mao, C., Feng, Y.,Wang, X., Ren, G., 2015. Review on research achievements of biogas from anaerobic digestion. *Renew. Sust. Energ. Rev.* 45, 540-555.

Meawad, A.S., Bojinova, D.Y., Pelovski, Y.G., 2010. An overview of metals recovery from thermal power plant solid wastes. *Waste Manag.* 30, 2548-2559.

Mehdi, Y., Hornick, J.-L., Istasse, L., Dufrasne, I., 2013. Selenium in the environment, metabolism and involvement in body functions. *Molecules* 18, 3292-3311.

Mohammandi, F., Littlejohn, P., West, A., Hall, A., 2014. Selen-IX™: Selenium removal from mining affected runoff using ion exchange based technology. *Hydrometallurgy* pp. 1-13.

MSE Technology Applications Inc., 2001. Selenium treatment/removal alternatives demonstration project: Mine waste technology program activity III, Project 20. Butte, Montana.

Muscatello, J.R., Janz, D.M., 2009. Selenium accumulation in aquatic biota downstream of a uranium mining and milling operation. *Sci. Total Environ.* 407, 1318-1325.

Muscatello, J.R., Belknap, A.M., Janz, D.M., 2008. Accumulation of selenium in aquatic systems downstream of a uraniummining operation in northern Saskatchewan, Canada. *Environ. Pollut.* 156, 387-393.

Nakamaru, Y., Tagami, K., Uchida, S., 2005. Distribution coefficient of seleniumin in Japanese agricultural soils. *Chemosphere* 58, 1347-1354.

Nakamaru, Y.M., Altansuvd, J., 2014. Speciation and bioavailability of selenium and antimony in non-flooded and wetland soils: A review. *Chemosphere* 111, 366-371.

Nancharaiah, Y.V., Lens, P.N.L., 2015a. Ecology and biotechnology of selenium-respiring bacteria. *Microbiol. Mol. Biol. Rev.* 79, 61-80.

Nancharaiah, Y.V., Lens, P.N.L., 2015b. Selenium biomineralization for biotechnological applications. *Trends Biotechnol.* 33, 323-330.

Nancharaiah, Y.V., Venkata Mohan, S., Lens, P.N.L., 2016. Biological and bioelectrochemical recovery of critical and scarce metals. *Trends Biotechnol.* 34, 127-155.

Narayanan, K.B., Sakthivel, N., 2010. Biological synthesis of metal nanoparticles by microbes. *Adv. Colloid Interf. Sci.* 156, 1-13.

Navarro-Alarcon, M., Cabrera-Vique, C., 2008. Selenium in food and the human body: A review. *Sci. Total Environ.* 1-3, 115-141.

Nnaji, C.C., 2014. A review of the upflow anaerobic sludge blanket reactor. *Desalin. Water Treat.* 52, 4122-4143.

Nogueira, C.W., Rocha, J.B., 2011. Toxicology and pharmacology of selenium: emphasis on synthetic organoselenium compounds. *Arch. Toxicol.* 85, 1313-1359.

NSMP, 2007. Identification and assessment of selenium and nitrogen treatment technologies and best management practices. (Available at) http://www.ocnsmp.com/library.asp.

Ohlendorf, H.M., 2002. The birds of Kesterson reservoir: a historical perspective. *Aquat. Toxicol.* 57, 1-10.

Opara, A., Peoples, M.J., Adams, D.J., Maehl, W.C., 2014. The Landusky mine biotreatment system: Comparison of conventional bioreactor performance with a new electro-biochemical reactor (EBR) technology. *Water Miner. Process.* 1-6.

Oremland, R.S., Blum, J.S., Culbertson, C.W., Visscher, P.T., Miller, L.G., Dowdle, P., Stromaier, F.E., 1994. Isolation, growth, and metabolism of an obligately anaerobic, selenate-respiring bacterium, strain SES-3. *Appl. Environ. Microbiol.* 60, 3011-3019.

Oremland, R.S., Blum, J.S., Bindi, A.B., Dowdle, P.R., Herbel, M., Stolz, J.F., 1999. Simultaneous reduction of nitrate and selenate by cell suspensions of selenium-respiring bacteria. *Appl. Environ. Microbiol.* 65, 4385-4392.

Oremland, R.S., Herbel, M.J., Blum, J.S., Langley, S., Beveridge, T.J., Ajayan, P.M., Sutto, T., Ellis, A.V., Curran, S., 2004. Structural and spectral features of selenium nanospheres produced by Se-respiring bacteria. *Appl. Environ. Microbiol.* 70, 52-60.

Pedrero, Z., Madrid, Y., 2009. Novel approaches for selenium speciation in foodstuffs and biological specimens: A review. *Anal. Chim. Acta* 634, 135-152.

Persico, J.L., Brookins, D.G., 1988. Selenium geochemistry at Bosque Del Apache National Wildlife Refuge. New Mex Geeological Soc Guidebook 39th F Conf. New Mexico: New Mexico Geological Society Guidebook, pp. 211-216.

Pickett, T., Harwood, J., 2014. Biofiltration systems answer selenium treatment challenge. WaterWorld. (Available at) http://www.waterworld.com/articles/iww/print/volume-12/issue-2/feature-editorial/biofiltration-systems-answer-selenium-treatmentchallenge.html.

Presser, T.S., Luoma, S.N., 2006. Forecasting selenium discharges to the San Francisco Bay-Delta estuary: Ecological effects of a proposed San Luis drain extension. US. Geol. Surv. Prof. Pap. 1646.

Purkerson, D.G., Doblin, M.A., Bollens, S.M., Luoma, S.N., Cutter, G.A., 2003. Selenium in San Francisco Bay zooplankton: Potential effects of hydrodynamics and food web interactions. *Estuaries* 26, 956-969.

Pyrzynska, K., 2002. Determination of selenium species in environmental samples. *Microchim. Acta* 140, 55-62.

Rayman, M.P., 2012. Selenium and human health. *Lancet* 379, 1256-1268.

Rech, S.A.,Macy, J.M., 1992. The terminal reductases for selenate and nitrate respiration in *Thauera selenatis* are two distinct enzymes. *J. Bacteriol.* 174, 7316-7320.

Rege, M.A., Yonge, D.R., Mendoza, D.P., Petersen, J.N., Bereded-Samuel, Y., Johnstone, D.L., Apel, W., Barnes, J.M., 1999. Selenium reduction by a denitrifying consortium. *Biotechnol. Bioeng.* 62, 479-484.

Rosen, B.P., Liu, Z., 2009. Transport pathways for arsenic and selenium: A minireview. *Environ. Int.* 35, 512-515.

Rutkowski, T., Hanson, R., Conroy, K., 2013. Mine water treatment options for meeting stringent selenium regulatory limits. Reliab Mine Water Technol. IMWA, Colorado, pp. 711-716.

Ryu, J.-H., Gao, S., Tanji, K.K., 2011. Accumulation and speciation of selenium in evaporation basins in California, USA. *J. Geochem. Explor.* 110, 216-224.

Saeki, K., Matsumoto, S., 1998. Mechanisms of ligand exchange reactions involving selenite sorption on goethite labeled with oxygen-stable isotope. *Commun. Soil Sci. Plant Anal.* 29, 3061-3072.

Sánchez-Andrea, I., Sanz, J.L., Bijmans, M.F., Stams, A.J., 2014. Sulfate reduction at low pH to remediate acid mine drainage. *J. Hazard. Mater.* 269, 98-109.

Santos, S., Ungureanu, G., Boaventura, R., Botelho, C., 2015. Selenium contaminated waters: An overview of analytical methods, treatment options and recent advances in sorption methods. *Sci. Total Environ.* 521-522, 246-260.

Schlekat, C.E., Lee, B.G., Luoma, S.N., 2002. Assimilation of selenium from phytoplankton by three benthic invertebrates: Effect of phytoplankton species. *Mar. Ecol. Prog. Ser.* 237, 79-85.

Schneider, L., Maher, W.A., Potts, J., Taylor, A.M., Batley, G.E., Krikowa, F., Chariton, A.A., Grubert, B., 2015. Modeling food web structure and selenium biomagnification in Lake Macquarie, New South Wales, Australia, using stable carbon and nitrogen isotopes. *Environ. Toxicol. Chem.* 34, 608-617.

Schrauzer, G.N., 2001. Nutritional seleniumsupplements: Product types, quality, and safety. *J. Am. Coll. Nutr.* 20, 1-4.

Schröder, I., Rech, S., Krafft, T., Macy, J.M., 1997. Purification and characterization of the selenate reductase from *Thauera selenatis*. *J. Biol. Chem.* 272, 23765-23768.

Sears, M.E., 2013. Chelation: Harnessing and enhancing heavy metal detoxification - A review. *Sci. World J.* 2013, 219840.

Seghezzo, L., Zeeman, G., van Lier, J.B., Hamelers, H.V.M., Lettinga, G., 1998. A review: The anaerobic treatment of sewage in UASB and EGSB reactors. *Bioresour. Technol.* 65, 175-190.

Sharma, V.K., McDonald, T.J., Sohn, M., Anquandah, G.A., Pettine, M., Zboril, R., 2015. Biogeochemistry of selenium: A review. *Environ. Chem. Lett.* 13, 49-58.

Sheoran, A.S., Sheoran, V., 2006. Heavy metal removal mechanism of acid mine drainage in wetlands: A critical review. *Miner. Eng.* 19, 105-116.

Siscar, R., Koenig, S., Torreblanca, A., Solé, M., 2013. The role of metallothionein and selenium in metal detoxification in the liver of deep-sea fish from the NW Mediterranean Sea. *Sci. Total Environ.* 466-467, 898-905.

Smith, K., Lau, A.O., Vance, F.W., 2009. Evaluation of treatment techniques for selenium removal. Engineers Society of Western Pennsylvania (Eds.), 70th Annu Int Water Conf. Curran Associates, Inc., Orlando, Florida, USA, pp. 75-92.

Sobolewski, A., 2005. Evaluation of treatment options to reduce water-borne selenium at coal mines in West-Central Alberta. Microbial Technologies Inc. (ISBN: 0-7785-4605-5).

Soda, S., Kashiwa, M., Kagami, T., Kuroda, M., Yamashita, M., Ike, M., 2011. Laboratory scale bioreactors for soluble selenium removal from selenium refinery wastewater using anaerobic sludge. *Desalination* 279, 433-438.

Sonstegard, J., Pickett, T., Harwood, J., Johnson, D., 2007. Full scale operation of GE ABMet® biological technology for the removal of selenium from FGD wastewaters. Engineers Society of Western Pennsylvania (Eds.), 69th Annu Int Water Conf. Curran Associates, Inc., Orlando, Florida, USA, p. 580.

Sonstegard, J., Harwood, J., Pickett, T., 2010. ABMet®: Setting the standard for selenium removal Editors: Engineers Society of Western Pennsylvania, Proc 71st Int Water Conf. Curran Associates

Inc., Texas, USA, p. 216.

Staicu, L.C., van Hullebusch, E.D., Lens, P.N.L., 2015a. Production, recovery and reuse of biogenic elemental selenium. *Environ. Chem. Lett.* 13, 89-96.

Staicu, L.C., van Hullebusch, E.D., Oturan, M.A., Ackerson, C.J., Lens, P.N.L., 2015b. Removal of colloidal biogenic selenium from wastewater. *Chemosphere* 125, 130-8.

Stewart, R., Grosell, M., Buchwalter, D., Fisher, N., Luoma, S., Matthews, T., Orr, P., Wang, W.-X., 2010. Bioaccumulation and trophic transfer of selenium. In: Chapman, P.M., Adams, W.J., Brooks, M.L., Delos, C.G., Luoma, S.N., Maher, W.A., Ohlendorf, H.M., Presser, T.S., Shaw, D.P. (Eds.), Ecol Assess Selenium Aquat Environ. Society of Environmental Toxicology and Chemistry, Florida, pp. 93-140.

Swift, M.C., 2002. Stream ecosystem response to, and recovery from, experimental exposure to selenium. *J. Aquat. Ecosyst. Stress. Recover.* 9, 159-184.

Taavitsainen, J., Lange, H., Laitinen, R.S., 1998. An ab initio MO study of selenium sulfide heterocycles Se_nS_{8-n}. *THEOCHEM J. Mol. Struct.* 453, 197-208.

Tabak, H.H., Scharp, R., Burckle, J., Kawahara, F.K., Govind, R., 2003. Advances in biotreatment of acid mine drainage and biorecovery of metals: I. Metal precipitation for recovery and recycle. *Biodegradation* 14, 423-436.

Takada, T., Hirata, M., Kokubu, S., Toorisaka, E., Ozaki, M., Hano, T., 2008. Kinetic study on biological reduction of selenium compounds. *Process Biochem.* 43, 1304-1307.

Tam, K., Ho, C.T., Lee, J.H., Lai, M., Chang, C.H., Rheem, Y., Chen,W., Hur, H.G., Myung, N.V., 2010. Growth mechanism of amorphous selenium nanoparticles synthesized by *Shewanella sp.* HN-41. *Biosci. Biotechnol. Biochem.* 74, 696-700.

The Free Press, 2013. BioteQ to deal with selenium in Elk Valley Waters. Free Press. (Available in) http://www.thefreepress.ca/news/209581551.html.

Tinggi, U., 2003. Essentiality and toxicity of selenium and its status in Australia: A review. *Toxicol. Lett.* 137, 103-110.

Tran, P.L., Lowry, N., Campbell, T., Reid, T.W.,Webster, D.R., Tobin, E., Mosley, T., Dertien, J., Colmer-Hamood, J.A., Hamod, A.N., 2012. An organoselenium compound inhibits *Staphylococcus aureus* biofilms on hemodialysis catheters in vivo. *Antimicrob. Agents Chemother.* 56, 972-978.

Tran, P., Patel, S., Hamood, A., Enos, T., Mosley, T., Jarvis, C., Desai, A., Lin, P., Reid, T.W., 2014. A novel organo-selenium bandage that inhibits biofilm development in a wound by gram-positive and gram-negative wound pathogens. *Antibiotics* 3, 435-449.

Turner, R.J., Borghese, R., Zannoni, D., 2012. Microbial processing of tellurium as a tool in biotechnology. *Biotechnol. Adv.* 30, 954-963.

Twidwell, L.G., McCloskey, J.M., Miranda, P., Gale, M., 1999. Technologies and potential technologies for removing selenium from process and mine wastewater. In: Gaballah, I., Hager, J., Solozaral, R. (Eds.), Glob Symp Recycl Waste Treat Clean Technol. San Sabastian, Spain, pp. 1645-1656.

Twidwell, L.G., McCloskey, J.M., Joyce, H., Dahlgren, E., Hadden, A., Keller, J.J., Free, M.L., 2005. Removal of selenium oxyanions from mine waters utilizing elemental iron and galvanically coupled metals. In: Young, C. (Ed.), Proceedings J.D. Miller Symposium, Innovations in Natural Resource Systems. Proc. J.D. Miller Sympo, Innov in Nat Res Sys. Littleton, CO, USA; SME, pp. 299-313.

USA Department of Health and Human Services, 2003. Toxicological profile for selenium. Georgia.

USEPA, 2014. External peer review draft - Aquatic life ambient water quality criterion for selenium - freshwater 2014. United States Environmental Protection Agency (EPA-820-F-14-005).

USGS, 2015. Mineral commodity summaries 2015. U.S. Geol. Surv. 196 (Available in) http://dx.doi.org/10.3133/70140094.

van Ginkel, S.W., Yang, Z., Kim, B.O., Sholin, M., Rittmann, B.E., 2011. The removal of selenate to low ppb levels from flue gas desulfurization brine using the H_2-based membrane biofilm reactor (MBfR). *Bioresour. Technol.* 102, 6360-6364.

Vercellino, T., Morse, A., Tran, P., Hamood, A., Reid, T., Song, L., Moseley, T., 2013a. The use of covalently attached organo-selenium to inhibit *S. aureus* and *E. coli* biofilms on RO membranes and feed spacers. *Desalination* 317, 142-151.

Vercellino, T., Morse, A., Tran, P., Song, L., Hamood, A., Reid, T., Moseley, T., 2013b. Attachment of organo-seleniumto polyamide composite reverse osmosis membranes to inhibit biofilm formation of *S. aureus* and *E. coli. Desalination* 309, 291-295.

Vesper, D.J., Roy, M., Rhoads, C.J., 2008. Selenium distribution and mode of occurrence in the Kanawha Formation, southern West Virginia, U.S. *Int. J. Coal Geol.* 73, 237-249.

Viamajala, S., Bereded-Samuel, Y., Apel, W.A., Petersen, J.N., 2006. Selenite reduction by a denitrifying culture: batch- and packed-bed reactor studies. *Appl. Microbiol. Biotechnol.* 71, 953-962.

Vinceti, M., Crespi, C.M., Bonvicini, F., Malagoli, C., Ferrante, M., Marmiroli, S., Stranges, S., 2013. The need for a reassessment of the safe upper limit of selenium in drinking water. *Sci. Total Environ.* 443, 633-642.

Wall, D.M., Allen, E., Straccialini, B., O'Kiely, P., Murphy, J.D., 2014. The effect of trace element addition to mono-digestion of grass silage at high organic loading rates. *Bioresour. Technol.* 172, 349-355.

Wang, Z., Gao, Y., 2001. Biogeochemical cycling of selenium in Chinese environments. *Appl. Geochem.* 16, 1345-1351.

Weeks, M.E., 1932. The discovery of the elements. VI. Tellurium and selenium. *J. Chem. Educ.* 9, 474-485.

Wen, H., Carignan, J., 2007. Reviews on atmospheric selenium: Emissions, speciation and fate. *Atmos. Environ.* 41, 7151-7165.

White, P.J., Broadley, M.R., Gregory, P.J., 2012. Managing the nutrition of plants and people. *Appl. Environ. Soil Sci.* 2012, 1-13.

WHO, 2011. Guidelines for drinking-water quality: Selenium in drinking-water. WHO Press, Switzerland.

Winkel, L.H., Johnson, C.A., Lenz, M., Grundl, T., Leupin, O.X., Amini, M., Grundl, T., Leupin, O.X., Amini,M., Charlet, L., 2012. Environmental selenium research: From microscopic processes to global understanding. *Environ. Sci. Technol.* 46, 571-579.

Winkel, L.H., Vriens, B., Jones, G., Schneider, L., Pilon-Smits, E., Bañuelos, G., 2015. Selenium cycling across soil-plant-atmosphere interfaces: A critical review. *Nutrients* 7, 4199-4239.

Wu, L., 2004. Review of 15 years of research on ecotoxicology and remediation of land contaminated by agricultural drainage sediment rich in selenium. *Ecotoxicol. Environ. Saf.* 57, 257-269.

Wu, L., Huang, Z.Z., 1991. Selenium accumulation and selenium tolerance of salt grass from soils with elevated concentrations of Se and salinity. *Ecotoxicol. Environ. Saf.* 22, 267-282.

Yanai, J., Mizuhara, S., Yamada, H., 2015. Soluble selenium content of agricultural soils in Japan and its determining factorswith reference to soil type, land use and region. *Soil Sci. Plant Nutr.* 61, 312-318.

Yang, G.Q., Wang, S.Z., Zhou, R.H., Sun, S.Z., 1983. Endemic selenium intoxication of humans in China. *Am. J. Clin. Nutr.* 37, 872-881.

Yee, N., Ma, J., Dalia, A., Boonfueng, T., Kobayashi, D.Y., 2007. Se(VI) reduction and the precipitation of Se(0) by the facultative bacterium *Enterobacter cloacae* SLD1a-1 are regulated by FNR. *Appl. Environ. Microbiol.* 73, 1914-1920.

Young, T.F., Finley, K., Adams, W., Besser, J., Hopkins, W.A., Jolley, D., McNaughton, E., Presser, T.S., Shaw, D.P., Unrine, J., 2010a. Selected case studies of ecosystem contamination by selenium. In: Chapman, P.M., Adams,W.J., Brooks, M.L., Delos, C.G., Luoma, S.N., Maher, W.A.,

Ohlendorf, H.M., Presser, T.S., Shaw, D.P. (Eds.), Ecol Assess Selenium Aquat Environ. Society of Environmental Toxicology and Chemistry, Florida, pp. 254-292.

Young, T.F., Finley, K., Adams, W., Besser, J., Hopkins, W.A., Jolley, D., McNaughton, E., Presser, T.S., Shaw, P.D., Unrine, J., 2010b. What you need to know about selenium. In: Chapman, P.M., Adams, W.J., Brooks, M.L., Delos, C.G., Luoma, S.N., Maher, W.A., Ohlendorf, H.M., Presser, T.S., Shaw, D.P. (Eds.), Ecol Assess Selenium Aquat Environ. Society of Environmental Toxicology and Chemistry, Florida, pp. 7-46.

Zahir, Z.A., Zhang, Y., Frankenberger Jr., W.T., 2003. Fate of selenate metabolized by *Enterobacter taylorae* isolated from rice straw. *J. Agric. Food Chem.* 51, 3609-3613.

Zannoni, D., Borsetti, F., Harrison, J.J., Turner, R.J., 2008. The bacterial response to the chalcogen metalloids Se and Te. *Adv. Microb. Physiol.* 531, 1-72.

Zehr, J.P., Oremland, R.S., 1987. Reduction of selenate to selenide by sulfate-respiring bacteria: Experiments with cell suspensions and estuarine sediments. *Appl. Environ. Microbiol.* 53, 1365-1369.

Zelmanov, G., Semiat, R., 2013. Selenium removal from water and its recovery using iron (Fe^{3+}) oxide/hydroxide-based nanoparticles sol (NanoFe) as an adsorbent. *Sep. Purif. Technol.* 103, 167-172.

Zhang, Y., Frankenberger Jr., W.T., 2003. Characterization of selenate removal from drainage water using rice straw. *J. Environ. Qual.* 32, 441-446.

Zhang, Y., Frankenberger Jr., W.T., 2005. Removal of selenium from river water by a microbial community enhanced with *Enterobacter taylorae* in organic carbon coated sand columns. *Sci. Total Environ.* 346, 280-285.

Zhang, Y., Zahir, Z.A., Frankenberger Jr., W.T., 2004. Fate of colloidal-particulate elemental selenium in aquatic systems. *J. Environ. Qual.* 33, 559-564.

Zhang, Y., Amrhein, C., Chang, A., Frankenberger Jr., W.T., 2008a. Effect of zero-valent iron and a redox mediator on removal of selenium in agricultural drainage water. *Sci. Total Environ.* 407, 89-96.

Zhang, Y., Okeke, B.C., Frankenberger Jr., W.T., 2008b. Bacterial reduction of selenate to elemental selenium utilizing molasses as a carbon source. *Bioresour. Technol.* 99, 1267-1273.

Zhang, J., Taylor, E.W., Wan, X., Peng, D., 2012. Impact of heat treatment on size, structure, and bioactivity of elemental selenium nanoparticles. *Int. J. Nanomedicine* 7, 815-825.

Zhang, Z., Zhang, J., Xiao, J., 2014. Selenoproteins and selenium status in bone physiology and pathology. *Biochim. Biophys. Acta* 1840, 3246-3256.

Ziemkiewicz, P.F., O'Neal, M., Lovett, R.J., 2011. Selenium leaching kinetics and in situ control. *Mine Water Environ.* 30, 141-150.

Zimmerman, M.T., Bayse, C.A., Ramoutar, R.R., Brumaghim, J.L., 2015. Sulfur and selenium antioxidants: Challenging radical scavenging mechanisms and developing structure-activity relationships based on metal binding. *J. Inorg. Biochem.* 145, 30-40.

CHAPTER 3

Effect of elevated nitrate and sulfate concentrations on selenate removal by mesophilic anaerobic granular bed reactors

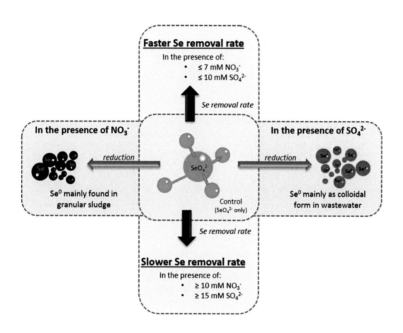

This chapter has been published in modified form:

Tan, L.C, Nancharaiah, Y.V., van Hullebusch, E., Lens, P.N.L. 2018. Effect of elevated nitrate and sulfate concentrations on selenate removal by mesophilic anaerobic granular sludge bed reactors. *Environ. Sci.: Wat. Res. Technol.* 4, 303-314. doi:10.1039/c7ew00307b.

Abstract

Simultaneous removal of selenate (SeO_4^{2-}), nitrate (NO_3^-) and sulfate (SO_4^{2-}), typically present in Se-contaminated wastewaters, by Eerbeek anaerobic granular sludge, was investigated in batch and continuous bioreactor experiments. Batch experiments showed SeO_4^{2-} removal was enhanced to 91% in simulated wastewater with $SeO_4^{2-} + NO_3^- + SO_4^{2-}$ (1:40:100 $SeO_4^{2-}:NO_3^-:SO_4^{2-}$ molar ratios) compared to simulated wastewater with SeO_4^{2-} alone (67%). SeO_4^{2-} removal was severely impacted by high concentrations of SO_4^{2-} ($SeO_4^{2-}:SO_4^{2-} > 1:300$). Removal of SeO_4^{2-}, NO_3^- and SO_4^{2-} at a 1:40:100 ratios was studied in a 2 L lab-scale upflow anaerobic sludge blanket (UASB) reactor operated at 20°C, 24 h hydraulic retention time and 2 g COD/L·d organic loading rate using lactate as electron donor. The removal efficiencies were stabilized at 100, 30 and 80% for NO_3^-, SO_4^{2-} and total Se, respectively, during 92 days of UASB operation. The total Se removal efficiencies dropped to 47% or even to a negative value when, respectively, SO_4^{2-} and NO_3^- were sequentially excluded from the influent. Speciation of Se, particularly the microbial production of colloidal Se^0 levels, was influenced by both SO_4^{2-} and NO_3^-. The results presented here demonstrate that UASB reactors are capable of removing SeO_4^{2-} in the presence of millimolar concentrations of NO_3^- and SO_4^{2-} typically found in Se-wastewaters.

Keywords: anaerobic granular sludge; co-contaminants; selenate removal; co-electron acceptors; simultaneous reduction

3.1   Introduction

Selenium (Se) is a ubiquitous metalloid element and a trace nutrient with a pivotal role in key metabolic functions for animals and humans (Brown and Arthur 2001). Se is used as raw material (i.e. glass, electronics and solar panels), supplement (i.e. vitamins, fertilizer, and wastewater) and in manufacturing products (i.e. shampoo). These anthropogenic activities account for almost 40% of the total Se emissions (10^9 g Se/yr) to the biosphere (Wen and Carignan 2007). Se concentrations in waste streams are typically found at < 12 mg Se/L (**Table 1**). Se has an unusual propensity to bioaccumulate over time in lakes, sediments, planktons, and fishes. This can lead to increase of Se concentrations up to 1.5 to 6-times relative to the background concentration over a long period (Lenz and Lens 2009). Lemly (2014) noted that Se pollution and bioaccumulation in the aquatic environment has caused the increase in mortality rate in fishes and thus impacting the livelihood of humans dependent on fishing with a calculated monetary loss of over $8.6 million annually. In order to avoid pollution and safeguard aquatic life, Se-containing wastewaters must be properly treated prior to discharge to the environment. Treatment of Se-laden wastewater must comply with the stringent discharge limit of 5 µg Se/L set by the United States Environmental Protection Agency (USEPA) (USEPA 2014).

Table 3.1 Selenium-laden wastewater composition (co-contaminants) from different industrial sources.

Wastewater	Se (mM)	SO_4^{2-} (mM)	NO_3^- (mM)	References
Mining-impacted	0-0.2	5-72	4	Santos et al. (2015); Smith et al. (2009)
Se refinery plant	up to 8	0-2	n.i.[*]	Soda et al. (2011)
Flue-gas desulfurization	0-2	31-209	0-7	Smith et al. (2009); Staicu et al. 2017
Agricultural	~0.01	6-106	0-4	Lenz et al. 2009; Smith et al. (2009)

[*]n.i. - no information

Although Se can exist in various forms in the aquatic environment, soluble Se anions such as selenate (SeO_4^{2-}) and selenite (SeO_3^{2-}) are the most common forms of Se encountered in waste streams (Lenz and Lens 2009; Nancharaiah and Lens 2015a). Physical (e.g. membrane filtration and adsorption using ferrihydrite) and chemical (e.g. metal precipitation and electro-coagulation) methods can effectively remove Se anions from wastewaters. But, these methods are costly in terms of energy and chemical consumption and also generate hazardous Se-bearing chemical sludge (Lenz and Lens 2009; Santos et al. 2015). Biological methods based on microbial reduction are effective for treating Se-laden wastewaters by converting soluble Se anions to insoluble and less toxic elemental selenium (Se^0), which can be present in the liquid phase as colloidal particles or immobilized in the biomass (Lenz and Lens 2009; Nancharaiah and Lens, 2015a, 2015b).

Soluble Se anions in wastewaters typically co-exist with other pollutants such as metals (Mal et al. 2016), dissolved solids (Lenz and Lens 2009) and anions (Tan et al. 2016), with nitrate (NO_3^-) and sulfate (SO_4^{2-}) as common constituents in most industrial wastewaters such as mining effluents and acid rock drainage (Johansson et al. 2015). Concentrations of NO_3^- and SO_4^{2-} are typically 100- to 1000-fold higher compared to those of the Se anion concentrations (**Table 3.1**). Therefore, these anions can affect SeO_4^{2-} reduction and decrease the Se removal efficiency of bioreactors (Lenz et al. 2009). Thus, careful monitoring of both the nitrate (Lortie et al. 1992; Takada et al. 2008) and sulfate (Lenz et al. 2008) concentrations are required to avoid inhibitory effects on SeO_4^{2-} reduction. Literature review shows that there is no consensus on the role of NO_3^- and/or SO_4^{2-} on biological SeO_4^{2-} removal (Tan et al. 2016).

Lai et al. (2014) showed that NO_3^- at a higher loading rate of 1.14 g NO_3^--N/m·d inhibited SeO_4^{2-} removal in a hydrogen-fed membrane biofilm reactor (MBfR). Contrary to this, Oremland et al. (1999) observed in batch experiments that the presence of NO_3^- actually stimulated the ability of *Sulfurospirillum barnesii* to reduce SeO_4^{2-} by keeping the cells in a high metabolic state. Zehr and Oremland (1987) observed an inhibition of SeO_4^{2-} removal by SO_4^{2-} at concentrations > 1 mM. In contrast, Chung et al. (2006) have found that biological SeO_4^{2-} reduction was unaffected by the presence SO_4^{2-}. In fact, it was the presence of SeO_4^{2-} that inhibited SO_4^{2-} removal in MBfR (Chung et al. 2006). SeO_4^{2-} and SO_4^{2-} are chemical analogs. Thus, SeO_4^{2-} can interfere with the SO_4^{2-} assimilatory and dissimilatory pathways in sulfate-reducers and inhibit SO_4^{2-} reduction (Hockin and Gadd, 2006). Likewise, Lenz et al. (2009) were able to simultaneously reduce SO_4^{2-} (2 mM) and SeO_4^{2-} (10 µM) in an upflow anaerobic sludge blanket (UASB) bioreactor bioaugmented with *S. barnesii*, but only after NO_3^- (15 mM) was completely denitrified.

As a result, there are only limited studies with contradicting results on biological SeO_4^{2-} removal in the presence of NO_3^- and SO_4^{2-} in bioreactors. Moreover, most studies like those mentioned above (Oremland et al, 2004; Chung et al. 2006; Lenz et al. 2009; Lai et al. 2014) focused on SeO_4^{2-} removal in the presence of either NO_3^- or SO_4^{2-}, whereas only a few studies have investigated SeO_4^{2-} removal in the presence of both NO_3^- and SO_4^{2-} (Tan et al. 2016).

More importantly, thus far, there are no studies that investigated the optimal ratio of NO_3^-, SO_4^{2-}, and SeO_4^{2-} for their simultaneous reduction and determined speciation of reduced Se in the presence of NO_3^- and SO_4^{2-}. Therefore, this study focused on investigating the effect of a range of NO_3^- and SO_4^{2-} concentrations as they occur in real wastewaters on SeO_4^{2-} removal using anaerobic granular sludge. Subsequently, a 2 L lab-scale continuous upflow anaerobic sludge blanket (UASB) reactor was operated with the optimal $SeO_4^{2-}:NO_3^-:SO_4^{2-}$ molar ratio to study the long-term stability of the biological SeO_4^{2-} removal process while treating synthetic mine-impacted wastewater.

3.2 Materials and methods

3.2.1 Source of biomass

Experiments were performed with methanogenic anaerobic granular sludge taken from a full-scale UASB reactor treating paper-mill wastewater (Eerbeek, The Netherlands), described in detail by Roest et al. (2005). Astratinei et al. (2006) showed that Eerbeek methanogenic anaerobic granular sludge is efficient in reducing SeO_4^{2-} when compared to other inoculum types (i.e. Nedalco methanogenic granular sludge, denitrifying biomass or selenogenic biomass). Granular sludge was supplied at 1% and 10% in wet weight per working for batch and bioreactor, respectively.

3.2.2 Synthetic wastewater

All experiments were conducted using synthetic mine-impacted wastewater composed of growth medium, electron donor and the three anions under study. The growth medium (**Table 3.2**) was prepared using the same method as in the study of Lenz et al. (2006), but excluding vitamins and yeast extract. Sodium lactate ($CH_3CH(OH)COONa$; 60%; VMR Chemicals, France) was used as electron donor. Potassium nitrate (KNO_3; Merck, Germany), potassium sulfate (K_2SO_4; Merck, Germany), and sodium selenate (Na_2SeO_4; Sigma-Aldrich, Germany) were the electron acceptors applied at different concentrations (**Table 3.3**).

3.2.3 Batch experiments for determining reduction profiles

Concentrations of NO_3^-, SO_4^{2-} and SeO_4^{2-} used were modeled after mine-impacted wastewater as shown in **Table 3.1**. SeO_4^{2-} reduction at different $NO_3^-:SeO_4^{2-}$ and $SO_4^{2-}:SeO_4^{2-}$ molar ratios were determined in batch tests using serum bottles as shown in **Table 3.3**. The initial SeO_4^{2-} concentration (0.1 mM) used was always at 0.1 mM for all conditions and all experiments were provided with excess lactate (20 mM). Assuming complete reduction of anions to Se^0, N_2 and HS^-, one mole of SeO_4^{2-}, NO_3^- and SO_4^{2-} requires 0.4, 0.5 and 0.7 moles of lactate, respectively. Batch tests were conducted using the following molar ratio: (a) SeO_4^{2-} :NO_3^- = 1:20, 1:40, 1:70, 1:100; (b) SeO_4^{2-}:SO_4^{2-} = 1:50, 1:100, 1:150, 1:200, 1:300; (c) SeO_4^{2-} :NO_3^-:SO_4^{2-} = 1:40:50, 1:40:100, 1:40:150, 1:40:200, 1:40:300; and (d) equimolar concentrations at 0.5 mM each for the three anions investigated.

Control incubations for individual anions (NO_3^--control, SO_4^{2-}-control, SeO_4^{2-}-control) were also performed to establish how much of each anion is individually reduced without competition, if any, from the other anions present in the system. Additional controls (biomass-control) with no substrate addition and autoclaved granular sludge exposed to the three anions combination were included to check for physiochemical processes occurring in the incubations for 5 days.

Table 3.2 Composition of the synthetic wastewater used

Compound	Batch bottles	UASB reactor
	Concentration (g/L)	
NH_4Cl	0.30	0.30
$CaCl_2 \cdot 2H_2O$	0.015	0.010
$MgCl_2 \cdot 6H_2O$	0.12	0.010
KCl	0.25	-
$Na_2HPO_4 \cdot 2H_2O$	0.29	0.053
KH_2PO_4	0.25	0.041
$NaHCO_3$	-	0.040
Acid trace metals solution[a]	0.1 mL/L	
Alkaline trace metals solution[b]	0.1 mL/L	

[a] *Acid trace metals solution (mM): 7.5 $FeCl_2$, 1 H_3BO_4, 0.5 $ZnCl_2$, 0.1 $CuCl_2$, 0.5 $MnCl_2$, 0.5 $CoCl_2$, 0.1 $NiCl_2$, 50 HCl*

[b] *Alkaline trace metals solution (mM): 0.1 Na_2WO_4, 0.1 Na_2MoO_4, 10 NaOH*

Prepared feed solutions were adjusted to neutral pH, poured into bottles that were sealed tightly and flushed with an excess of N_2 gas for 5-10 min to make the medium anaerobic. The batch experiments were performed in a 30°C room, placed on an orbital shaker (180 rpm) for homogeneous mixing and incubated for 5 days (t = 120 h). The bioconversion was monitored through sampling at 24 h time intervals.

The influence of NO_3^- and SO_4^{2-} concentrations on the end-product of the SeO_4^{2-} conversions were investigated in a separate experiment by making a Se mass balance. All batch tests were performed in triplicates. The incubation conditions for the Se mass balance experiment were the same as described above (batch experiments), except that sampling was only conducted at the end of the run (t = 120 h). The Se mass balance was determined for the total Se in the liquid, total dissolved Se in the liquid, unaccounted dissolved Se, Se anions (SeO_4^{2-} and SeO_3^{2-}), colloidal Se^0 in the liquid, Se in the biomass and Se in the gas phase.

$$Total\ Se\ in\ the\ liquid = total\ dissolved\ Se + colloidal\ Se^0 \qquad \text{Eq. (3.1)}$$

$$Unaccounted\ dissolved\ Se = total\ dissolved\ Se - SeO_4^{2-} - SeO_3^{2-} \qquad \text{Eq. (3.2)}$$

$$Se\ in\ gas\ phase = initial\ Se - total\ Se\ in\ liquid - Se\ in\ the\ biomass \qquad \text{Eq. (3.3)}$$

Table 3.3 Experimental design for reduction profile experiments in batch bottles and the UASB reactor

Parameters	Serum bottles	UASB reactor
Total volume (mL)	300	2500
Working volume (mL)	200	2000
Biomass, wet weight (g)	2	200
Mixing	180 rpm	re-circulation + upflow velocity
pH	7.0	7.0 - 7.5
Temperature (°C)	30	20 (room temp)
Lactate (mM)	20	20
SeO_4^{2-} (mM)	0.1	0.1
NO_3^- (mM)	2, 4, 7, 10	4, 10, 18
SO_4^{2-} (mM)	5, 10, 15, 20, 30	15 - 20
Molar ratio		
NO_3^-:SeO_4^{2-}	20 - 100	40, 100, 180
SO_4^{2-}:SeO_4^{2-}	50 - 300	150 - 200

3.2.4   UASB reactor operation

A 2 L continuous UASB bioreactor (**Fig. 3.1**) was used to investigate the reactor performance in removing SeO_4^{2-} when exposed simultaneously to anion co-contaminants. The operating parameters and experimental design for the UASB reactor are summarized in **Fig. 3.1c** and **Table 3.3**, respectively. The pH was maintained at 7.0 using a phosphate buffer and

NaHCO$_3$ mixed in the feed solution. The reactor was operated in a thermostatic room of 25 ($\pm$ 2) °C, while the reactor bulk liquid had an average temperature of 20°C. The superficial liquid upflow velocity of the reactor was set to 1.5 m/h and the hydraulic retention time (HRT) to 24 h.

Table 3.4 Operating conditions of the continuously operated UASB reactor used

| Experimental Periods | Days | Influent parameters and concentration | | | |
| | | COD | NO$_3^-$ | SO$_4^{2-}$ | SeO$_4^{2-}$ |
		mg/L (mM)	mg/L (mM)	mg/L (mM)	mg Se/L (mM)
I[a]	0-29	1780 (20)	248 (4)	1440 (15)	8 (0.1)
II	30-92	1780 (20)	248 (4)	1440 (15)	8 (0.1)
III[b]	93-168	1780 (20)	248 (4)	1440 (15)	8 (0.1)
IV(a)	169-184	1780 (20)	248 (4)	0	8 (0.1)
IV(b)	185-202	1780 (20)	620 (10)	0	8 (0.1)
IV(c)	203-211	1780 (20)	248 (4)	0	8 (0.1)
IV(d)	212-231	1780 (20)	1116 (18)	0	8 (0.1)
V	232-266	1780 (20)	0	0	8 (0.1)

[a] start-up period of the reactor; [b] reactor crashed due to influent pump malfunction (day 93-110); attempted to recover reactor performance was done by increasing pH from 7.0 to 8.5

The UASB reactor operating conditions are summarized in **Table 3.4**. The reactor was operated at an organic loading rate (OLR) of 2 g COD/L·d using lactate as the electron donor. The start-up operation (period I) lasted for 29 days; after which, the succeeding operation (period II) was considered to be under pseudo-steady-state condition. The influent pump malfunction for 15 days which lead to the reactor failing and attempts to recover the reactor performance was done by increasing the influent pH from 7.0 to 8.5 (phase III). From periods I to III, the UASB reactor was operated under simultaneous exposure to the three anions under investigation. To study the effect of the NO$_3^-$ concentration on the Se removal efficiency, period IV was operated under NO$_3^-$ + SeO$_4^{2-}$ conditions (SO$_4^{2-}$ withdrawn from the feed). Period IV was divided into 4 parts in which the NO$_3^-$ concentration varied from 4 mM to 18 mM. For period V, NO$_3^-$ was removed from the influent and the UASB reactor operated under SeO$_4^{2-}$ only fed conditions.

3.2.5 Analytical methods

Effluent samples were analyzed for the concentration of residual Se as total Se (Se$_{tot}$) and dissolved Se (Se$_{diss}$). Se$_{tot}$ includes dissolved Se and colloidal Se^0 present in the liquid phase, while Se$_{diss}$ includes only Se forms dissolved in the liquid phase. For Se$_{tot}$ measurements, liquid samples were acidified with concentrated HNO_3 and analyzed using an atomic absorption spectroscopy graphite furnace (AAS-GF, ThermoElemental Solaar MQZe GF95, Se lamp at 196.0 nm). For Se_{diss} measurements, liquid samples were first centrifuged at 37,000 g for 15 min to remove suspended cells and colloidal Se^0 particles, followed by filtration through a 0.45 μm membrane filter (Jain et al. 2016). The supernatant was used to measure Se_{diss} using AAS-GF. The Se content of the granular sludge (0.5 g dry weight) was also measured after acid digestion (MARS 5 pKo Temp CEM Microwave) and Se analysis of the digestate via AAS-GF.

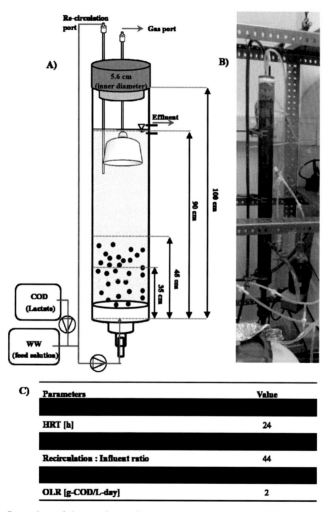

Fig. 3.1 (A) Configuration of the continuously operated UASB reactor, (B) digital photograph of the actual set-up, and (C) reactor operating parameters used.

Lactate, NO_3^-, SO_4^{2-} and SeO_4^{2-} were measured using ion chromatography (IC, Dionex ICS 1000 AS4A column) at retention times of approximately 1.3, 3.9, 7.5, and 8.0 min, respectively.

However, at high concentrations, the SO_4^{2-} peak interfered with the SeO_4^{2-} peak. Due to this, SeO_4^{2-} was analyzed as Se_{diss} (subtracting SeO_3^{2-} concentration) using AAS-GF whenever SO_4^{2-} was present in the wastewater. When appropriate, SeO_3^{2-} was measured as described by Li et al. (2014) using a UV/Vis spectrophotometer (Lambda 365, Perkin-Elmer). All liquid samples were filtered through 0.45 μm cellulose acetate syringe filters (Sigma-Aldrich, USA) to remove any particulate matter prior to analysis. Chemical oxygen demand (COD), nitrite (NO_2^-), total dissolved sulfide (TDS) and pH were measured using standard methods (APHA/AWWA/WEF 2005).

3.2.6   FISH imaging

Granules remained within the sludge bed at the bottom of the reactor and biomass loss was negligible during the entire reactor operation. No additional biomass was added during the entire duration of the reactor run. Specific withdrawal of samples of granular sludge (150 mL) at the end of each period was done (taken from the sampling port **Fig. 3.1**) before shifting to the next period. The microbial communities present in the sludge samples were qualitatively analyzed using fluorescence in situ hybridization (FISH). Vigorous biomass mixing within the reactor was done first by increasing the liquid upflow velocity to ensure a good representation of biomass sampling.

All microbial cells were visualized by DAPI straining with specific oligonucleotide probes ARCH915 (Sekiguchi et al. 1999) and EUB I-III (Daims et al. 1999) for archaeal and bacterial cells, respectively. Two subgroups of the proteobacteria (beta-proteobacteria and gamma-proteobacteria) were included using BET42a and GAMMA42a probes (Lee et al. 2008). Sulfate-reducing bacteria (SRB) were visualized using the *Desulfobulbaceae* DBB60 (Sekiguchi et al. 1999) and *Desulfovibrionales* SRB385 (Amann et al. 1990) probes. The inoculum anaerobic granular sludge was used as the control.

3.3   Results

3.3.1.   Reduction profiles of NO_3^-, SO_4^{2-} and SeO_4^{2-} by anaerobic granular sludge

Simultaneous reduction and the optimal molar ratio of $SeO_4^{2-}:NO_3^-:SO_4^{2-}$ for SeO_4^{2-} removal were investigated in batch experiments at 30°C and near-neutral pH. The percentage removal of Se_{diss} in the absence and presence of NO_3^-, SO_4^{2-} and both NO_3^- and SO_4^{2-} at different molar ratios are shown in **Fig. 3.2**, while the removal of lactate, NO_3^- and SO_4^{2-} under different experimental conditions are shown in **Appendix 1, Table S3.1**.

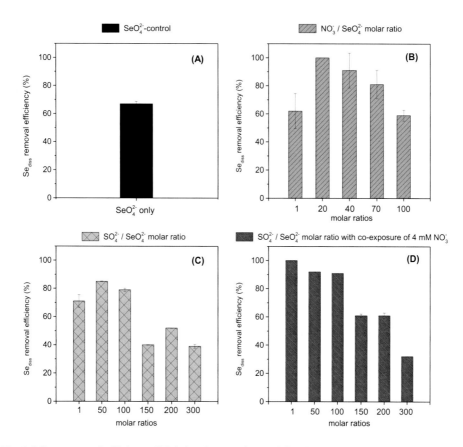

Fig. 3.2 Se$_{diss}$ removal efficiency (%) in batch tests after 120h incubation for different molar ratios: (A) SeO$_4^{2-}$, (B) NO$_3^-$/SeO$_4^{2-}$, (C) SO$_4^{2-}$/SeO$_4^{2-}$ and (C) SO$_4^{2-}$/SeO$_4^{2-}$ with 4 mM NO$_3^-$. At equimolar conditions, 0.5 mM was used for all oxyanions while 0.2 ($\pm$ 0.1) mM SeO$_4^{2-}$ was used for the rest of the molar ratio conditions.

Removal of individual oxyanions

Removal of the three anions was observed when supplied individually along with lactate at 100% for NO$_3^-$, 67 ($\pm$ 2)% for SeO$_4^{2-}$, and 15 ($\pm$ 0.4)% for SO$_4^{2-}$. No reduction of NO$_3^-$, SO$_4^{2-}$ or SeO$_4^{2-}$ was observed with autoclaved biomass or in the absence of lactate (**Appendix 1, Table S3.1**).

Table 3.5 Se mass balance experiment accounting for reduced Se speciation after 5 days of incubation under different batch conditions

Batch conditions	Molar ratio	Initial Se (µM)	Unaccounted Se$_{diss}$	Se content in different phases (%)				
				SeO$_4^{2-}$	SeO$_3^{2-}$	Colloidal Se^0	Se in biomass	Gaseous Se
SeO$_4^{2-}$ only	n/a	230 (±26)	27 (±1)	27 (±2)	7 (±4)	9 (±1)	29 (±1)	2.0 (±0.1)
SeO$_4^{2-}$ + NO$_3^-$	1:40	250 (±5)	19 (±2)	24 (±1)	5 (±1)	19 (±1)	32 (±4)	1.0 (±0.6)
	1:100	228 (±5)	15 (±1)	31(±6)	8 (±2)	19 (±4)	26 (±5)	1.9 (±0.1)
SeO$_4^{2-}$ + SO$_4^{2-}$	1:100	226 (±4)	0 (±0)	n/a	15 (±3)	66 (±7)	17 (±5)	2.1 (±0.5)
	1:200	223 (±28)	33 (±3)	n/a	15 (±2)	39 (±1)	11 (±1)	1.8 (±0.2)
SeO$_4^{2-}$ + NO$_3^-$ + SO$_4^{2-}$	1:40:100	226 (±30)	0 (±0)	n/a	15 (±2)	69 (±1)	14 (±3)	2.0 (±0.4)
	1:40:200	220 (±6)	30 (±4)	n/a	11 (±1)	46 (±3)	12 (±1)	1.7 (±0.4)

SeO_4^{2-} removal in the presence of NO_3^-

SeO_4^{2-} removal was determined at different SeO_4^{2-}:NO_3^- molar ratios of 1 to 100. SeO_4^{2-} removal was unaffected by NO_3^- at equimolar concentrations and at a high SeO_4^{2-}:NO_3^- molar ratio of 1:100. Increased SeO_4^{2-} removal efficiencies ranging from 81 to 100% were observed for SeO_4^{2-}:NO_3^- molar ratios of 1:20 to 1:70 (**Fig. 3.2 B**). At intermediate molar ratios, NO_3^- had a stimulating effect on the SeO_4^{2-} removal (**Fig. 3.2 B**). No SeO_3^{2-} accumulation was detected throughout the batch tests and complete NO_3^- removal was observed within 48 h for all molar ratios tested (**Appendix 1, Table S3.1**). Transient NO_2^- accumulation (maximum at 0.13 ($\pm$ 0.02) mM during 21 h) and its removal (below detection limit after 52 h) was noticed (**Appendix 1, Fig. S3.1**).

Acclimatization of the granular sludge to conditions of NO_3^- + SeO_4^{2-} or SeO_4^{2-} only were investigated at concentrations of 4 mM NO_3^- and 0.1 mM SeO_4^{2-} (**Appendix 1, Fig. S3.2**). Granular sludges from the SeO_4^{2-}-control and for a molar ratio NO_3^-/SeO_4^{2-} of 40 at week 0 experiments were re-incubated for 2 more weeks. The NO_3^- + SeO_4^{2-} incubation did not show differences on SeO_4^{2-} removal, attaining a consistent average of 90 ($\pm$ 2)% Se removal efficiency after 120 h for all 3 weeks (**Appendix 1, Fig. S3.2b**). The SeO_4^{2-}-control, on the other hand, slightly increased from 67 ($\pm$ 2)% to 85 ($\pm$ 1)% in the third week of re-incubation (**Appendix 1, Fig. S1a**). This suggests the enrichment of SeO_4^{2-} reducing bacteria after repeated exposure, contributing to a higher SeO_4^{2-} reduction efficiency. Overall, there was no inhibition of the Se removal at high NO_3^- concentrations. In contrast, NO_3^- enhanced the SeO_4^{2-} removal under low NO_3^- concentrations (NO_3^-/SeO_4^{2-} molar ratio < 70).

Se speciation and mass balance analysis showed comparable results for NO_3^- + SeO_4^{2-} and SeO_4^{2-} only conditions (**Table 3.5**). Most of the Se in the NO_3^- + SeO_4^{2-} batch incubations could be accounted for in the form of SeO_4^{2-}, whereas Se in granular sludge amounted to 71 ($\pm$ 8), unaccounted Se_{diss} to 42 ($\pm$ 4) and colloidal Se^0 to 47 ($\pm$ 1) μM Se. Unaccounted Se_{diss} includes Se compounds other than SeO_4^{2-} and SeO_3^{2-}, such as organo-Se compounds, which were not measured due to analytical limitations. Notably, the colloidal Se^0 fraction had almost doubled when SeO_4^{2-} was supplied along with NO_3^-. Higher (1:100) or lower (1:40) SeO_4^{2-}:NO_3^- molar ratios showed no marked differences in the Se speciation.

SeO_4^{2-} removal in the presence of SO_4^{2-}

Similar to what was observed with NO_3^-, Se_{diss} showed increased removal efficiency when the $SeO_4^{2-}:SO_4^{2-}$ molar ratios were $\leq 1:100$ (**Fig. 3.2C**). At both $SeO_4^{2-}:SO_4^{2-}$ molar ratios of 1:50 and 1:100, the Se removal efficiency amounted to an average of 82 ($\pm$ 4)%, which is 23% higher than the Se removal efficiency in the SeO_4^{2-} only batch incubations. At $SeO_4^{2-}:SO_4^{2-}$ molar ratios of $\geq 1:150$ to 1:300, the Se removal efficiency was further reduced to an average of 44 ($\pm$ 7)%.

At SO_4^{2-} supplemented conditions, a SO_4^{2-} removal efficiency of only about 15 ($\pm$ 0.4)% (SO_4^{2-} only) and 6 ($\pm$ 2)% ($SO_4^{2-}+NO_3^-$) was attained (**Appendix 1, Table S3.1**). Higher total dissolved sulfide (TDS) concentrations were produced after 120 h in serum bottles without SeO_4^{2-}, an average of 4 ($\pm$ 1) mM at both SO_4^{2-} only and $SO_4^{2-} + NO_3^-$ conditions. In contrast, TDS measured in the presence of SeO_4^{2-} for all conditions was low at an average concentration of 0.7 ($\pm$ 0.3) mM. An average SO_4^{2-} removal efficiency of 26 ($\pm$ 8)% was obtained for molar ratios $SeO_4^{2-}:SO_4^{2-} > 1:1$, while at the ratio $SeO_4^{2-}:SO_4^{2-} = 1:1$, the SO_4^{2-} removal efficiency was 63 ($\pm$ 10)%.

In contrast to the Se speciation mass balance results for $SeO_4^{2-} + NO_3^-$ (**Table 3.5**), there was a notable difference in the Se mass balance when SO_4^{2-} was present in batch incubations. Compared to the SeO_4^{2-} only and $NO_3^- + SeO_4^{2-}$ conditions, a higher percentage of colloidal Se^0 was observed for both $SeO_4^{2-}:SO_4^{2-}$ molar

SeO_4^{2-} removal in the presence of both SO_4^{2-} and NO_3^-

An increased Se removal efficiency was observed when all three anions were present in the solution at $SeO_4^{2-}:NO_3^-:SO_4^{2-}$ molar ratios of 1:40:50 and 1:40:100 (**Fig. 3.2D**), amounting to 92 ($\pm$ 1)% Se removal efficiency, which is 25% higher than the SeO_4^{2-} only condition. Initially, this increase in Se removal efficiency was considered to be the effect of NO_3^-. However, under the same conditions without NO_3^- (**Fig. 3.2C**), the Se percentage removal difference was marginally at only 10% lower. Increasing the $SeO_4^{2-}:NO_3^-:SO_4^{2-}$ molar ratios to 1:40:150 and 1:40:200 showed similar Se removal efficiencies to the SeO_4^{2-} only batch incubation. Further increasing the $SeO_4^{2-}:NO_3^-:SO_4^{2-}$ molar ratio to 1:40:300 lowered the Se removal efficiency to an average of 16 ($\pm$ 1)%. Complete NO_3^- removal was attained within 48 h, while SO_4^{2-} removal was only marginal at 24 ($\pm$ 4)% for all $SeO_4^{2-}:NO_3^-:SO_4^{2-}$ molar ratios investigated (**Appendix 1, Table S3.1**). On the other hand, no major differences were noted for the Se speciation mass balance with or without $SO_4^{2-} + SeO_4^{2-}$ or with NO_3^- (**Table 3.5**).

3.3.2 Treatment of synthetic mine-impacted wastewater in a UASB reactor

Based on the batch experiments, the highest possible $SeO_4^{2-}:NO_3^-:SO_4^{2-}$ molar ratio to observe an increase in SeO_4^{2-} removal was 1:40:100 (**Fig. 3.2**). Therefore, a concentration of 0.1 mM SeO_4^{2-}, 4 mM NO_3^- and 10 mM SO_4^{2-} was chosen for the study of Se removal in a continuous UASB reactor.

UASB reactor treating wastewater with co-electron acceptors

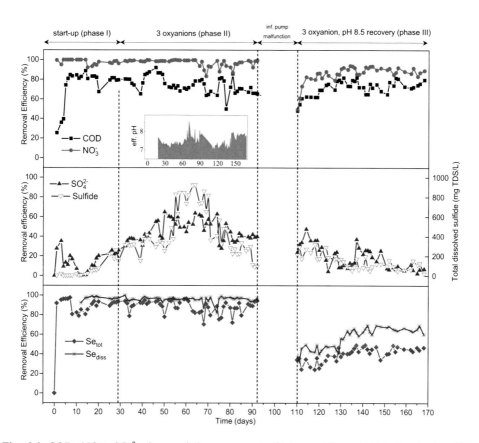

Fig. 3.3 COD, NO_3^-, SO_4^{2-}, Se_{tot} and Se_{diss} removal efficiency (%) and total dissolved sulfide concentration for the UASB reactor operated for 92 days (phase I-II). From day 93-110, influent pump malfunction occurred, causing the reactor crash. Day 110-168, reactor was operated at pH 8.5 for recovery period (phase III).

Fig. 3.3 shows the COD, NO_3^-, SO_4^{2-} and SeO_4^{2-} removal efficiencies during phase I-II of UASB reactor operation. The effluent pH varied from 7-8 and the COD consumption was about

70 ($\pm$ 11)%. NO_3^- removal was similar to what was observed in batch tests, attaining complete removal within 24 h. Se_{tot} and Se_{diss} removal efficiencies were also high, achieving an average of 78 ($\pm$ 11)% and 93($\pm$ 5)%, respectively. Only SO_4^{2-} showed a poor removal efficiency, achieving an average of 38 ($\pm$ 16)%, but producing a high TDS concentration of 12 ($\pm$ 6) mM (383 $\pm$ 204 mg TDS/L). The presence of NO_3^- and operation at 20°C most likely contributed to the poor SO_4^{2-} removal of the UASB reactor.

On day 93, the influent pump malfunctioned for 15 days causing for the reactor to crash. An attempt to recover the bioreactor by increasing the influent pH from 7.0 to 8.5 was conducted from day 110 until day 168 (58 days, Phase III, **Fig. 3.3**) There was an observed 30-40% dropped in reactor performance during initial re-startup period particularly for Se removal efficiencies. Phase III reactor performance showed gradual improvement for both NO_3^- and Se removal efficiencies but did not restore the reactor performance to same level achieved during phase II. An average removal efficiency of 85 ($\pm$ 19)% NO_3^-, 17 ($\pm$ 25)% SO_4^{2-}, 42 ($\pm$ 21)% Se_{tot}, and 64 ($\pm$ 20)% Se_{diss} was achieved. It was observed that SO_4^{2-} did not recovery at all during phase III. Therefore, SO_4^{2-} was removed for the succeeding phases and reactor was operated at $NO_3^- + SeO_4^{2-}$ condition while influent pH was lowered back to 7.0.

UASB reactor treating NO_3^- and SeO_4^{2-} contaminated wastewater

Phase IV (**Fig. 3.4**) was conducted under $NO_3^- + SeO_4^{2-}$ conditions (SO_4^{2-} was omitted) with progressively increasing NO_3^- concentrations (**Table 3.4**). There was no notable change in COD consumption (72 $\pm$ 4%) for phase IV compared to the previous phases (I-III). The effluent pH fluctuated, from 7 and then reaching 8-9, most likely due to the reduction of NO_3^-. NO_3^- removal was completed for the entire phase IV, attaining an average of 97 ($\pm$ 1)%. NO_2^- (maximum at 22 mg NO_2^--N/L) was present in the effluent, but did not accumulate in the UASB reactor. There was a distinct change in the Se removal efficiency during phase IVb (10 mM), which increased to 55 ($\pm$ 17)% Se_{tot} and 79 ($\pm$ 4)% Se_{diss} removal. Increasing the NO_3^- concentration to 18 mM (IVd) resulted in a decrease in the Se_{tot} and Se_{diss} removal efficiencies to 31 ($\pm$ 14)% and 53 ($\pm$ 12)%, respectively. This could indicate that increasing the NO_3^- concentration to 10 mM improved reactor recovery in terms of re-establishing the SeO_4^{2-} removal. However, increasing NO_3^- concentrations to 18 mM (phase IVd) negatively impacted the SeO_4^{2-} removal performance in the reactor (**Fig. 3.4**).

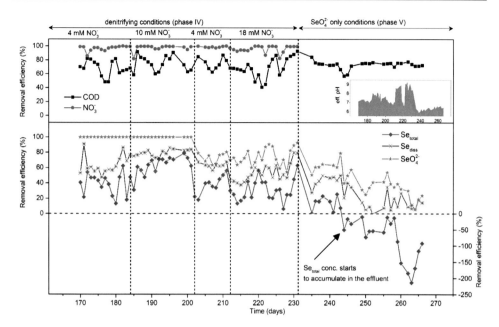

Fig. 3.5 COD, NO_3^-, Se_{tot} and Se_{diss} removal efficiency (%) for the UASB reactor operated under NO_3^- + SeO_4^{2-} condition with varying NO_3^- concentrations (168-231 days, phase IV) and under SeO_4^{2-}-only reducing conditions (232-266 days, phase V).

UASB reactor treating only SeO_4^{2-} contaminated wastewater

Phase V (**Fig. 3.4**) was operated from day 232 to 266 under solely SeO_4^{2-} reducing conditions to compare with phases II and IV. The effluent pH was consistently below 7, despite no changes made in the feed composition. This supports the role of NO_3^- reduction in achieving an effluent pH of 7 or higher during phase IV. COD consumption in phase V was at 73 ($\pm$ 5)%. Conversely, Se removal during phase V was very poor, with increasing Se_{tot} (12 $\pm$ 4 mg Se_{tot}/L) concentrations present in the effluent. The Se_{diss} removal efficiency was only 25 ($\pm$ 17)%, while the SeO_4^{2-} removal efficiency was 42 ($\pm$ 18)%. This suggests that the microbial community developed in the UASB reactor during phase IV required NO_3^- to achieve Se removal. Low SeO_3^{2-} concentrations at < 1.7 mg Se/L were detected in the effluent throughout the reactor run. Se concentrations trapped in the biomass sampled on day 266 (**Appendix 1, Fig. S3.3**) averaged 62 ($\pm$ 39) mg Se/g TS.

3.3.3   Microbial community analysis through FISH imaging

FISH image analysis of the inoculum granular sludge (**Fig. 3.5a**) showed both archaeal (green) and bacterial (blue) communities, with only a few SRB present (red). In contrast, a good

89

mix of archaea, bacteria and SRB were observed at the end of phase II (**Fig. 3.5b**), when fed with three oxyanions. In phase III (**Fig. 3.5c**), with reactor failure, fewer bacteria were observed, while no archaea or SRB communities were detected anymore. Archaea repopulated in phase IVb (10 mM NO_3^-, **Fig. 3.5d**), but were not observed further on during operation under phase IVd (18 mM NO_3^-, **Fig. 3.5e**) and V (SeO_4^{2-} reducing condition, **Fig. 3.5f**). The proteobacteria subgroups beta-proteobacteria and gamma-proteobacteria showed shifts along the different operational phases. At phase II, an even mixture of beta-proteobacteria and gamma-proteobacteria was observed (**Fig. 3.5b**); however, the gamma-proteobacteria community dominated from phase III onwards until phase V (**Fig. 3.5c and 3.5f**).

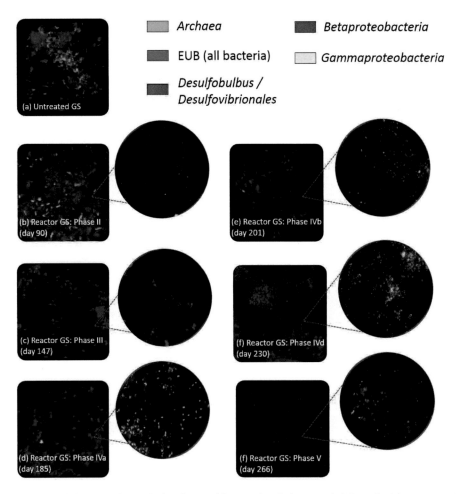

Fig. 3.5 Microbial community analysis of anaerobic granular sludge sampled from the laboratory-scale UASB reactor using *fluorescence in situ hybridization* (FISH) technique. FISH images are grouped into EUB-SRB-Archaea (box shape) and BETA-GAMMA (circle shape).

3.4 Discussion

3.4.1 SeO$_4^{2-}$ removal with co-electron acceptors in batch and continuous systems

This study shows that both NO$_3^-$ and SO$_4^{2-}$ concentrations strongly influence SeO$_4^{2-}$ removal by anaerobic granular sludge. At SO$_4^{2-}$ + SeO$_4^{2-}$ + NO$_3^-$ conditions, as in real mine-impacted wastewater or drainage water, a slightly higher Se removal efficiency (10% higher) was observed compared to SO$_4^{2-}$ + SeO$_4^{2-}$ incubations. This observation suggests that the presence of both SO$_4^{2-}$ and NO$_3^-$ actually has a stimulating effect on SeO$_4^{2-}$ removal. However, the SeO$_4^{2-}$ removal percentage decreased when NO$_3^-$ and SO$_4^{2-}$ concentrations exceeded 7 mM and 10 mM, respectively (**Fig. 3.2**). Based on batch experiments, the highest possible SeO$_4^{2-}$:NO$_3^-$:SO$_4^{2-}$ molar ratio showing increased SeO$_4^{2-}$ removal (compared to control condition) was at 1:40:100, achieving 90% SeO$_4^{2-}$ removal (**Fig. 3.2**). Increasing the SeO$_4^{2-}$:NO$_3^-$:SO$_4^{2-}$ molar ratio to more than 1:40:100 did not show any positive effect on SeO$_4^{2-}$ removal (**Fig. 3.2**).

The optimal concentration ratios of SeO$_4^{2-}$:NO$_3^-$:SO$_4^{2-}$ molar ratio was applied to the bioreactor to determine long-term performance in a continuous system. One of the aims of the reactor study was to investigate whether the anaerobic granular sludge was able to simultaneously reduce NO$_3^-$, SO$_4^{2-}$ and SeO$_4^{2-}$ from the beginning without biomass modification or adaptation. Reactor studies considering simultaneous removal of NO$_3^-$, SO$_4^{2-}$ and SeO$_4^{2-}$ are few between. One study by Lenz et al. (2009) operated an UASB reactor to treat synthetic agricultural drainage wastewater containing 21 mM nitrate, 2 mM SO$_4^{2-}$ and 0.01 mM SeO$_4^{2-}$ using Eerbeek granular sludge bioaugmented with Se-reducing organism *S. barnesii*. Although the use of bioaugmented *S. barnesii* was able to improve Se removal efficiencies, their study was only able to achieve high SeO$_4^{2-}$ and total Se removal when NO$_3^-$ was omitted from the feed, while no SO$_4^{2-}$ removal was achieved when NO$_3^-$ was present.

In contrast to Lenz et al. (2009), this work has demonstrated simultaneously high removal efficiency of NO$_3^-$ and SeO$_4^{2-}$ with minimal SO$_4^{2-}$ removal without biomass modification or adaptation. SeO$_4^{2-}$ was reduced efficiently (> 80%) from 10 ($\pm$ 2) to 0.9 ($\pm$ 0.1) mg Se/L with a total Se effluent concentration of 2.0 ($\pm$ 0.6) mg Se/L in the UASB bioreactor fed with both high concentrations of NO$_3^-$ and SO$_4^{2-}$. SeO$_4^{2-}$ removal was most likely mediated through microbial reduction of SeO$_4^{2-}$ to Se^0 as evident from the typical red color of the Se^0 (**Appendix 1, Fig. S3.4**) and continual increase of Se content in the biomass (**Appendix 1, Fig. S3.3**). UASB reactor was operated with a relatively higher HRT of 24 h to establish high removal efficiencies of anions. SeO$_4^{2-}$ was efficiently removed and the majority of the biotransformed

Se was retained in the granular sludge. But the total Se leaving the treated water (~2 mg Se/L) from UASB reactor is far from the current water quality criterion of fresh water discharge. Therefore, additional post-treatment techniques like simple slow sand filtration or dissolved air flotation can be used to further reduce the total Se concentration prior to discharge (CH2M HILL 2010).

Influence of NO_3^- on SeO_4^{2-} removal

The influence of NO_3^- on the SeO_4^{2-} removal has been previously reported, i.e. different loading rates of NO_3^- can influence Se removal (Chung et al. 2006; Takada et al. 2008). Takada et al. (2008) observed that the reduction of SeO_4^{2-} to SeO_3^{2-} was not affected by < 500 mg NO_3^--N/L (36 mM NO_3^-). However, further reduction of SeO_3^{2-} to Se^0 was found to be highly inhibited by NO_3^- at concentrations as low as 5 mg NO_3^--N/L (0.4 mM NO_3^-). Chung et al. (2006) observed that a small amount of NO_3^- (5 mg N/L) improved the SeO_4^{2-} removal in a MBfR, provided that electron donor (H_2) supply was not limiting. Dessì et al. (2016) operated UASB reactors at under 30 and 55°C under SeO_4^{2-}-reducing condition for biomass adaptation to Se before introducing NO_3^- into the feed. The authors observed no increase in Se removal in the presence of NO_3^- at a NO_3^-:SeO_4^{2-} molar ratio of 2 and 10. However, at a NO_3^-:SeO_4^{2-} molar ratio of 100:1, a decrease of about 10-20% in Se (Se_{tot} and Se_{diss}) removal was observed particularly in mesophilic conditions. In contrast, in this study, the reactor operation under denitrifying condition improved Se removal (from 60% to 80% Se_{diss} removal) at a NO_3^-:SeO_4^{2-} molar ratio of 100:1 (**Fig. 3.4** phase IVb). However, increasing the NO_3^-:SeO_4^{2-} molar ratio to 180:1 (**Fig. 3.4** phase V) led to a decrease in Se removal; while omitting NO_3^- completely impacted Se removal (**Fig. 3.4** phase V). Both the batch and continuous system showed that when NO_3^- concentrations were too high, there was a decrease in Se removal, possibly linked to inhibition of SeO_4^{2-} reductase.

The effect of NO_3^- in improving SeO_4^{2-} removal has been linked to specific microbial pathways and metabolic activities (DeMoll-decker and Macy 1993; Oremland et al. 1999; Yee et al. 2007). One plausible explanation for the influence of NO_3^- on SeO_4^{2-} removal lies in the reductases available for the conversion of both NO_3^- and SeO_4^{2-} in facultative bacteria. SeO_4^{2-} reduction can be achieved by the action of either SeO_4^{2-} reductase (*Ser*ABC) or NO_3^- reductase (*Nar* and *Nap*), which is then enzymatically reduced to Se^0 by periplasmic NO_2^- reductase (*Nir*) (Nancharaiah and Lens 2015a). Depending on the microorganism, a higher preference to reduce NO_3^- in the same reduction pathway with SeO_4^{2-} or a separate reduction pathway for both NO_3^-

and SeO_4^{2-} can be present, this could indicate whether competition between the two anions will occur or not (DeMoll-decker and Macy 1993).

Another explanation for the enhanced SeO_4^{2-} removal in the presence of NO_3^- could be the state of reduction activity of the microorganism. Oremland et al. (1999) showed that the SeO_4^{2-} removal activity of *S. barnesii* was dependent on the growth medium. Authors observed that when pre-grown in the presence of NO_3^- (< 1 mM), SeO_4^{2-} reduction was faster (7 h) as compared to the cells grown in the absence of NO_3^- (12 h). Oremland et al. (1999) hypothesized that the presence of NO_3^- promoted a higher state of metabolic activity and, therefore, achieved faster SeO_4^{2-} reduction rates. The acclimatization experiment indicates a similar result wherein under $NO_3^- + SeO_4^{2-}$ condition, SeO_4^{2-} was consistently removed at ~90% removal efficiency from initial incubation until week 2 of re-incubation (**Appendix 1, Fig. S3.2b**). On the other hand, SeO_4^{2-} only batch condition showed a rise in SeO_4^{2-} removal from ~60 to ~80% after repeated batch exposure indicating that Se-reducers needed further time to develop and be metabolically active (**Appendix 1, Fig. S3.2a**).

An in-depth study of the formed microbial community and subsequent enzymatic factors could potentially give more information on the influence of NO_3^- and SO_4^{2-} on SeO_4^{2-} removal. Though phylogenetically diverse, many denitrifying bacteria belong to the gamma subclass *Proteobacteria* phylum (Lai et al. 2014). FISH images (**Fig. 3.5**) indicated that the induced gamma-proteobacteria dominate the community until the end of the operation run. Both studies by Dessì et al. (2016) and Lai et al. (2014) showed that the SeO_4^{2-} reducing community was largely shaped by the presence of NO_3^- in the bioreactor influent. The studies showed that the microbial community significantly changed upon the first addition of NO_3^- in the system. Afterward, the microbial community structure remained the same in further UASB reactor operation, whether NO_3^- was present or removed from the influent. This indicates that denitrifying bacteria could be a controlling factor for SeO_4^{2-} removal. Future work needs to focus on understanding the exact mechanism involved in the SeO_4^{2-} reduction in the presence of co-electron acceptors, particularly at the stimulating $NO_3^-:SeO_4^{2-}$ and $SO_4^{2-}:SeO_4^{2-}$ molar ratios. Additional studies can be done using pure cultures to investigate changes in specific growth kinetics and enzymatic reductions under different NO_3^- to SeO_4^{2-} molar ratios in order to establish the controlling factors involved in the increased SeO_4^{2-} removal in the presence of NO_3^-.

Influence of SO_4^{2-} on SeO_4^{2-} removal

In contrast to NO_3^-, there are only limited reports on the effect of SO_4^{2-} on SeO_4^{2-} removal. Hockin and Gadd (2006) reported greater SeO_4^{2-} removal by a *Desulfomicrobium* sp. biofilm when under excess SO_4^{2-} conditions (28 mM) as compared to under SO_4^{2-} limited conditions (5 mM). Lenz et al. (2008) investigated the effect of two different SO_4^{2-} concentrations (26 and 1.3 mM) for the removal of 0.01 mM SeO_4^{2-} in a UASB reactor and recommended that the $SeO_4^{2-}:SO_4^{2-}$ molar ratio should not exceed 1:521 (SO_4^{2-} at 1.3 mM) in order to achieve complete SeO_4^{2-} removal (89% Se_{diss} removal). The batch tests showed that at a $SeO_4^{2-}:SO_4^{2-}$ molar ratio of $\leq$ 1:100, SO_4^{2-} removal was larger, while a negative impact on the SeO_4^{2-} removal was observed when the $SeO_4^{2-}:SO_4^{2-}$ molar ratio exceeds $>$ 1:100 (**Fig. 2**). Additionally, this study attained a similar Se removal efficiency as reported by Lenz et al. (2008) at a $SeO_4^{2-}:NO_3^-:SO_4^{2-}$ molar ratio of 1:40:100. Although the molar ratios in this study were lower than the one recommended by Lenz et al. (2008), the SO_4^{2-} and SeO_4^{2-} concentrations used were higher (10 mM and 0.1 mM, respectively). This indicates that not only the $SeO_4^{2-}:SO_4^{2-}$ ratio should be carefully monitored, but the anion concentrations should be controlled as well. It was reported by Chung et al. (2006) that a SeO_4^{2-} threshold concentration of ≤ 1 mM should be observed and increasing beyond this concentration, SeO_4^{2-} would be inhibitory to sulfate-reducing bacteria. The concentrations and $SeO_4^{2-}:NO_3^-:SO_4^{2-}$ molar ratios investigated in this study were based on typical SeO_4^{2-} and SO_4^{2-} concentrations found in mining-impacted wastewater (**Table 3.1**) and therefore more applicable for scale-up operation.

Enhanced SeO_4^{2-} removal by SO_4^{2-} has no clear link to the microbial reduction pathway. Though it has been reported that SRB (along with denitrifying bacteria) are capable of reducing SeO_4^{2-}, it is unclear to what extent they were involved in the SeO_4^{2-} reduction. There is a possibility that sulfide toxicity (Lenz et al. 2008) caused the differences in SeO_4^{2-} removal. However, the sulfide concentrations produced during batch tests were low (**Appendix 1, Table S3.1**), while in the reactor, SeO_4^{2-} removal was not affected by the sulfide concentration in phases I and II (**Fig. 3.3**). Abiotic reactions within the biofilm between Se and S compounds could have contributed to the increased SeO_4^{2-} removal in the presence of SO_4^{2-}. The reduction of SeO_4^{2-} to Se^0 is generally a two-step process in which SeO_3^{2-} is an intermediate anion. Some microorganisms are capable of reducing both SeO_4^{2-} and SeO_3^{2-} to Se^0 while others can reduce only either SeO_4^{2-} to SeO_3^{2-} or SeO_3^{2-} to Se^0. In the batch tests, the sulfide concentrations were lower when SeO_4^{2-} was present (**Appendix 1, Table S3.1**), suggesting the possibility of sulfide utilization by SeO_3^{2-} (intermediate product of SeO_4^{2-} reduction) to form Se^0. In the experiment by Hockin and Gadd (2003) using a *Desulfomicrobium norvegicum* biofilm grown at 30°C and pH 7.0, the formation of Se^0 was noted as a result of the abiotic reduction of SeO_3^{2-} (200 μM) by biogenic sulfide (10 mM). The reduction of SeO_4^{2-} to SeO_3^{2-} and subsequent abiotic reaction with biogenic sulfide could thus contribute to an enhanced SeO_4^{2-} removal when SO_4^{2-} is present in the wastewater system.

Further research should be conducted to determine how both NO_3^- and SO_4^{2-}, at certain molar ratios with SeO_4^{2-}, promote or inhibit SeO_4^{2-} removal. Possible high-throughput techniques such as NanoSIMS (isotope tracking) coupled with secondary X-ray fluorescence (S-XRF) can visualize and map the Se, N and S distribution in and uptake by the microorganisms and anaerobic granules at high resolution (Moore et al. 2010). By visualization of how and where SeO_4^{2-}, NO_3^- and SO_4^{2-} are taken up, the controlling metabolic interactions within a syntrophic system can be identified.

3.4.2   Effect of NO_3^- and SO_4^{2-} on Se species mass balance

Apart from the operational conditions, the presence of anions in the wastewater can also potentially control how Se transformation or speciation occur. Dessì et al. (2016) demonstrated that the presence of NO_3^- in the influent of UASB reactor led to an excess release of colloidal Se^0 in the effluent. Contrary to this, the Se species mass balance (**Table 3.5**) indicated that more Se was incorporated into the granular sludge when grown in the presence of NO_3^-. The UASB reactor operated in this study showed negative Se_{tot} removal efficiencies once the NO_3^- was omitted from the synthetic influent (**Fig. 3.4**). Furthermore, in the UASB reactor, there was evident dropped in Se concentration detected in the biomass once NO_3^- was omitted from the feed (**Appendix 1, Fig. S3.3**) and even showing an increase in total Se (higher than inlet concentration) release in the effluent (**Fig. 3.4**). However, the increase in NO_3^- concentration to 18 mM during phase IVd could also have triggered elevated production of extracellularly Se^0 that was carried over to phase V. On the other hand, when grown in the presence of SO_4^{2-} (whether NO_3^- was present or absent), more colloidal Se^0 (red formation) was observed. Hockin and Gadd (2006) reported similar results for a sulfate-reducing biofilm, where under excess SO_4^{2-} concentration (28 mM), SeO_4^{2-} was bioreduced to Se^0, while under sulfate-limiting conditions (5 mM), SeO_4^{2-} was bioreduced to selenide.

3.4.3 Practical implications

The characteristics of Se-laden wastewater are site-specific and depend on the industrial wastewater processing sector (i.e. coal mining, mineral ores mining and agricultural farming) (Lenz and Lens 2009; Tan et al. 2016). Therefore, it is expected that real Se-laden wastewaters will have different molar ratios that can be below or above the recommended molar ratios of $SeO_4^{2-}:NO_3^-:SO_4^{2-}$ observed in this study. By having knowledge on a concentration range of SO_4^{2-} and NO_3^- that can actually promote SeO_4^{2-} removal, different operational process conditions can be applied depending on the real wastewater scenario.

In order to utilize the advantage of a higher SeO_4^{2-} removal with NO_3^- and SO_4^{2-}, improvement on the biological operation process can be made through wastewater process line integration/mixing or through reactor effluent recycling. An in-depth knowledge of the mining processing routes (e.g. smelting or acidification) (Norgate et al. 2007) and resulting wastewater characteristics produced for each process would be required in order to carry out wastewater stream mixing. For example, in copper refinery slimes, SeO_4^{2-} wastewater streams are produced after the leaching process from the slug slime (Hoffmann 1989) and can be mixed with the wastewater stream from the slag tailing or sulfuric acid treatment (Norgate et al. 2007) containing SO_4^{2-}. Another example that can be applied is from Se-refining plants, where concentrations of Se anions can reach up to 30 mg Se/L after the kiln powder leaching process (Ike et al. 2017). Ike et al. (2017) employed an aerobic reactor inoculated with *Pseudomonas stutzeri* NT-I for the treatment of wastewater from the Se refinery plant to reduce the soluble Se anions concentration by > 90%. However, maintaining the survival of the pure cultures is challenging in a long-term operation and aeration would add additional cost. Therefore, an alternative operation can be suggested by operating an anaerobic reactor and add NO_3^- and SO_4^{2-} at the appropriate molar ratios using stringent bioprocess control (e.g. proportional-integral-derivative (PID) controllers) to achieve high soluble Se removal efficiencies. Overall, this study gave insight on the effect of co-electron acceptors on SeO_4^{2-} removal in batch and continuous systems and can help in the advancement and application of biological processes such as predicting or improving reactor performance.

3.5 Conclusions

This study showed that methanogenic granular sludge can simultaneously remove SeO_4^{2-}, NO_3^- and SO_4^{2-}, even in the presence of high NO_3^- and SO_4^{2-} concentrations, albeit at different efficiencies. A $SeO_4^{2-}:NO_3^-:SO_4^-$ molar ratio at 1:40:100 was observed to be the optimal condition, showing an increase in SeO_4^{2-} removal efficiency by 37% when compared to the control condition (SeO_4^{2-} only). Long-term UASB reactor operation at high NO_3^- (4 mM) and SO_4^{2-} (15 mM) concentrations for 92 days achieved 100% NO_3^-, 30% SO_4^{2-} and 80% total Se removal efficiencies. Speciation analysis in batch experiments showed an overall increase in colloidal Se^0 concentration in the liquid phase when NO_3^-, SO_4^{2-} or both were included. In contrast, an increase in total Se concentration in the effluent was observed in the UASB reactor when SO_4^{2-} and NO_3^- were sequentially removed from the influent. Overall, the molar ratio of NO_3^-:SeO_4^{2-} and SO_4^{2-}:SeO_4^{2-} showed to be an important factor in the reduction of SeO_4^{2-}. At the proper molar ratios, both SO_4^{2-} and NO_3^- had a positive effect on the SeO_4^{2-} removal rate in both batch incubations and a continuous UASB reactor. As such, controlling the NO_3^- and SO_4^{2-} concentration could direct possibilities in the improvement of UASB reactors for the treatment of SeO_4^{2-} containing wastewaters.

3.6 References

Amann, R.I., Binder, B.J., Olson, R.J., Chisholm, S.W., Devereux, R., Stahl, D.A., 1990. Combination of 16S rRNA-targeted oligonucleotide probes with flow cytometry for analyzing mixed microbial populations. *Appl Env. Microbiol.* 56, 1919-1925.

Association, A.P.H., Association A.W.W., Federation, W.E., 2005. Standard methods for examination of water and wastewater, 5th ed. American Public Health Association, Washington, DC, USA.

Astratinei, V., van Hullebusch, E., Lens, P.N.L., 2006. Bioconversion of selenate in methanogenic anaerobic granular sludge. *J. Environ. Qual.* 35, 1873-1883.

Brown, K.M., Arthur, J.R., 2007. Selenium, selenoproteins and human health: A review. *Public Health Nutr.* 4, 593-599.

CH2M HILL, 2010. Review of available technologies for the removal of selenium from water - Final report prepared for North American Metal Council.

Chung, J., Nerenberg, R., Rittmann, B.E., 2006a. Bioreduction of selenate using a hydrogen based membrane biofilm reactor. *Environ. Sci. Technol.* 40, 1664-1671.

Daims, H., Brühl, A., Amann, R., Schleifer, K.-H., Wagner, M., 1999. The domain-specific probe EUB338 is insufficient for the detection of all bacteria: Development and evaluation of a more comprehensive probe set. *Syst. Appl. Microbiol.* 22, 434-444.

DeMoll-decker, H., Macy, J.M., 1993. The periplasmic nitrite reductase of *Thauera selenatis* may catalyze the reduction of selenite to elemental selenium. *Arch. Microbiol.* 160, 241-247.

Dessì, P., Jain, R., Singh, S., Seder-Colomina, M., van Hullebusch, E.D., Rene, E.R., Ahammad, S.Z., Carucci, A., Lens, P.N.L., 2016. Effect of temperature on selenium removal from wastewater by UASB reactors. *Wat. Res.* 94, 146-154.

Hockin, S., Gadd, G.M., 2006. Removal of selenate from sulfate-containing media by sulfate-reducing bacterial biofilms. *Environ. Microbiol.* 8, 816-826.

Hockin, S.L., Gadd, G.M., 2003. Linked redox precipitation of sulfur and selenium under anaerobic conditions by sulfate-reducing bacterial biofilms. *Appl. Enivironmental Microbiol.* 69, 7063-7072.

Hoffmann, J.E., 1989. Recovering selenium and tellurium from copper refinery slimes. *JOM* 41, 33-38.

Ike, M., Soda, S., Kuroda, M., 2017. Bioprocess approaches for the removal of selenium from industrial waste and wastewater by *Pseudomonas stutzeri* NT-I, In: Hullebusch, E.D. van (Ed.), Bioremediation of Selenium Contaminated Wastewater. Springer International Publishing, pp. 57-73.

Jain, R., Seder-Colomina, M., Jordan, N., Dessi, P., Cosmidis, J., van Hullebusch, E.D.,Weiss, S., Farges, F., Lens, P.N.L., 2015b. Entrapped elemental selenium nanoparticles affect physicochemical properties of selenium fed activated sludge. *J. Hazard.Mater.* 295, 193-200.

Johansson, C.L., Paul, N.A., de Nys, R., Roberts, D.A., 2015. The complexity of biosorption treatments for oxyanions in a multi-element mine effluent. *J. Environ. Manage.* 151, 386-392.

Lai, C.-Y., Wen, L.-L, Shi, L.-D., Zhao, K.-K., Wang, Y.-Q., Yang, X., Rittman, B.E., Zhou, C., Tang, Y., Zheng, P., Zhao, H.-P., 2016. Selenate and nitrate bioreductions using methane as the electron donor in a membrane biofilm reactor. *Environ. Sci. Technol.* 50, 10179-10186.

Lai, C.Y., Yang, X., Tang, Y., Rittmann, B.E., Zhao, H.P., 2014. Nitrate shaped the selenate-reducing microbial community in a hydrogen-based biofilm reactor. *Environ. Sci. Technol.* 48, 3395-3402.

Lee, H., Park, Y., Choi, E., Lee, J., 2008. Bacterial community and biological nitrate removal: Comparisons of autotrophic and heterotrophic reactors for denitrification with raw sewage. *J. Microbiol. Biotechnol.* 18, 1826-1835.

Lemly, A.D., 2014. Teratogenic effects and monetary cost of selenium poisoning of fish in Lake Sutton, North Carolina. *Ecotoxicol. Environ. Saf.* 104, 160-167.

Lenz, M., Enright, A.M., O'Flaherty, V., van Aelst, A.C., Lens, P.N.L., 2009. Bioaugmentation of UASB reactors with immobilized *Sulfurospirillum barnesii* for simultaneous selenate and nitrate removal. *Appl. Microbiol. Biotechnol.* 83, 377-388.

Lenz, M., Gmerek, A., Lens, P.N.L., 2006. Selenium speciation in anaerobic granular sludge. *Int. J. Environ. Anal. Chem.* 86, 615-627.

Lenz, M., Lens, P.N.L., 2009. The essential toxin: The changing perception of selenium in environmental sciences. *Sci. Total Environ.* 407, 3620-3633.

Lenz, M., van Hullebusch, E.D., Hommes, G., Corvini, P.F., Lens, P.N.L., 2008. Selenate removal in methanogenic and sulfate-reducing upflow anaerobic sludge bed reactors. *Water Res.* 42, 2184-2194.

Li, D.-B., Cheng, Y.-Y., Wu, C., Li, W.-W., Li, N., Yang, Z.-C., Tong, Z.-H., Yu, H.-Q., 2014. Selenite reduction by *Shewanella oneidensis* MR-1 is mediated by fumarate reductase in periplasm. *Sci. Rep.* 4, 3735.

Lortie, L., Gould, W.D., Rajan, S., McCready, R.G.L., Cheng, K.-J., 1992. Reduction of selenate and selenite to elemental selenium by a *Pseudomonas stutzeri* isolate. *Appl. Environ. Microbiol.* 58, 4042-4044.

Mal, J., Nancharaiah, Y.V., van Hullebusch, E.D., Lens, P.N.L., 2016. Effect of heavy metal co-contaminants on selenite bioreduction by anaerobic granular sludge. *Bioresour. Technol.* 206, 1-8.

Moore, K.L., Schröder, M., Lombi, E., Zhao, F.J., McGrath, S.P., Hawkesford, M.J., Shewry, P.R., Grovenor, C.R.M., 2010. NanoSIMS analysis of arsenic and selenium in cereal grain. *New Phytol.* 185, 434-445.

Nancharaiah, Y.V., Lens, P.N.L., 2015a. Ecology and biotechnology of selenium-respiring bacteria. *Microbiol. Mol. Biol. Rev.* 79, 61-80.

Nancharaiah, Y.V., Lens, P.N.L., 2015b. Selenium biomineralization for biotechnological applications. *Trends Biotechnol.* 33, 323-330.

Norgate, T.E., Jahanshahi, S., Rankin, W.J., 2007. Assessing the environmental impact of metal production processes. *J. Clean. Prod.* 15, 838-848.

Oremland, R.S., Blum, J.S., Bindi, A.B., Dowdle, P.R., Herbel, M., Stolz, J.F., 1999. Simultaneous reduction of nitrate and selenate by cell suspensions of selenium-respiring bacteria. *Appl. Environ. Microbiol.* 65, 4385-4392.

Roest, K., Heilig, H.G.H.., Smidt, H., De Vos, W.M., Stams, A.J.M., Akkermans, A.D.L., 2005. Community analysis of a full-scale anaerobic bioreactor treating paper mill wastewater. *Syst. Appl. Microbiol.* 28, 175-85.

Santos, S., Ungureanu, G., Boaventura, R., Botelho, C., 2015. Selenium contaminated waters: An overview of analytical methods, treatment options and recent advances in sorption methods. *Sci. Total Environ.* 521-522, 246-260.

Sekiguchi, Y., Kamagata, Y., Nakamura, K., Ohashi, A., Harada, H., 1999. Fluorescence in situ hybridization using 16S rRNA-targeted oligonucleotides reveals localization of methanogens and selected uncultured bacteria in mesophilic and thermophilic sludge granules. *Appl. Environ. Microbiol.* 65, 1280-1288.

Smith, K., Lau, A.O., Vance, F.W., 2009. Evaluation of treatment techniques for selenium removal. Engineers Society of Western Pennsylvania (Eds.), 70th Annu Int Water Conf. Curran Associates, Inc., Orlando, Florida, USA, pp. 75-92.

Soda, S., Kashiwa, M., Kagami, T., Kuroda, M., Yamashita, M., Ike, M., 2011. Laboratory scale bioreactors for soluble selenium removal from selenium refinery wastewater using anaerobic sludge. *Desalination* 279, 433-438.

Staicu, L.C., Morin-Crini, N., Crini, G., 2017. Desulfurization: Critical step towards enhanced selenium removal from industrial effluents. *Chemosphere* 172, 111-119.

Takada, T., Hirata, M., Kokubu, S., Toorisaka, E., Ozaki, M., Hano, T., 2008. Kinetic study on biological reduction of selenium compounds. *Process Biochem.* 43, 1304-1307.

Tan, L.C., Nancharaiah, Y.V., van Hullebusch, E.D., Lens, P.N.L., 2016. Selenium: Environmental significance, pollution, and biological treatment technologies. *Biotechnol. Adv.* 34, 886-907.

USEPA, 2014. External peer review draft - Aquatic life ambient water quality criterion for selenium - freshwater 2014. United States Environmental Protection Agency (EPA-820-F-14-005).

Wen, H., Carignan, J., 2007. Reviews on atmospheric selenium: Emissions, speciation and fate. *Atmos. Environ.* 41, 7151-7165.

Yee, N., Ma, J., Dalia, A., Boonfueng, T., Kobayashi, D.Y., 2007. Se(VI) reduction and the precipitation of Se(0) by the facultative bacterium *Enterobacter cloacae* SLD1a-1 are regulated by FNR. *Appl. Environ. Microbiol.* 73, 1914-1920.

Zehr, J.P., Oremland, R.S., 1987. Reduction of selenate to selenide by sulfate-respiring bacteria: Experiments with cell suspensions and estuarine sediments. *Appl. Environ. Microbiol.* 53, 1365-1369.

CHAPTER 4

Selenate removal in biofilm systems: effect of nitrate and sulfate on selenium removal efficiency, biofilm structure, and microbial community

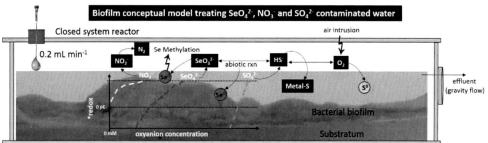

This chapter has been submitted in modified form:

Tan, L.C, Espinosa-Ortiz E.J., Nancharaiah, Y.V., van Hullebusch, E., Gerlach, R., Lens, P.N.L. 2018. Selenate removal in biofilm systems: effect of nitrate and sulfate on selenium removal efficiency, biofilm structure, and microbial community. *J. Ind. Microbiol. Biotechnol.* doi:10.1002/jctb.5586.

Abstract

Selenium (Se) discharged into natural waterbodies can accumulate over time and have negative impacts on the environment. Se-laden wastewater streams can be treated using biological processes. However, the presence of other electron acceptors in wastewater, such as nitrate (NO_3^-) and sulfate (SO_4^{2-}), can influence selenate (SeO_4^{2-}) reduction and impact the efficiency of biological treatment systems. SeO_4^{2-} removal by biofilms formed from an anaerobic sludge inoculum was investigated in the presence of NO_3^- and SO_4^{2-} using drip flow reactors operated continuously for 10 days at pH 7.0 and 30°C. The highest total Se (~60%) and SeO_4^{2-} (~80%) removal efficiencies were observed when the artificial wastewater contained SO_4^{2-}. A maximum amount of 68 µmol Se/cm^2 was recovered from the biofilm matrix in SO_4^{2-} +SeO_4^{2-} exposed biofilms and biofilm mass was 2.7 fold increased for biofilms grown in the presence of SO_4^{2-}. When SeO_4^{2-} was the only electron acceptor, biofilms were thin and compact. In the simultaneous presence of NO_3^- or SO_4^{2-}, biofilms were thicker (> 0.6 mm), less compact and exhibited gas pockets. The presence of SO_4^{2-} had a beneficial effect on biofilm growth and the SeO_4^{2-} removal efficiency, while the presence of NO_3^- did not have a significant effect on SeO_4^{2-} removal by the biofilms.

Keywords: biofilm; co-electron acceptors; selenium removal; biofilm characterization; selenate; nitrate; sulfate

4.1 Introduction

Selenium (Se) is a trace metalloid that is vital, but at the same time harmful to living organisms at an unusually small concentration range of only 5- to 10-fold difference between essential and toxic concentrations (Lenz and Lens 2009). Apart from naturally occurring processes that mobilize Se from minerals and volcanic rocks, anthropogenic activities (i.e. mining and agriculture) are the major contributors of Se mobilization and release into the environment (Wen and Carignan 2007). Se oxyanions such as selenite (SeO_3^{2-}) and selenate (SeO_4^{2-}) are typically found at low levels (< 12 mg Se/L) in drainage, acid rock drainage and mining wastewaters compared to other contaminants like nitrate (NO_3^-) and sulfate (SO_4^{2-}) which are typically present at 250 mg/L and 3000 mg/L, respectively (Tan et al. 2016). Se is considered a problematic pollutant because it has the propensity to bioaccumulate in organisms (Hockin and Gadd 2003). Se release and accumulation into the aquatic environment has posed serious threats to egg-laying vertebrates, causing reproductive failures and teratogenic effects, which subsequently resulted in monetary losses of millions of dollars to fisheries (Lemly, 2014). Due to the environmental impacts of Se release into the aquatic environment, Se removal from contaminated waters before discharge is essential for the protection of living organisms and the environment.

Membrane filtration, ion exchange, adsorption and chemical reduction methods are typically used for Se removal from contaminated waters (Tan et al. 2016). However, the use of biological methods is becoming attractive since these methods offer an environmentally friendly and potentially cost-effective alternative, might be suitable for large scale application and might allow for biogenic elemental Se (Se^0) recovery (Nancharaiah and Lens 2015a). The biological process converts the soluble Se oxyanions to insoluble and less toxic Se^0 (Mal et al. 2017) which can potentially be recovered and reused for various industrial applications, such as fertilizers for Se-deficient soils or use in photovoltaic cells (Nancharaiah and Lens 2015b; Nguyen et al. 2016). Despite the recent advances in the biological treatment of Se-laden wastewaters, there are still many knowledge gaps in the application of the treatment, particularly when considering the complexity of Se-laden wastewaters. One of the challenges of Se-laden wastewaters, such as mining effluents, is the presence of other oxyanions, such as NO_3^- and SO_4^{2-}, that are typically present in concentrations more than 100 to 1000 times greater than Se (Tan et al. 2016).

NO_3^- and SO_4^{2-} are terminal electron acceptors for denitrifying and sulfate-reducing bacteria, respectively, and can either inhibit or enhance microbial SeO_4^{2-} removal by microorganisms (Tan et al. 2016). Oremland et al. (1999) observed that the presence of NO_3^- (0.1 mM) promoted the ability of *Sulfospirillum barnesii* to reduce SeO_4^{2-} (0.1 mM) by keeping the cells at a constant state of high metabolic activity. In contrast, Lai et al. (2014) reported a 30% decrease in SeO_4^{2-} (0.02 mM) removal in a hydrogen-based membrane biofilm reactor (MBfR) in the presence of NO_3^- (<0.9 mM), attributed to competition for H_2 (electron donor) and possible suppression of the SeO_4^{2-} reductases. A study by Hockin and Gadd (2006) using a sulfate-reducing biofilm composed of *Desulfomicrobium norvegicum* carried out SeO_4^{2-} removal and suggested greater SeO_4^{2-} reduction in the presence of excess SO_4^{2-} (28 mM) compared to SO_4^{2-} limiting conditions (5 mM). In contrast, Ontiveros-Valencia et al. (2016) did not observe any change in the removal efficiency of SeO_4^{2-} (< 0.08 mM) by an anaerobic biofilm growing in a hydrogen-fed membrane reactor in the presence or absence of SO_4^{2-}.

Most studies have focused on SeO_4^{2-} reduction in the presence of either NO_3^- or SO_4^{2-}, while only a few studies have conducted an in-depth analysis of the effect of the concurrent presence of NO_3^- and SO_4^{2-} on SeO_4^{2-} reduction in biological systems. In a previous study using anaerobic granular sludge (**chapter 3**), there was an observable impact of NO_3^- and SO_4^{2-} on SeO_4^{2-} removal efficiencies. In addition, the effect Se might have on biomass growth, Se speciation and Se^0 fate in the presence of NO_3^- and/or SO_4^{2-} have yet to be reported, particularly in a biofilm system. There are only limited studies on SeO_4^{2-} removal and biofilm-Se interactions with co-electron acceptors, most of which were investigated using MBfRs with H_2 as electron donor (Ontiveros-Valencia et al. 2016; Van Ginkel et al. 2011). Therefore, this study investigates the responses of lactate-fed anaerobic biofilms growing in drip flow reactors (DFRs) to SeO_4^{2-} and other electron acceptors (SO_4^{2-} and NO_3^-) by determining: i) the impact on the SeO_4^{2-} removal efficiency and ii) changes in biofilm characteristics, including biofilm architecture and microbial community composition.

4.2 Materials and methods

4.2.1 Inoculum and biofilm growth conditions

Anaerobic granular sludge taken from a full-scale upflow anaerobic sludge bed (UASB) reactor treating paper-mill wastewater (Eerbeek, The Netherlands) was used as the inoculum (Roest et al. 2005) for biofilm development. About 1 g wet granular sludge was homogenized using a homogenizer potter tube and used as the seed inoculum (0.25 g dry weight). All experiments were carried out using synthetic mining wastewater composed of growth medium, electron donor and the three oxyanions under study.

The composition of the growth medium, according to **chapter 3**, was as follows (in g/L): NH_4Cl (0.30), $CaCl_2 \cdot 2H_2O$ (0.10), $MgCl_2 \cdot 6H_2O$ (0.01) and $NaHCO_3$ (0.04). Phosphate buffer (0.053 g/L Na_2HPO_4 and 0.041 g/L KH_2PO_4) was included in the medium to maintain near neutral pH (7-8) conditions. Acid and alkaline trace metal solutions (0.1 mL each, as described in Stams et al. (1992)) were added to 1 L of synthetic wastewater. Sodium lactate was used as the electron donor while NO_3^-, SO_4^{2-} and SeO_4^{2-} (provided as KNO_3, K_2SO_4 and Na_2SeO_4) were provided as electron acceptors. Assuming complete oxidation of lactate and reduction of oxyanions to Se^0, N_2 and HS^-, 1 mole of SeO_4^{2-}, NO_3^- and SO_4^{2-} requires 0.4, 0.6 and 0.8 moles of COD, respectively. All feed solutions were purged with nitrogen gas to remove oxygen from the artificial wastewater. It should be noted that the synthetic mine wastewater used in this study simulates, as close as possible, the composition of Se-laden wastewater in the mining industry as described in Stover et al. (2006).

4.2.2   Reactor configuration

Drip flow reactors (DFRs) provide plug flow-like systems under low-shear/laminar flow and are flexible and adaptable to a variety of conditions (Goeres et al. 2009). Multi-panel DFRs (15.24 cm × 12.70 cm × 2.54 cm), as shown in **Fig. 4.1**, were used with silicon coupons (0.3 cm × 7.5 cm × 2.5 cm) as the substratum for biofilm growth. Each DFR was assembled according to Goeres et al. (2009). Biofilms were cultivated under anoxic conditions at 30°C for a total period of 12 days. All DFR incubations were performed at least in duplicate. Effluent liquid samples were collected daily and analyzed for lactate, NO_3^-, SO_4^{2-} and Se concentrations (total Se, SeO_4^{2-} and SeO_3^{2-}).

DFRs were initially purged with N_2 while all influent solutions were purged with N_2 for 30 minutes and changed every 2 days. Each DFR was initially inoculated with 1 g homogenized wet granules in 10 mL with only lactate and growth medium (without oxyanions). Each DFR was operated in batch mode for 2 days to allow for the attachment and growth of cells on the substratum. After 2 days, the DFR was inclined at a 10° angle and operated continuously at a flow rate of 0.2 mL/min (0.288 L/day) for 10 days. During continuous mode, biofilms were exposed to 20 mM lactate and different conditions: (A) SeO_4^{2-}, (B) $NO_3^- + SeO_4^{2-}$, (C) $SO_4^{2-} + SeO_4^{2-}$, and (D) $NO_3^- + SO_4^{2-} + SeO_4^{2-}$, (F) $NO_3^- + SO_4^{2-}$ and (E) no electron acceptors. SeO_4^{2-},

NO_3^- and SO_4^{2-} were provided at an average concentration of 0.13 ($\pm$ 0.02) mM, 4.8 ($\pm$ 0.3) mM and 12.9 ($\pm$ 0.3) mM, respectively.

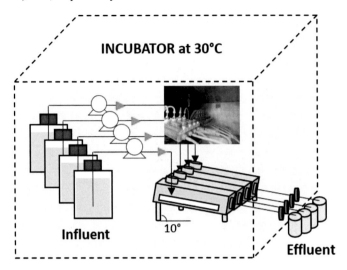

Fig. 4.1 Schematic diagram of multi-panel drip flow reactors (DFR) used for biofilm experiments.

4.2.3 Biofilm characterization

Biofilms grown under different conditions on the silicon coupons were cut into sections for various analyses at the end of each experimental run. Biofilm samples were scraped from the silicon coupon into a micro-centrifuge tube (pre-weighed) for measurement of wet biomass weight. The biofilm dry weight, ash-free dry weight and total Se in the biofilm per area (cm^2) was determined. Dry weight was reported as total solids (TS), while ash-free dry weight was reported as volatile solids (VS).

Biofilm dry weight and ash-free dry weight were measured following standard protocols. Biomass was first dried in an incubator at 100°C for 3 days (dry weight) followed by heating in a furnace at 500°C for 4 hours (ash-free weight). Biofilm samples (2 samples per coupon) for total Se content were first digested using concentrated nitric acid (HNO_3) for 2 days, centrifuged and diluted using 5% HNO_3 (modified from Jain et al. 2015). Total Se was then analyzed using inductively coupled plasma mass spectrometry (ICP-MS, Agilent 7500ce). Cell viability was analyzed using the Live/Dead® *Bac*Light™ Bacterial Viability kit containing SYTO9 and propidium iodide (PI) (Manteca et al. 2005); green ('live') and red ('dead') cells were counted using epifluorescence microscopy (Nikon Eclipse E800).

4.2.4   Biofilm imaging

The architecture of the biofilms formed under different operating conditions was visualized using a Leica TCS-SP2 AOBS confocal laser scanning microscope (CLSM). Biofilms formed on the silicon coupons were first stained using the Live/Dead® BacLight^{TM} kit for 20 min at 30°C in the dark. Excess dye was removed by washing with pure water and images were taken at 100× and 630× magnifications. Imaris software (Bitplane Scientific Software) was used for processing the CLSM images.

Biofilm thickness was estimated using a cryo-section method. Briefly, stained biofilm samples were frozen by placing them on dry ice and covering with a tissue embedding medium (OCT, optimum cutting temperature, Tissue-Tek). Samples were sliced into 5 μm sections using a Leica CM1850 cryostat at -20°C. Images were acquired using a Nikon Eclipse E800 microscope in fluorescence and transmission modes using differential interference contrast optics and processed for thickness measurements using MetaMorph (Molecular Devices). Five samples were analyzed per reactor for thickness measurements.

The elemental composition of the biofilm matrix was analyzed using a scanning electron microscope (SEM, Zeiss Supra^{TM} 55VP) equipped with an energy dispersive X-ray spectroscopy unit (EDX, Princeton Gamma-Tech) using secondary electron and back-scattering mode. For SEM analysis, the biofilm samples were gently washed with Milli-Q water and dried at ambient temperature and pressure. Dried samples were deposited onto carbon tape and coated with iridium before imaging.

4.2.5   Microbial community analysis

Genomic DNA was extracted from the samples following the protocol of Lueders et al. (2004) and quantified using a NanoDrop-1000 spectrophotometer. Extracted DNA was amplified by PCR using primers Pro 341 forward and Pro 805 reverse targeting the V4 region of the 16S rRNA gene of bacteria and archaea (Takahashi et al. 2014). Amplicons were checked by agarose gel electrophoresis and sequenced using the Illumina MiSeq standard protocol sequencing platform "16S Metagenomic Sequencing Library Preparation". Sequences produced were analyzed using the standard operating procedures of the bioinformatics platform Mothur (Schloss et al. 2009). Detailed description of the microbial community analysis procedure can be found in **Appendix 7**.

4.2.6 Analytical methods

Effluent samples were collected daily and filtered using 0.2 µm cellulose acetate membranes. Filtered samples were analyzed for lactate, NO_3^-, SO_4^{2-} and SeO_4^{2-} using ion chromatography (IC, Dionex ICS-1100). The IC column used was a Dionex Ion Pac^{TM} AS22, operation was set at 1.2 mL/min flow rate, using a 25 µL sample loop with 4.5 mM Na_2CO_3 and 1.4 mM $NaHCO_3$ as eluent. The SeO_3^{2-} concentrations were determined using a spectrophotometric method (Li et al. 2014). Briefly, 1 mL of sample was mixed with 0.5 mL of 4 M HCl and 1 mL of 1 M ascorbic acid; after 10 min, absorbance was measured at 500 nm using a Genesys 10UV scanning spectrophotometer (Thermo Scientific) and compared to the absorbance of equivalently treated standards with known SeO_3^{2-} concentrations. Total dissolved sulfides were determined colorimetrically as well (at 480 nm) after the formation of a colloidal copper sulfide precipitate as described by Cord-Ruwisch (1985). Unfiltered liquid samples were acidified with HNO_3 and measured for total Se using ICP-MS. At the end of each experiment, effluent samples (50 mL) were collected and dried at room temperature. Dried particles were analyzed for elemental components using SEM-EDX as described above.

4.2.7 Reactor performance parameters and statistical analysis

Average daily removal rates and removal efficiencies for lactate, NO_3^-, SO_4^{2-}, SeO_4^{2-} and total Se were calculated. The removal rates were calculated according to Eq. (4.1):

$$\text{Removal rate (mmol/day)} = Q \times (C_{in} - C_{out}) \qquad \text{Eq. (4.1)}$$

Where Q is flow rate in L/day while C_{in} and C_{out} are, respectively, influent and effluent concentration in mM. Statistical differences were evaluated using analysis of variance (ANOVA) with Tukey test method for multiple comparison; a $P_{value} \leq 0.05$ was considered statistically significantly different.

4.3 Results

4.3.1 Effect of co-electron acceptors on Se removal

The pH values in the DFRs remained close to neutral for all experimental runs with the measured effluent pH ranging from 7.0-8.0 for all DFR incubations. **Table 4.1** shows the removal performance of the DFRs under different incubation conditions. Total Se and SeO_4^{2-} removal rates were the highest for the treatments receiving SO_4^{2-} ($SO_4^{2-} + SeO_4^{2-}$ and $SO_4^{2-} +$

NO_3^- + SeO_4^{2-}), attaining an average of 61 ($\pm$ 2)% total Se and 77 ($\pm$ 1)% SeO_4^{2-} removal efficiency. The highest SeO_3^{2-} concentration detected in the effluent amounted to 0.04 ($\pm$ 0.01) mM for biofilms exposed to both SeO_4^{2-} and SO_4^{2-}, while the other incubations had an average SeO_3^{2-} concentration of 0.02 ($\pm$ 0.01) mM. Significantly lower SeO_4^{2-} removal ($P_{value} \leq 0.0001$) was observed by the DFR biofilms grown in the absence of SO_4^{2-} (SeO_4^{2-} and NO_3^- + SeO_4^{2-}). DFR incubations with SeO_4^{2-} alone attained 17 ($\pm$ 9)% total Se and 30 ($\pm$ 6)% SeO_4^{2-} removal efficiencies, while NO_3^-+SeO_4^{2-} incubations attained 17 ($\pm$2)% total Se and 37 ($\pm$ 12)% SeO_4^{2-} removal efficiencies.

NO$_3^-$ removal in the DFR biofilm systems was successfully achieved (~97%) when grown with SO_4^{2-} (**Table 4.1**). Low NO_3^- removal (56 $\pm$ 17%) was observed in the absence of SO_4^{2-} (**Table 4.1**). SO_4^{2-} removal was low for all experimental conditions, reaching only a 7 mmol/day removal rate as compared to the influent SO_4^{2-} mass flow rate of 44 mmol/day. However, there was a significant increase ($P_{value} = 0.009$) in the SO_4^{2-} removal efficiency from 5% to 15% in the absence of SeO_4^{2-}.

4.3.2   Biofilm characterization under different growth conditions

Biofilm structure

Images of biofilms grown under different incubation conditions are shown in **Fig. 4.2**. In the absence of oxyanions, biofilms on the silicone coupons were whitish and fairly transparent. Biofilms exposed to SeO_4^{2-} or NO_3^-+SeO_4^{2-} developed a light orange coloration, while biofilms exposed to SO_4^{2-}+SeO_4^{2-} (with or without the presence of NO_3^-) showed intense red coloration with interspersed whitish to yellowish deposits. CLSM images did not reveal significant morphological or architectural differences among the biofilms grown under the different conditions (**Appendix 2, Figure S4.1**).

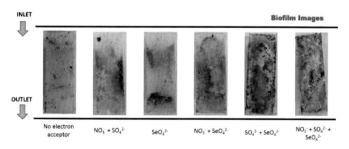

Fig. 4.2 Representative images of biofilms formed under different incubations taken using a digital camera. Silicon coupons were 2.5 cm wide and 7.5 cm long.

Table 4.1 Average removal performance (daily removal rate and removal efficiency) in the drip flow reactor for lactate, NO_3^-, SO_4^{2-}, and Se (Se_{tot} and SeO_4^{2-}) in the effluent. Lactate, NO_3^-, SO_4^{2-} and SeO_4^{2-} were provided at the following rates in mmol/day: 61 ($\pm$ 6), 16 ($\pm$ 2), 44 ($\pm$ 3) and 0.50 ($\pm$ 0.03), respectively.

	Incubation	A[*]	B[**]	C[*]	D[**]	E[**]	F[*]
	units	No electron donor	NO_3^- + SO_4^{2-}	SeO_4^{2-}	NO_3^- + SeO_4^{2-}	SO_4^{2-} + SeO_4^{2-}	NO_3^- + SO_4^{2-} + SeO_4^{2-}
Daily removal rate:							
Lactate	mmol/day	38 ($\pm$ 19)	58 ($\pm$ 5)	32 ($\pm$ 15)	29 ($\pm$ 3)	57 ($\pm$ 3)	42 ($\pm$ 12)
NO_3^-	mmol/day	-	18 ($\pm$ 1)	-	9 ($\pm$ 2)	-	13 ($\pm$ 1)
SO_4^{2-}	mmol/day	-	7 ($\pm$ 5)	-	-	4 ($\pm$ 2)	4 ($\pm$ 1)
Se_{tot}	mmol/day	-	-	0.07 ($\pm$ 0.04)	0.08 ($\pm$ 0.06)	0.30 ($\pm$ 0.05)	0.30 ($\pm$ 0.08)
SeO_4^{2-}	mmol/day	-	-	0.10 ($\pm$ 0.03)	0.20 ($\pm$ 0.07)	0.40 ($\pm$ 0.09)	0.40 ($\pm$ 0.07)
Removal efficiency:							
Lactate	% consumed	62 ($\pm$ 4)	90 ($\pm$ 12)	54 ($\pm$ 9)	42 ($\pm$ 14)	89 ($\pm$ 12)	79 ($\pm$ 10)
NO_3^-	% removed	-	99 ($\pm$ 1)	-	56 ($\pm$ 17)	-	95 ($\pm$ 9)
SO_4^{2-}	% removed	-	15 ($\pm$ 11)	-	-	5 ($\pm$ 3)	5 ($\pm$ 2)
Se_{tot}	% removed	-	-	17 ($\pm$ 9)	17 ($\pm$ 13)	59 ($\pm$ 11)	62 ($\pm$ 18)
SeO_4^{2-}	% removed	-	-	30 ($\pm$ 6)	37 ($\pm$ 12)	77 ($\pm$ 15)	76 ($\pm$ 13)

Note: [*]n = 4 experimental replicates, [**]n = 2 experimental replicates

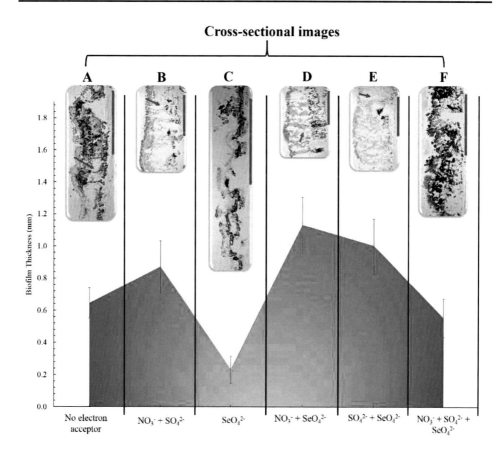

Fig. 4.3 Cross-sectional images of biofilms stained with SYTO 9 (green) and propidium iodide (red) with corresponding thickness measurement in mm (area graph). All images were visualized using fluorescence microscopy under 40× magnification. Each scale bar (red line) represents 1 mm. Red arrows indicate presumed gas pockets observed in the cross-sectional images.

Biofilm thicknesses were determined from cross-section images as shown in **Fig. 4.3**. In general, all biofilms had a thickness greater than 0.6 mm, except those grown with SeO_4^{2-} only. Empty spaces were observed in the biofilms, presumably voids or gas pockets within the biofilm matrix (indicated by red arrows in **Fig. 4.3**). Even though analyses of these gases was not performed in this study, these gas pockets in the biofilms are likely due to the microbial production of gases such as hydrogen sulfide, nitrogen gas, hydrogen or carbon dioxide. Gas bubbles were also visible along some of the biofilms (**Fig. 4.2**), particularly those biofilms grown in the presence of SO_4^{2-}. Biofilms exposed to SeO_4^{2-} showed a more compact structure

(**Fig. 4.3c**), with a significantly lower average biofilm thickness of 0.20 ($\pm$ 0.08) mm ($P_{value} \leq$ 0.0001).

Biomass weight and biofilm activity

Dry weight of biofilms varied significantly among different incubations as shown in **Table 4.2**. Incubations exposed to SO_4^{2-} resulted in increased dry weights (6.6 $\pm$ 0.4 mg TS/cm^2) compared to incubations without SO_4^{2-} (2.5 $\pm$ 1.2 mg TS/cm^2). Ash-free dry weight results showed that the biofilms were mainly composed of organic materials (about $>$ 70% were volatile matter) for all incubations (**Table 4.2**).

Table 4.2 Biomass dry weight and ash-free dry weight of biofilms from each incubation. Dry weight was reported as total solid (TS) while ash-free dry weight was reported as volatile solid (VS).

Incubation		Biofilm biomass	
		Dry weight (mg TS/cm^2)	Ash-free dry weight (mg VS/cm^2)
A[*]	No electron acceptor	1.8 ($\pm$ 0.7)	1.4 ($\pm$ 0.4)
B[**]	$NO_3^- + SO_4^{2-}$	6.6 ($\pm$ 1.8)	6.0 ($\pm$ 1.4)
C[*]	SeO_4^{2-}	1.9 ($\pm$ 0.6)	1.9 ($\pm$ 0.9)
D[**]	$NO_3^- + SeO_4^{2-}$	3.9 ($\pm$ 1.3)	3.6 ($\pm$ 1.2)
E[**]	$SO_4^{2-} + SeO_4^{2-}$	7.1 ($\pm$ 1.2)	6.4 ($\pm$ 1.4)
F[*]	$NO_3^- + SO_4^{2-} + SeO_4^{2-}$	6.3 ($\pm$ 1.1)	4.0 ($\pm$ 1.5)

Note: [*]n = 4 experimental replicates, [**]n = 2 experimental replicates

Biofilm activity was assessed through lactate consumption (**Table 4.1**) and cell viability (**Appendix 2, Fig. S4.2**). Lactate was supplied as the sole external electron donor for all biofilm experiments and was provided in excess at an average of 18.5 ($\pm$ 0.9) mM (at a rate of 61 mmol/day). Under conditions where no electron acceptor was present, about 62 ($\pm$ 4)% of the lactate was consumed. Incubations grown with SeO_4^{2-} alone and $NO_3^- + SeO_4^{2-}$ showed similar lactate consumption, averaging 54 ($\pm$ 9)% and 42 ($\pm$ 14)%, respectively. In contrast, biofilms grown in the presence of SO_4^{2-} ($NO_3^- + SO_4^{2-}$ and $SO_4^{2-} + SeO_4^{2-}$) exhibited significantly higher lactate consumption rates of 52 ($\pm$ 9) mmol/day, corresponding to a removal efficiency of 86 ($\pm$ 6)%. Cell viability analysis showed that biofilms exposed to SeO_4^{2-} alone had the highest viable cell counts per unit area (3.9$\times$10^8 $\pm$ 1.3$\times$10^8 viable cell counts/cm^2, $P_{value} \leq$ 0.004) compared to the other DFR incubations (**Appendix 2, Fig. S4.2**). The $SO_4^{2-} + SeO_4^{2-}$ exhibited the lowest cell viability at 2.9$\times$10^7 ($\pm$ 8.5$\times$10^6) viable cell counts/cm^2, which was 11 times lower than for the SeO_4^{2-} incubations.

112

Microbial community composition

Relative abundances (RA) at the bacterial family level were determined for each incubation (**Fig. 4.4**). Biofilms that developed in the absence of electron acceptors (fed with lactate alone) showed a similar relative abundance of *Firmicutes* (38.3 ± 9.8 % RA) and *Proteobacteria* (43.1 ± 10.0 % RA). On the other hand, biofilms grown with external electron acceptors were mainly composed of *Bacteroidetes*, *Firmicutes* and *Proteobacteria* (98.8 ±0.5 % RA) of which *Proteobacteria* accounted for 65.2 (± 7.4)% RA.

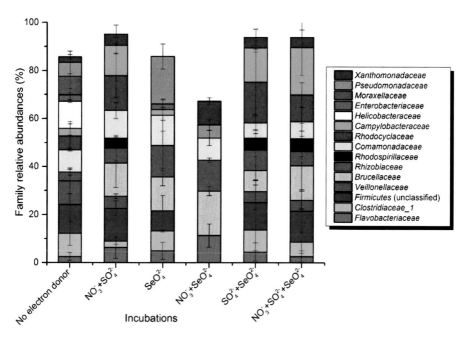

Fig. 4.4 Relative abundances of microbial groups (as family phylotypes) in biofilm samples of each incubation. Relative abundances < 1% are not displayed.

Biofilms incubated with SeO_4^{2-} alone developed *Pseudomonadaceae* (19.8 ±5.3 % RA) as the dominant bacterial family of which 90% were associated with *Pseudomonas* spp. Common phylotypes of *Comamonadaceae* (~10.9% RA), *Rhizobiaceae* (~18.5% RA) and *Brucellaceae* (~12.9% RA) were observed in the biofilms grown with SeO_4^{2-} alone and $NO_3^-+SeO_4^{2-}$ DFR incubations. Biofilms incubated in the presence of SO_4^{2-} developed similar communities dominated by *Campylobacteraceae*, *Rhodocyclaceae*, *Brucellaceae* and unclassified members of *Firmicutes* at an average of 15.5 (± 3.8)%, 14.3 (± 3.0)%, 12.3 (± 3.1)% and 12.7 (± 1.1)% RA, respectively.

Se concentration in the biofilm

Biofilms formed in the presence of $SO_4^{2-}+SeO_4^{2-}$ showed the highest amount of Se entrapped in the biofilm matrix at 68 ($\pm$ 32) μmol Se/cm^2, while 25 ($\pm$ 18) μmol Se/cm^2 was obtained for biofilms grown with $NO_3^-+SO_4^{2-}+SeO_4^{2-}$. Biofilms formed in conditions without SO_4^{2-} showed < 5 μmol Se/cm^2 entrapped in the biofilm matrix. An overall Se mass balance for the biofilm system was calculated (**Appendix 2, Table S4.1**). For 10 days continuous operation, an average of 0.38 ($\pm$ 0.03) mmol Se (30 $\pm$ 2 mg Se) passed through the DFR. For SeO_4^{2-} and $NO_3^-+SeO_4^{2-}$ incubations, roughly 88 ($\pm$ 1)% of Se was detected in the effluent while only 2 ($\pm$ 0.8)% was entrapped in the biofilm (**Appendix 2, Table S4.1**). In contrast, $SO_4^{2-}+SeO_4^{2-}$ and $NO_3^-+SO_4^{2-}+SeO_4^{2-}$ incubations were able to trap 17 ($\pm$ 1)% and 6 ($\pm$ 0.4)% of Se in the biofilm matrix with only 59 ($\pm$ 1)% and 48 ($\pm$ 5)% being released in the effluent, respectively (**Appendix 2, Table S4.1**); a Se effluent mass balance is presented in **Fig. 4.5**. Results indicated that total Se detected in the effluent samples was mainly due to incomplete reduction of Se oxyanions (SeO_4^{2-} and SeO_3^{2-}). It should be considered that the biomass samples were randomly selected and biofilm heterogeneity might introduce some discrepancy in closing the overall Se mass balance. Additionally, effluent sampling was conducted only once a day and considered representative; possible fluctuations in the effluent concentrations within the day might have occurred. Therefore, discrepancies in the Se mass balance are likely due to small sample sizes and random selection.

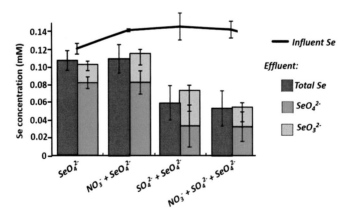

Fig. 4.5 Effluent Se mass balance averaged from 10 days of continuous growth of biofilms in DFRs fed with Se contaminated wastewater and different co-electron acceptors.

EDX results are shown in **Fig. 4.6a** while SEM images of biofilms grown under different growth conditions are shown in **Fig. 4.6 b-e**. Apart from carbon and oxygen signals emanating from the organic biofilm matrix, no other elements were present in detectable quantities for biofilms developed in the absence of external electron acceptors (**Appendix 2, Fig. 4.3**). Biofilms formed in the presence of SeO_4^{2-} contained spherical nanoparticles and the spheres were identified to contain Se (**Fig. 4.6a**). Biofilms incubated with SeO_4^{2-} alone, $NO_3^-+SeO_4^{2-}$ and $NO_3^-+SO_4^{2-}+SeO_4^{2-}$ showed similar EDX spectra, composed of mainly carbon and oxygen peaks with a significant Se signal. EDX spectra of $SO_4^{2-}+SeO_4^{2-}$ incubations indicated the presence of both Se and S in the ratio of approximately 3.5:1.

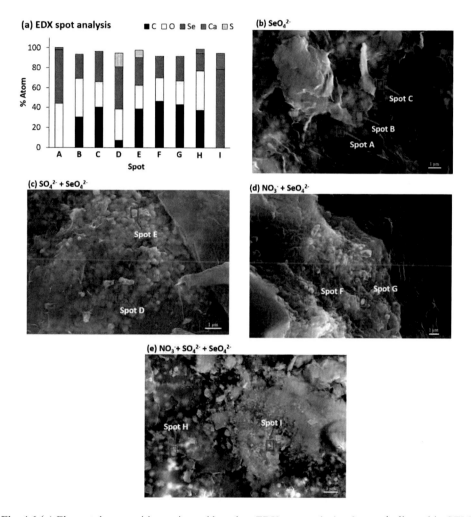

Fig. 4.6 (a) Elemental composition estimated based on EDX spot analysis of areas indicated in SEM images (b-e). SEM images were acquired in secondary electron mode for each treatment: (b) SeO_4^{2-} only, (c) $SO_4^{2-}+SeO_4^{2-}$, (d) $NO_3^-+SeO_4^{2-}$, and (e) $NO_3^-+SO_4^{2-}+SeO_4^{2-}$. Magnification is 25000×.

4.4 Discussion

4.4.1 Influence of co-electron acceptors on Se removal and microbial community structure

Effect of SO_4^{2-}

Removal of Se was greatly influenced by the presence of SO_4^{2-} (**Table 4.1**) showing significantly higher removal efficiencies for both total Se and SeO_4^{2-} compared to SO_4^{2-}-free treatments. Promotion of Se removal in the presence of SO_4^{2-} could be due to the microbial community developed under different incubation conditions (**Fig. 4.4**), leading to different biomass accumulation (**Table 4.2**), or increased biofilm activity (assessed via lactate consumption, **Table 4.1**).

In biofilms grown only with SeO_4^{2-}, a large percentage of the community consisted of *Pseudomonadaceae*. A number of *Pseudomonas* spp. are known to be capable of both SeO_4^{2-} and SeO_3^{2-} reduction to Se^0 (Gonzalez-Gil et al. 2016). A low relative abundance of *Pseudomonas spp.* was detected in biofilms formed in the other incubations, particularly those grown with SO_4^{2-}. This might indicate the development of a microbial community well-adapted to the reduction of SeO_4^{2-} in the biofilms grown only with SeO_4^{2-}. Biofilm communities grown in the presence of SO_4^{2-} ($NO_3^- + SO_4^{2-}$, $SO_4^{2-} + SeO_4^{2-}$ and $NO_3^- + SO_4^{2-} + SeO_4^{2-}$) contained an abundance of both sulfate reducing bacteria (SRB, e.g. *Firmicutes*) and denitrifying bacteria (e.g. *Campylobacteraceae*, *Brucellaceae* and *Rhodocyclaceae*). *Sulfurospirillum* spp. represented about 88% of the *Campylobacteraceae* and have been described as SeO_4^{2-} respiring bacteria. *Sulfurospirillum* spp. are also able to grow under microaerobic conditions and most species can respire NO_3^- (Nancharaiah and Lens 2015a; Oremland et al. 1999). Similarly, various members of the *Rhodocyclaceae* (i.e. *Zoogloea* sp. and *Azospira* sp.) can utilize both NO_3^- and SeO_3^{2-} as electron acceptor and have been found in biofilms growing in Se containing medium (Fakra et al. 2015; Hunter 2007). A number of unclassified members of the *Firmicutes* (containing the second largest known group of SRB, i.e. *Desulfovibrio*) were also detected in the presence of SO_4^{2-} (**Fig. 4.4**).

SeO_4^{2-} removal does not necessarily require Se-specific bacteria such as Se-respiring bacterial strains that can use SeO_4^{2-} as terminal electron donor to support growth. SeO_4^{2-} can be reduced to Se^0 through nonspecific mechanisms in the presence of SRB and denitrifying bacteria using assimilatory or dissimilatory pathways (Eswayah et al. 2016). However, for some microorganisms (e.g. *Thauera selenatis* and *Sulfospirillum barnesii*) dissimilatory reduction of

SeO_4^{2-} can support growth via anaerobic respiration, while in other cases (e.g. SRB) reduction of Se oxyanions can be part of a detoxification response or may be an adventitious reaction catalyzed by enzymes with a principally different function (Eswayah et al. 2016; Muyzer and Stams 2008). This is evident in the SeO_4^{2-} only biofilm where, although a Se-reducing community developed, biomass accumulation and lactate consumption was poor, possibility leading to poor Se removal. In contrast, biofilms grown with SO_4^{2-} showed substantially higher biomass accumulation and lactate consumption compared to incubations without SO_4^{2-} and could indicate that biofilms grown with SO_4^{2-} were more metabolically active. It is possible that sulfide precipitates contributed to the higher dry weight biomass of the incubations with SO_4^{2-}. However, it should be noted that in the SO_4^{2-} incubation with SeO_4^{2-}, only 4 mmol/day SO_4^{2-} was removed from the supplied 44 mmol/day. Additionally, even with the increase in SO_4^{2-} removal rate in the $NO_3^- + SO_4^{2-}$ incubations of 7 mmol/day, the dry weight biomass was the same with the $SO_4^{2-} + SeO_4^{2-}$ and the $NO_3^- + SO_4^{2-} + SeO_4^{2-}$ incubations. Therefore, the contribution of sulfide deposits to the increase to biomass dry weight was likely negligible.

Tomei et al. (1995) reported similar results when studying the effects of SO_4^{2-} and SeO_4^{2-} on a pure culture of *Desulfovibrio desulfuricans*; higher biomass accumulation was observed when *D. desulfuricans* was grown in the presence of both SO_4^{2-} and SeO_4^{2-}, while lower biomass accumulation was observed when the *D. desulfuricans* biofilm was grown without SO_4^{2-} and only in the presence of SeO_4^{2-}. Lower cell growth of SRB cells due to SeO_4^{2-} can be attributed to SeO_4^{2-} interference with SO_4^{2-} activation and SeO_4^{2-} metabolism in SRB cells. The authors postulated that the inhibition of *D. desulfuricans* growth in the absence of SO_4^{2-} is due to the structural similarity between the Se and S atoms. Se oxyanions can act as S analogues, stopping SO_4^{2-} activation and disrupting both assimilatory and dissimilatory SO_4^{2-} reducing pathways (Bulska et al. 2006; Hockin and Gadd 2006), thereby reducing SRB cell growth (Malagoli et al. 2015; Tomei et al. 1995).

Another reason for the increase in Se removal in the presence of SO_4^{2-} could be related to abiotic reactions possibly occurring between Se oxyanions and S compounds within the biofilm matrix. In the experiment by Hockin and Gadd (2003) using *Desulfomicrobium norvegicum* biofilms grown at 30°C and pH 7.0, formation of Se^0 was noted as a result of abiotic reduction of SeO_3^{2-} (200 µM) by biogenic sulfide (10 mM). The reduction of SeO_3^{2-}, the intermediate product between SeO_4^{2-} and Se^0, by biogenic sulfide could have contributed to the more complete removal of SeO_4^{2-}. In the work presented here, there was no measurable dissolved sulfide in the effluent detected for any of the incubations (detection limit approximately 0.1

mM). This could indicate either usage of sulfide for abiotic reduction or sulfur entrapment within the biofilm. Though it is difficult to properly quantify the abiotic reactions that were occurring within the biofilm, it is possible that interactions between SeO_3^{2-} and biogenic sulfide occurred when the biofilms were grown in the presence of $SO_4^{2-} + SeO_4^{2-}$.

Effect of NO_3^-

In contrast to SO_4^{2-}, NO_3^- did not have a stimulating effect on the removal of Se in the biofilm systems showing a similar Se removal efficiency, biomass accumulation and lactate consumption as biofilms grown in the presence of solely SeO_4^{2-}. Additionally, NO_3^- removal was poor in the incubations void of SO_4^{2-}. SO_4^{2-} does not typically have a negative effect on NO_3^- reduction, considering the highly thermodynamically favorable denitrification reactions compared to reduction of the other oxyanions investigated. In most biological wastewater treatment studies, the presence of other oxyanions has not been identified as a detrimental factor for NO_3^- removal (**chapter 3**; Lai et al. 2014; Lenz et al. 2009; Oremland et al. 1999).

Microbial communities in biofilms grown in the presence of $NO_3^- + SeO_4^{2-}$ showed three dominant bacterial families: *Comamonadaceae*, *Rhizobiaceae* and *Brucellaceae*, all known for their denitrifying bacterial members, which are often found in freshwater or soil and can play an important role in the nitrogen cycle. Additionally, although *Pseudomonadaceae* are also known for their NO_3^- respiring bacterial members, there was only a small relative abundance percentage found in biofilms grown under $NO_3^- + SeO_4^{2-}$ conditions ($5.4 \pm 2.1\%$ RA) compared to biofilms grown with SeO_4^{2-} alone ($19.8 \pm 2.1\%$ RA). Some members of the *Comamonadaceae* (e.g. *Comamonas sp.*) are described to be SeO_3^{2-} reducers (Fakra et al. 2015), while *Ochrobactrum* spp. is capable of reducing SeO_3^{2-} and tellurite to Se^0 and Te^0 nanoparticles, respectively (Zonaro et al. 2015). However, there is little information available on the potential role of these organisms in the reduction of SeO_4^{2-}. On the other hand, *Rhizobium* spp. (constituting ~79% of the *Rhizobiaceae*) have been studied by Hunter and Manter (2007) under denitrifying conditions and observed that under anoxic conditions, *Rhizobium* sp. growth was dependent on the presence of NO_3^- and that NO_3^- was necessary to facilitate SeO_4^{2-} or SeO_3^{2-} reduction.

While the microbial community analyses showed the presence of denitrifiers (**Fig. 4.4**), the low NO_3^- and SeO_4^{2-} removal efficiencies (**Table 4.1**) could possibly be due to an overall decrease in microbial activity in the absence of SO_4^{2-}, which is supported by the low lactate consumption (average removal efficiency $48 \pm 9 \%$) and low biomass production (averaged 2.9

$\pm$ 1 mg TS/cm^2) in the SeO$_4^{2-}$ and NO$_3^-$ + SeO$_4^{2-}$ incubations. Another explanation for the low NO$_3^-$ and SeO$_4^{2-}$ removal could be that the inoculum used for the study, identified to be dominated by methanogens and SRB (Roest et al. 2005), contained a low denitrifying population at the initial stage of incubation. Though microbial communities were characterized, this was conducted at the end of a 12-day experimental run (2 days of batch incubation and 10 days of continuous operation) and a denitrifying population might not have developed well during these relatively short term experiments. This could indicate that the inoculum was indeed predisposed to SO$_4^{2-}$ reducing conditions and would have required more time to establish a denitrifying microbial community. Hence, longer term monitoring of the microbial community, the evaluation of different inocula and their ability to adapt to the treatment conditions should be considered in future studies.

Alternatively, competition for reductases and/or intermediates of SeO$_4^{2-}$ reduction might have occurred. In the case of SRBs, Hockin and Gadd (2006), who also observed increased Se removal efficiencies in the presence of high SO$_4^{2-}$ concentrations (28 mM), proposed that SeO$_4^{2-}$ reduction takes place in the periplasm of SRBs, allowing for simultaneous reduction without competition between SeO$_4^{2-}$ and SO$_4^{2-}$. For certain denitrifying or Se-reducing bacteria (i.e. *Enterobacter cloacae*), membrane-bound NO$_3^-$ reductases (*Nar*), periplasmic NO$_3^-$ reductases (*Nap*) and SeO$_4^{2-}$ reductases (*Ser*) have all been shown to catalyze the reduction of SeO$_4^{2-}$ to SeO$_3^{2-}$ (Ridley et al. 2006). Furthermore, some denitrifiers (i.e. *Rhizobium sullae* and *Sulfurospirillum barnesii*) are only capable of reducing SeO$_4^{2-}$ to SeO$_3^{2-}$ and require another intermediate, such as nitrite, to further reduce SeO$_3^{2-}$ to Se^0 via nitrite reductases, while others can only reduce either SeO$_4^{2-}$ to SeO$_3^{2-}$ or SeO$_3^{2-}$ to Se^0 (Oremland et al. 1999; Yang et al. 2011). As such, it is possible that low SeO$_4^{2-}$ removal under NO$_3^-$ + SeO$_4^{2-}$ incubation could be due to the lack of accumulated intermediate, i.e. nitrite, or specific organisms to properly facilitate the full reduction process from SeO$_4^{2-}$ to Se^0.

4.4.2 Se immobilization and biofilm response to Se exposure

The biofilms in the DFR appeared to be the thinnest when only SeO$_4^{2-}$ was fed into the system, while those grown with SeO$_4^{2-}$ plus co-electron acceptors were generally thicker (**Fig. 4.3**). Thinner biofilms could be a result of stress from or toxicity of SeO$_4^{2-}$ to the biomass, e.g. production of intermediate SeO$_3^{2-}$ or entrapment of biogenic Se^0. Densification of fungal biofilms upon exposure to SeO$_3^{2-}$ has been described for *Phanerochaete chrysosporium* pellets (Espinosa-Ortiz et al. 2016). The authors attributed this densification to a stress response due to the presence of SeO_3^{2-} as well as the deposition of biogenic Se^0 within the pellets. Biogenic Se^0 nanoparticles have been described to harbor antimicrobial and antibiofilm activities due to the production of reactive oxygen species (Zonaro et al. 2015). However, this study did not monitor whether the dense biofilms indeed produced reactive oxygen species.

One of the main issues in biological treatment of Se-laden wastewaters is the release of colloidal Se^0 into the aqueous phase, which not only potentially compromises discharge criteria but also raises questions regarding the fate and stability of Se^0 in the environment (Buchs et al. 2013). Additional chemical post-treatment methods such as electrocoagulation and precipitation might be required to ensure the removal of Se^0 colloidal particles (Nancharaiah and Lens 2015b; Staicu et al.. 2015). Therefore, this study investigated SeO_4^{2-} transformation, retention and immobilization of biogenic Se^0 in the biofilms.

This study provides strong evidence of Se immobilization within the biofilm matrix: (1) visual red deposits inside the biofilms (**Fig. 4.2**), (2) detection of Se concentration within the biofilm matrix and (3) Se-containing spherical particles in the biofilms revealed by SEM-EDX analysis (**Fig. 4.6**). Biofilm images revealed an orange-red coloration of biofilms (**Fig. 4.2**), particular those grown in the presence of SO_4^{2-}, which is a strong indication of a transformation of SeO_4^{2-} to biogenic Se^0 and its retention within the biofilm.(Dessì et al. 2016) Additionally, effluent Se mass balance calculations showed a negligible release of colloidal Se^0 (estimated as the difference between total Se and Se oxyanion concentration) in the effluent of the biofilm reactors (**Fig. 4.5**). HPLC-ICP-MS analyses of effluent samples revealed no peaks for other dissolved Se compounds aside from SeO_3^{2-} and SeO_4^{2-} (**Appendix 2, Fig. S4.4**). Visually, no reddish color (indication of colloidal Se^0) was observed in the effluent for any of the incubations. Instead, incubations with SO_4^{2-} showed a yellowish coloration in the effluent, while the effluents of treatments without SO_4^{2-} remained mostly clear to slightly whitish (**Appendix 2, Fig. S4.5a**). Yellowish coloration could be an indication of elemental sulfur (S^0). This was confirmed by EDX analysis of dried effluent samples. The EDX spectra revealed S-associated peaks from effluent samples with yellowish color (**Appendix 2, Fig. S4.5c**), while no S signal was detected in the clear-to-whitish effluent samples from incubations without SO_4^{2-} (**Appendix 2, Fig. S4.5b**). Possible formation of Se-S particles inside the biofilms was indicated by the EDX spectra (**Appendix 2, Fig. S4.5a**). Detection of Se-S particles were similarly observed by Davis et al. (2007) during *in situ* SeO_4^{2-} bioprecipitation in a SO_4^{2-} reducing groundwater zone and by Hockin and Gadd (2003) under SO_4^{2-} reducing conditions in batch experiments investigating SeO_3^{2-} reduction.

This study indicates that the biofilms formed in the DFR retained Se^0 formed by microbial reduction within the biofilm matrix. In particular, biofilms formed in the presence of SO_4^{2-} + SeO_4^{2-} accumulated more total Se in the biofilm matrix while exhibiting greater biomass growth and better SeO_4^{2-} removal. Treatment of Se-laden wastewaters coupled with SO_4^{2-} removal could have potential applications in the reuse of Se^0 through biological reduction of Se-oxyanions (Jain et al. 2015; Nancharaiah and Lens 2015b). However, Se recovery and purification techniques from biological systems are still largely unexplored and will require further research before actual large-scale application. However, this study did not evaluate the effect of other heavy metals, which might be present in mining wastewaters; this should be the focus of future studies.

Alternative nondestructive biofilm imaging techniques such as cryo-slicing coupled with low-temperature FESEM (Hockin and Gadd 2003) or synchrotron techniques such as X-ray fluorescence imaging and scanning transmission X-ray microscopy (Yang et al. 2016, 2011) could be used in the future for the investigation of the localization of the precipitated Se and S granules. This would allow further insights into the fate and behavior of Se within the biofilm as influenced by the presence of NO_3^- and/or SO_4^{2-}.

4.5 Conclusions

Effective SeO_4^{2-} removal through microbial reduction to elemental Se^0 and retention of biogenic elemental Se^0 within the biofilm matrix was demonstrated in drip flow biofilm reactors over a 10-day period. The experimental work conducted in this study has practical implications in industrial system management and the treatment of Se-laden wastewater containing high amounts of co-electron acceptors, i.e. NO_3^- and SO_4^{2-}. This study demonstrates that biofilms can be established in simulated mining wastewater and that SeO_4^{2-} removal efficiency by lactate-fed anaerobic biofilms can be affected by the presence of specific oxyanions. The presence of SO_4^{2-} in the synthetic wastewater significantly improved biofilm growth and the SeO_4^{2-} removal efficiency, whereas the presence of NO_3^- did not have a significant effect on SeO_4^{2-} removal by the biofilms. Thus, the presence of co-electron acceptors in Se-laden wastewater -and possibly even their addition- should be carefully considered, as it can be beneficial for Se removal. Electron acceptors also affected the morphology of the biofilms. Relatively thin biofilms were formed in the presence of SeO_4^{2-} alone, while thicker biofilms accumulated in the presence of NO_3^- or SO_4^{2-}, which could have implications in the management of biofilms in scaled-up operations. Additionally, biogenic Se^0 was mainly retained within the biofilm matrix, which minimized the discharge of colloidal Se^0 in the treated water leaving the reactor. This indicates, overall, that anaerobic treatment using biofilm systems can be directly applied and is suitable for treating Se-laden wastewater with co-electron acceptors.

4.6   References

Buchs, B., Evangelou, M.W.H., Winkel, L.H.E., Lenz, M., 2013. Colloidal properties of nanoparticular biogenic selenium govern environmental fate and bioremediation effectiveness. *Environ. Sci. Technol.* 47, 2401-2407.

Bulska, E., Wysocka, I.A., Wierzbicka, M.H., Proost, K., Janssens, K., Falkenberg, G., 2006. In vivo investigation of the distribution and the local speciation of selenium in *Allium cepa L.* by means of microscopic X-ray absorption near-edge structure spectroscopy and confocal microscopic X-ray fluorescence analysis. *Anal. Chem.* 78, 7616-7624.

Cord-Ruwisch, R., 1985. A quick method for the determination of dissolved and precipitated sulfides in cultures of sulfate-reducing bacteria. *J. Microbiol. Methods* 4, 33-36.

Davis, A.C., Patterson, B.M., Grassi, M.E., Robertson, B.S., Prommer, H., Mckinley, A.J., 2007. Effects of increasing acidity on metal(loid) bioprecipitation in groundwater: Column studies. *Environ. Sci. Technol.* 41, 7131-7137.

Dessì, P., Jain, R., Singh, S., Seder-Colomina, M., van Hullebusch, E.D., Rene, E.R., Ahammad, S.Z., Carucci, A., Lens, P.N.L., 2016. Effect of temperature on selenium removal from wastewater by UASB reactors. *Water Res.* 94, 146-154.

Espinosa-Ortiz, E.J., Pechaud, Y., Lauchnor, E., Rene, E.R., Gerlach, R., Peyton, B.M., van Hullebusch, E.D., Lens, P.N.L., 2016. Effect of selenite on the morphology and respiratory activity of *Phanerochaete chrysosporium* biofilms. *Bioresour. Technol.* 210, 138-145.

Eswayah, A.S., Smith, T.J., Gardiner, P.H.E., 2016. Microbial transformations of selenium species of relevance to bioremediation. *Appl. Environ. Microbiol.* 82, 4848-4859.

Fakra, S.C., Luef, B., Castelle, C.J., Mullin, S.W., Williams, K.H., Marcus, M.A., Schichnes, D., Banfield, J.F., 2015. Correlative cryogenic spectro-microscopy to investigate selenium bioreduction products. *Environ. Sci. Technol.* doi: 10.1021/acs.est.5b01409.

Goeres, D.M., Hamilton, M.A., Beck, N.A., Buckingham-Meyer, K., Hilyard, J.D., Loetterle, L.R., Lorenz, L.A., Walker, D.K., Stewart, P.S., 2009. A method for growing a biofilm under low shear at the air-liquid interface using the drip flow biofilm reactor. *Nat. Protoc.* 4, 783-788.

Gonzalez-Gil, G., Lens, P.N.L., Saikaly, P.E., 2016. Selenite reduction by anaerobic microbial aggregates: Microbial community structure, and proteins associated to the produced selenium spheres. *Front. Microbiol.* 7, 571.

Hockin, S., Gadd, G.M., 2006. Removal of selenate from sulfate-containing media by sulfate-reducing bacterial biofilms. *Environ. Microbiol.* 8, 816-826.

Hockin, S.L., Gadd, G.M., 2003. Linked redox precipitation of sulfur and selenium under anaerobic conditions by sulfate-reducing bacterial biofilms. *Appl. Enivironmental Microbiol.* 69, 7063-7072.

Hunter, W.J., 2007. An *Azospira oryzae* (*syn Dechlorosoma suillum*) strain that reduces selenate and selenite to elemental red selenium. *Curr. Microbiol.* 54, 376-381.

Hunter, W.J., Manter, D.K., 2007. Reduction of selenite to elemental red selenium by *Rhizobium sp.* strain B1. *Curr. Microbiol.* 55, 344-349.

Jain, R., Seder-Colomina, M., Jordan, N., Dessi, P., Cosmidis, J., van Hullebusch, E.D.,Weiss, S., Farges, F., Lens, P.N.L., 2015b. Entrapped elemental selenium nanoparticles affect physicochemical properties of selenium fed activated sludge. *J. Hazard.Mater.* 295, 193-200.

Lai, C.Y., Yang, X., Tang, Y., Rittmann, B.E., Zhao, H.P., 2014. Nitrate shaped the selenate-reducing microbial community in a hydrogen-based biofilm reactor. *Environ. Sci. Technol.* 48, 3395-3402.

Lemly, A.D., 2014. Teratogenic effects and monetary cost of selenium poisoning of fish in Lake Sutton, North Carolina. *Ecotoxicol. Environ. Saf.* 104, 160-167.

Lenz, M., Enright, A.M., O'Flaherty, V, van Aelst, A.C., Lens, P.N.L., 2009. Bioaugmentation of UASB reactors with immobilized *Sulfurospirillum barnesii* for simultaneous selenate and nitrate removal. *Appl. Microbiol. Biotechnol.* 83, 377-388.

Lenz, M., Lens, P.N.L., 2009. The essential toxin: The changing perception of selenium in environmental sciences. *Sci. Total Environ.* 407, 3620-3633.

Li, D.-B., Cheng, Y.-Y.,Wu, C., Li,W.-W., Li, N., Yang, Z.-C., Tong, Z.-H., Yu, H.-Q., 2014. Selenite reduction by *Shewanella oneidensis* MR-1 is mediated by fumarate reductase in periplasm. *Sci. Rep.* 4, 3735.

Lueders, T., Manefield, M., Friedrich, M.W., 2004. Enhanced sensitivity of DNA- and rRNA-based stable isotope probing by fractionation and quantitative analysis of isopycnic centrifugation gradients. *Environ. Microbiol.* 6, 73-78.

Mal, J., Veneman, W.J., Nancharaiah, Y.V., van Hullebusch, E.D., Peijnenburg, W.J.G.M., Vijver, M.G., Lens, P.N.L., 2016. A comparison of fate and toxicity of selenite, biogenically and chemically synthesized selenium nanoparticles to the zebrafish (*Danio rerio*) embryogenesis. *Nanotoxicology* 11, 87-97.

Malagoli, M., Schiavon, M., dall'Acqua, S., Pilon-Smits, E.A.H., 2015. Effects of selenium biofortification on crop nutritional quality. *Front. Plant Sci.* 6, 280.

Manteca, Á., Fernández, M., Sánchez, J., 2005. A death round affecting a young compartmentalized mycelium precedes aerial mycelium dismantling in confluent surface cultures of *Streptomyces antibioticus*. Microbiology 151, 3689-3697.

Muyzer, G., Stams, A.J.M., 2008. The ecology and biotechnology of sulphate-reducing bacteria. *Nat. Rev. Microbiol.* 6, 441-454.

Nancharaiah, Y.V., Lens, P.N.L., 2015a. Ecology and biotechnology of selenium-respiring bacteria. *Microbiol. Mol. Biol. Rev.* 79, 61-80.

Nancharaiah, Y.V., Lens, P.N.L., 2015b. Selenium biomineralization for biotechnological applications. *Trends Biotechnol.* 33, 323-330.

Nguyen, V.K., Park, Y., Yu, J., Lee, T., 2016. Microbial selenite reduction with organic carbon and electrode as sole electron donor by a bacterium isolated from domestic wastewater. B*ioresour. Technol.* 212, 182-189.

Ontiveros-Valencia, A., Penton, C.R., Krajmalnik-Brown, R., Rittmann, B.E., 2016. Hydrogen-fed biofilm reactors reducing selenate and sulfate: Community structure and capture of elemental selenium within the biofilm. *Biotechnol. Bioeng.* 113, 1736-44.

Oremland, R.S., Blum, J.S., Bindi, A.B., Dowdle, P.R., Herbel, M., Stolz, J.F., 1999. Simultaneous reduction of nitrate and selenate by cell suspensions of selenium-respiring bacteria. *Appl. Environ. Microbiol.* 65, 4385-4392.

Ridley, H., Watts, C.A., Richardson, D.J., Butler, C.S., 2006. Resolution of distinct membrane-bound enzymes from *Enterobacter cloacae* SLD1a-1 that are responsible for selective reduction of nitrate and selenate oxyanions. *Appl. Environ. Microbiol.* 72, 5173-5180.

Roest, K., Heilig, H.G.H.., Smidt, H., De Vos, W.M., Stams, A.J.M., Akkermans, A.D.L., 2005. Community analysis of a full-scale anaerobic bioreactor treating paper mill wastewater. *Syst. Appl. Microbiol.* 28, 175-85.

Schloss, P.D., Westcott, S.L., Ryabin, T., Hall, J.R., Hartmann, M., Hollister, E.B., Lesniewski, R.A., Oakley, B.B., Parks, D.H., Robinson, C.J., Sahl, J.W., Stres, B., Thallinger, G.G., Van Horn, D.J., Weber, C.F., 2009. Introducing mothur: Open-source, platform-independent, community-supported software for describing and comparing microbial communities. Appl. Environ. Microbiol. 75, 7537-7541. doi:10.1128/AEM.01541-09

Staicu, L.C., van Hullebusch, E.D., Lens, P.N.L., 2015. Production, recovery and reuse of biogenic elemental selenium. *Environ. Chem. Lett.* 13, 89-96.

Stams, A.J.M., Grolle, K.C.F., Frijters, C.T.M., van Lier, J.B., 1992. Enrichment of thermophilic propionate-oxidizing bacteria in syntrophy with *Methanobacterium thermoautotrophicum* or *Methanobacterium thermoformicicum*. *Appl. Environ. Microbiol.* 58, 346-352.

Stover, E.L., Pudvay, M., Kelly, R.F., Lau, A.O., 2006. Biological treatment of flue gas desulfurization wastewater. Proceeding Engineer's Society of Western Pennsylvannia: IWC-07-49.

Takahashi, S., Tomita, J., Nishioka, K., Hisada, T., Nishijima, M., 2014. Development of a prokaryotic universal primer for simultaneous analysis of bacteria and archaea using next-generation sequencing. *PLOS One* 9, e105592.

Tan, L.C., Nancharaiah, Y.V., van Hullebusch, E.D., Lens, P.N.L., 2016. Selenium: Environmental significance, pollution, and biological treatment technologies. *Biotechnol. Adv.* 34, 886-907.

Tomei, F.A., Barton, L.L., Lemanski, C.L., Zocco, T.G., Fink, N.H., Sillerud, L.O., 1995. Transformation of selenate and selenite to elemental selenium by *Desulfovibrio desulfuricans*. *J. Ind. Microbiol. Biotechnol.* 14, 329-336.

van Ginkel, S.W., Yang, Z., Kim, B.O., Sholin, M., Rittmann, B.E., 2011. The removal of selenate to low ppb levels from flue gas desulfurization brine using the H_2-based membrane biofilm reactor (MBfR). *Bioresour. Technol.* 102, 6360-6364.

Wen, H., Carignan, J., 2007. Reviews on atmospheric selenium: Emissions, speciation and fate. *Atmos. Environ.* 41, 7151-7165.

Yang, S.I., George, G.N., Lawrence, J.R., Kaminskyj, S.G.W., Dynes, J.J., Lai, B., Pickering, I.J., 2016. Multispecies biofilms transform selenium oxyanions into elemental selenium particles: Studies using combined synchrotron X-ray fluorescence imaging and scanning transmission X-ray microscopy. *Environ. Sci. Technol.* 50, 10343-10350.

Yang, S.I., Lawrence, J.R., Swerhone, G.D.W., Pickering, I.J., 2011. Biotransformation of selenium and arsenic in multi-species biofilm. *Environ. Chem.* 8, 543-551.

Zonaro, E., Lampis, S., Turner, R.J., Junaid, S., Vallini, G., 2015. Biogenic selenium and tellurium nanoparticles synthesized by environmental microbial isolates efficaciously inhibit bacterial planktonic cultures and biofilms. *Front. Microbiol.* 6, 658.

CHAPTER 5

Biological treatment of selenium-laden wastewater containing nitrate and sulfate in an upflow anaerobic sludge bed reactor at pH 5.0

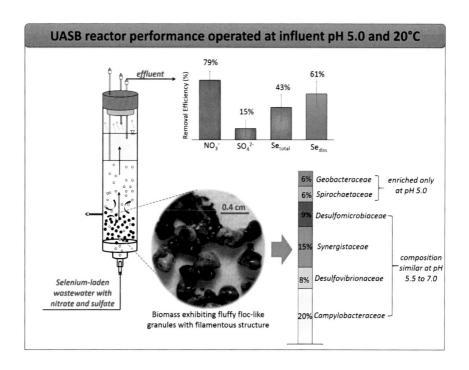

This chapter has been submitted in modified form:

Tan, L.C, Nancharaiah, Y.V., Lu, S., van Hullebusch, E., Gerlach, G., Lens, P.N.L. Biological treatment of selenium-laden wastewater containing nitrate and sulfate in an upflow anaerobic sludge bed reactor at pH 5.0. *Chemosphere*. (*under review*)

Abstract

This study investigated the removal of selenate (SeO_4^{2-}), sulfate (SO_4^{2-}) and nitrate (NO_3^-) at different influent pH values ranging from 7.0 to 5.0 and 20°C in an upflow anaerobic sludge blanket (UASB) reactor using lactate as an electron donor. At pH 5.0, the UASB reactor showed a 20 to 30% decrease in reactor performance within a standard error of 2.5% as compared to operation at pH 5.5 to 7.0, reaching removal efficiencies of 79%, 15%, 43% and 61% for NO_3^-, SO_4^{2-}, Se_{tot} and Se_{diss}, respectively. The sludge formed during low pH operation had a fluffy, floc-like appearance with filamentous structure, possibly due to the low polysaccharide (PS) to protein (PN) ratio (0.01 PS/PN) in the soluble extracellular polymeric substances (EPS) matrix of the biomass. Scanning electron microscopy with energy dispersive X-ray spectroscopy (SEM-EDX) analysis of the sludge confirmed Se oxyanion reduction and deposition of Se^0 particles inside the biomass. Microbial community analysis using Illumina MiSeq sequencing revealed that the families of *Campylobacteraceae* and *Desulfomicrobiaceae* were the dominant phylotypes throughout the reactor operation at approximately 23% and 10% relative abundance, respectively. Furthermore, approximately 10% relative abundance of both the *Geobacteraceae* and *Spirochaetaceae* was observed in the granular sludge during the pH 5.0 operation. Overall, this study demonstrated the feasibility of UASB operation at pH values ranging from 7.0 to 5.0 for removing Se and other oxyanions from wastewaters.

Keywords: Selenate bioreduction; UASB reactor; anaerobic granular sludge; microbial diversity; acid mine drainage

128

5.1 Introduction

Selenium (Se) is both an essential and toxic trace element for living cells, with only 5- to 10-fold difference between essential and toxic concentrations (Nancharaiah and Lens 2015). Se accumulation and intoxication can cause mutations, reproductive failures and teratogenic effects, particularly with aquatic egg-laying organisms upon exposure to toxic Se levels (Lemly 2014). As such, it is now widely recognized that removal of Se from contaminated waters is essential for the protection of animal, human and ecosystem health (Tan et al. 2016).

Compared to physical and chemical technologies, biological treatment has emerged in recent years as a promising Se removal technology due to the possibility of simultaneous treatment of various oxyanions as well as yielding elemental Se (Se^0) nanoparticles that can potentially be recovered and reused (Tan et al. 2016). Despite these recent advances, there are still many challenges in applying this bioprocess to real wastewaters. One such challenge is the low pH of some Se-laden wastewaters, e.g. acid mine drainage (AMD) wastewater. AMD wastewater is characterized by low pH values (varying between pH 2 to 5) and varying heavy metal concentrations depending on the mining site (Koschorreck and Tittel 2007). Se speciation depends on pH, ionic strength and redox properties of a system (Erosa et al. 2009). Adjustment of the initial pH entails additional costs when the process is transferred into full-scale applications treating real wastewaters. Espinosa-Ortiz et al. (2015) reported selenite (SeO_3^{2-}) removal from pH 4.5 medium using *Phanerochaete chrysosporium* fungal pellets and achieved a total soluble selenium (SeO_3^{2-}) removal efficiency of 70%. However, the fungal pellets were incapable of reducing and removing selenate (SeO_4^{2-}).

Table 5.1 gives an overview of bioreactor studies related to the treatment of Se-laden wastewaters. It shows that there are very limited studies investigating the effect of pH on Se removal. The majority of the studies, as summarized in **Table 5.1**, adjusted the initial pH to a circum-neutral pH during laboratory testing or used a buffer medium before performing Se bioremoval to avoid large pH fluctuations and associated impact on the biological system. Therefore, bioprocesses capable of Se removal from wastewaters with low pH without prior neutralization need to be developed.

This study investigated the effect of decreasing influent pH (from 7.0 to pH 5.0) on the performance of a continuously operated upflow anaerobic sludge blanket (UASB) reactor (referred to as low pH reactor) treating SeO_4^{2-}, SO_4^{2-} and NO_3^- using anaerobic granular sludge at room temperature. The protein, polysaccharide, and Se content of the UASB granules were determined at various time points and compared to the inoculum and granular sludge from a control UASB reactor (referred to as control reactor, UASB operated in **chapter 3**) operated at pH 7.0. Low pH reactor biomass morphology was also analyzed at the end of the reactor run and compared to inoculum and control reactor. The microbial community composition of the inoculum and sludge sampled from the low pH reactor was analyzed and related to low pH UASB reactor performance and operational strategies.

5.2   Materials and methods

5.2.1   Source of biomass

Experiments were performed with anaerobic granular sludge taken from a full-scale UASB reactor treating paper-mill wastewater in Eerbeek (The Netherlands). Biomass was supplied at 10 g (wet weight) to 100 mL medium as described in Lenz et al. (2008). No additional biomass was added during the entire reactor run. Prior to each operational change (see below), a small portion of the biomass was withdrawn from the UASB reactor for analysis of microbial community, protein and polysaccharide content.

5.2.2   Synthetic wastewater

All experiments were carried out using synthetic mining wastewater containing (in g/L): NH_4Cl (0.30), $CaCl_2 \cdot 2H_2O$ (0.01), $MgCl_2 \cdot 6H_2O$ (0.01), and $NaHCO_3$ (0.04). Acid and alkaline trace metal solutions (0.1 mL each) were added to 1 L of synthetic wastewater (composition of trace metal solutions is described in Stams et al. (1992). No phosphate or pH buffer was included in the medium and the pH was adjusted to the desired pH using 0.5 M HCl. Analytical grade sodium lactate was used as the electron donor and potassium nitrate, potassium sulfate and sodium selenate were added as electron acceptors. Their concentrations depended on the operational period, as described below.

5.2.3   UASB reactor

Low pH bioreactor configuration and operating conditions

The low pH UASB reactor had a working volume of 2.0 L, an inner diameter of 5.6 cm and a length of 1 m (**Fig. 5.1**). It was operated at room temperature (20 ± 2°C) at an average hydraulic residence time (HRT) of 24 h. The influent flow rate was set at 1.4 mL/min and the upflow velocity at 1.5 m/h with a re-circulation flow of 62 mL/min. The organic loading rate was 2 g COD/L·d using lactate as the electron donor.

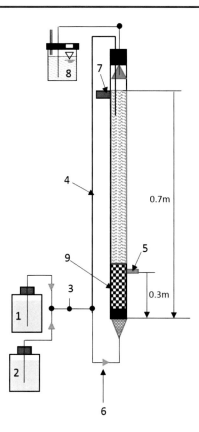

Fig. 5.1 Reactor schematic diagram and operating parameters of the UASB reactors used in this study. (1) electron donor tank, (2) growth medium with oxyanion tank, (3) influent sampling point, (4) recirculation system, (5) sludge sampling point, (6) upflow velocity pump, (7) effluent sampling point, (8) gas outlet and trap, and (9) expansion bed for the anaerobic granular sludge.

The low pH UASB reactor was provided with excess electron donor and exposed to three oxyanions, i.e. SeO_4^{2-}, NO_3^- and SO_4^{2-}, for the entire duration of the study. An average of 48 ($\pm$ 5) mg NO_3^--N/L and 2 ($\pm$ 0.5) g SO_4^{2-}/L were fed into the low pH reactor for the entire UASB reactor operation. The low pH UASB reactor was operated in three different operational periods (Phase I to III). The first phase of the low pH reactor operation was divided into 5 parts. Phase Ia used approximately 10 mg Se/L as SeO_4^{2-}, 42 mg NO_3^--N/L and 2 g SO_4^{2-}/L at pH 7.0 and was considered the start-up phase, lasting for 29 days. Subsequent phases were maintained long enough to reach a pseudo-steady-state. The influent pH was gradually decreased from pH 7.0 (phase Ib, day 30-61) to pH 6.0 (phase Ic, day 62-90), to pH 5.5 (phase Id, day 91-129) and finally to pH 5.0 (phase Ie, day 130-168). Each phase was operated for approximately 32 days before the pH was changed. After phase Ie, the incoming feed pH was maintained at pH 5.0 and

131

the removal efficiency stability of the bioreactor was assessed by changing the SeO_4^{2-} concentration from ~12 mg Se/L (phase I) to a range of 1-5 mg Se/L (phase IIa, day 169-202) and then back to ~12 mg Se/L (phase IIb, day 203-231). Finally, the effect of SeO_3^{2-} addition at pH 5.0 was investigated during phase III (day 232-266), where the total Se concentration (~10 mg Se/L) in the influent consisted of 50% SeO_3^{2-} and 50% SeO_4^{2-}. The average influent Se concentrations (as SeO_4^{2-}) during phase I was 12 ($\pm$ 0.8) mg Se/L·d, 3 ($\pm$ 0.8) mg Se/L·d (phase IIa) and 12 ($\pm$ 0.8) mg Se/L·d during phase IIb. In phase III, the concentration of total Se was split into 50% SeO_4^{2-} and 50% SeO_3^{2-} in mg Se/L at a concentration of 5 ($\pm$ 1) mg Se/L and 4 ($\pm$ 0.5) mg Se/L, respectively.

Control reactor set-up

A control reactor was operated alongside the low pH reactor with an influent stream pH of 7.0 and was presented in **chapter 3**. The control reactor set-up was operated similarly to phases Ia and Ib of the low pH reactor with phase I of the control reactor (pH 7.0) lasting for 92 days. The sludge characteristics of the control reactor (i.e. polysaccharide/protein ratio and biomass morphology) were reported and compared with sludge characteristics of the low pH reactor.

Analytical methods

Liquid samples were collected and centrifuged at 37,000 g for 15 min to remove suspended cells and Se^0 particles (Jain et al. 2015) before the supernatant was used for analysis. Chemical oxygen demand (COD), total dissolved sulfide (TDS), and total solids (TS) measurements were carried out using standard methods (APHA/AWWA/WEF 2005). Ion chromatography (IC, Dionex ICS 1000 AS4A column) was used to measure the concentration of NO_3^- and SO_4^{2-} at retention times of 3.9 and 7.5 min, respectively. Due to SO_4^{2-} peak interference with the SeO_4^{2-} peak, SeO_4^{2-} was not determined by IC. Instead, all influent and effluent Se concentrations were analyzed as Se using a graphite furnace atomic absorption spectrometer (GFAAS, ThermoElemental Solaar MQZe GF95, Se lamp at 196.0 nm) and reported as total Se (Se_{tot}) and dissolved Se (Se_{diss}). Se_{tot} includes both dissolved Se and colloidal Se present in the collected liquid sample, while Se_{diss} includes only dissolved Se remaining in the effluent after centrifugation (37,000 g) and filtration (0.45 µm, acetate membrane, Sigma-Aldrich, USA). The SeO_3^{2-} concentration in the reactor was measured using a spectrophotometric method as described by (Li et al. 2014).

Statistical analysis

Averages and standard deviations were calculated for the oxyanion removal efficiencies. Statistical differences in lactate, NO_3^-, SO_4^{2-}, Se_{tot} and Se_{diss} removal efficiencies between the low pH reactor operation phases were evaluated using analysis of variance (ANOVA) with Tukey tests, where a *P-value* ≤ 0.05 was considered statistically significant. Minitab 17 Statistical (Minitab Inc., US) and OriginPro 9.0 (OriginLab Corporation, US) software were used for statistical analysis and for presenting results in the form of graphs, respectively.

5.2.4 Anaerobic granular sludge characterization

Microbial community analysis

Granular sludge samples from the UASB reactor at the end of each period and from the inoculum were analyzed for microbial community composition. Prior to sampling, vigorous biomass mixing within the reactor was performed by briefly increasing the upflow velocity. Genomic DNA was extracted from the samples following the protocol of Lueders et al. (2004). Extracted DNA was quantified using a NanoDrop-1000 spectrophotometer and amplified by PCR using the primers Pro 341 forward and Pro 805 reverse targeting the V4 region of the 16S rRNA gene of bacteria and archaea (Takahashi et al. 2014). Amplicons were checked by agarose gel electrophoresis and sequenced using the Illumina MiSeq sequencing platform following the standard protocol. Sequences produced were analyzed using the standard operating procedure of the bioinformatics platform Mothur (Schloss et al. 2009). Further details on the microbial community analysis procedure can be seen in **Appendix 7**.

Protein, polysaccharides, and selenium content

On days 168 (end of phase Ie) and 266 (end of phase III), granular sludge was taken from low pH reactor for analysis of protein, polysaccharide and Se content. Liquid suspension with biomass (100 mL) was taken and the granular sludge was homogenized using a glass potter homogenizer. Protein and polysaccharide concentrations were determined after extracting soluble extracellular polymeric substances (EPS) following the protocol of Geyik and Çeçen (2014). Briefly, a known weight of sludge was homogenized using a potter homogenizer and centrifuged at 12,000 g for 10 min. The supernatant was considered the 1st soluble EPS fraction. The pellet was re-suspended in PBS and centrifuged again at 12,000 g for 10 min. This supernatant was taken as the 2nd soluble EPS fraction. The two EPS fractions were pooled, filtered using a 0.45 µm membrane and stored at -20°C until analysis.

Analyses of protein and polysaccharide content were carried out using the Pierce™ BSA protein assay kit (ThermoScientific) and phenol-sulfuric acid method (Albalasmeh et al. 2013), respectively. The total Se in the granular sludge was measured after acid digestion with 65% HNO_3 (MARS 5 pKo Temp CEM Microwave) using GFAAS.

The morphology of granular sludge from the inoculum, control reactor and low pH reactor was determined using stereomicroscopy and scanning electron microscopy (SEM, Zeiss Supra™ 55VP) equipped with energy dispersive X-ray spectroscopy (EDX, Princeton Gamma-Tech). EDX was used for detection of Se and S on the surface of the granular sludge. For SEM-EDX analysis, the sludge samples were prepared by gentle rinsing with Milli-Q water (18 MΩ/cm) and leaving them to dry on a dish at ambient temperature. Dried samples were deposited on carbon tape and coated with iridium before imaging.

5.3   Results

5.3.1   Low pH reactor at decreasing pH

The low pH reactor performance in terms of COD, NO_3^-, SO_4^{2-} and Se removal efficiencies is shown in **Fig. 5.2**, while the averaged effluent concentrations are reported in **Table 5.1**. Statistical groupings of the different phases during the low pH reactor operation were developed for comparison of the removal performances (**Appendix 3**, **Table S5.1**). Phases that did not exhibit statistically significantly different removal efficiencies were grouped together. The pH of the feed was gradually lowered from 7.0 to 5.0 to avoid a sudden pH shock on the biomass. Although the influent pH was lowered from 7.0 to 5.0, the effluent pH values were always between pH 7 and 8, with an average pH of 7.4 ($\pm$ 0.3).

The low pH reactor demonstrated relatively stable COD, NO_3^- and Se removal for influent pH values between 7 and 5.5. The COD removal efficiency was, on average, 74 ($\pm$ 3)% from start-up until the end of phase Id. However, the COD removal efficiency diminished to 58 ($\pm$ 9)% when the influent pH was decreased to 5.0 in phase Ie (*P-value* $\leq$ 0.0001). The NO_3^- removal efficiency during phase Ia to Ic (influent pH 7.0 to 6.0) was on average 97 ($\pm$ 2)% and decreased slightly to 90 ($\pm$ 3)% during phases Id and Ie (influent pH 5.5 and 5.0). Overall, NO_3^- removal efficiencies were not statistically significantly different during phases Ia through Ie at an average of 94 ($\pm$ 5)% (0.30 $\leq$ *P-value* $\leq$ 0.52).

Table 5.1 Comparison of studies on Se removal in different bioreactors in the presence of co-oxyanions.

Reactor type	$Se_{inf.}$	Other Ions	Inoculum	Operational Conditions	Removal Efficiency	References
H_2-MBfR	0.55 mg SeO_4^{2-}-Se/L	5 mg NO_3^--N/L 79 mg SO_4^{2-}/L	Biofilm from pilot plant treating groundwater	pH 7.0-9.0; H_2 gas	90% Se; 100% NO_3^-; 75% SO_4^{2-}	Chung et al. (2006)
MBfR	0.55 mg SeO_4^{2-}-Se/L	0-10 mg NO_3^--N/L	Denitrifying anaerobic methane oxidation (DAMO) culture	pH 7.0; 30°C; CH_4 gas	94% SeO_4^{2-} (without NO_3^-); 60-75% SeO_4^{2-}; 75% NO_3^-	Lai et al. (2016)
H_2-MBfR	0.55-6.6 mg SeO_4^{2-}-Se/L	40-90 mg SO_4^{2-}/L	Anoxic sludge from wastewater treatment plant	pH 7.0-9.0; HRT 3.3 h; 20°C; H_2 gas	> 90% SeO_4^{2-}; 25% SO_4^{2-}	Ontiveros-Valencia et al. (2016)
Upflow fungal pelleted reactor	0.79 mg SeO_3^{2-}-Se/L	No co-contaminants	*Phanerochaete chrysosporium*	pH 4.5; HRT 24 h; 30°C; Glucose	70% SeO_3^{2-}	Espinosa-Ortiz et al. (2015)
UASB	0.79 mg SeO_4^{2-}-Se/L	< 2500 mg SO_4^{2-}/L	Eerbeek anaerobic granular sludge	pH 7.0; HRT 6 h; 30°C; lactate	99% Se (methanogenic condition); 97% (SO_4^{2-} reducing condition); 20-90% SO_4^{2-}	Lenz et al. (2008)
UASB	0.79 mg SeO_4^{2-}-Se/L	930 mg NO_3^--N/L 192 mg SO_4^{2-}/L	Eerbeek anaerobic granular sludge with immobilized *Sulfurospirillum barnesii*	pH 7.0; HRT 6 h; 30°C; lactate	97% Se; 100% NO_3^-; 90% SO_4^{2-} (after NO_3^- had been removed)	Lenz et al. (2009)
UASB	13 mg Se/L[1]	1-10 g/L heavy metals 35.6 S g/L	Anaerobic sludge from mesophilic UASB reactor treating beverage wastewater	pH <1.0 adjusted to 6.0-9.0; HRT 24 h; 30°C; ethanol	97% Se_{tot}	Soda et al. (2011)
UASB	5-12 mg SeO_4^{2-}-Se/L	47 mg NO_3^--N/L 1500 mg SO_4^{2-}/L	Eerbeek anaerobic granular sludge	pH 5.0-7.0; HRT 24 h; 20°C; lactate	43% Se_{tot}; 61% Se_{diss}; 85% SeO_3^{2-}; 92% NO_3^-; 15% SO_4^{2-}	This study

*H_2-MBfR – hydrogen-based membrane biofilm reactors; UASB – upflow anaerobic sludge bed

Table 5.2 Summary of reactor performance of the UASB reactor operating at pH 5.0 (average concentration of constituents in the effluent and percent removal)

Phases	Description	Start-up pH 7.0, 10 mg Se/L	pH 7.0, 12 mg Se/L	pH 6.0, 12 mg Se/L	pH 5.5, 12 mg Se/L	pH 5.0, 12 mg Se/L	pH 5.0, <5 mg Se/L	pH 5.0, 12 mg Se/L	pH 5.0, 50% SeO$_3$$^{2-}$
	Experimental Periods	Phase Ia	Phase Ib	Phase Ic	Phase Id	Phase Ie	Phase IIa	Phase IIb	Phase III
	Days	0-29	30-61	62-90	91-129	130-168	169-202	203-231	232-266
Se$_{tot}$ conc. (removal %)	mg Se/L	1.1 ± 0.7 (91 ± 6%)	1.0 ± 0.5 (91 ± 5%)	1.9 ± 0.8 (84 ± 6%)	2.7 ± 1.7 (78 ± 14%)	6.7 ± 1.4 (43 ± 13%)	1.3 ± 0.3 (52 ± 13%)	7.3 ± 1.8 (39 ± 18%)	6.7 ± 0.8 (27 ± 9%)
Se$_{diss}$ conc. (removal %)	mg Se/L	0.3 (± 0.2) (97 ± 2%)	0.4 ± 0.1 (96 ± 1%)	0.5 ± 0.4 (96 ± 4%)	1.2 ± 1.1 (90 ± 9%)	4.7 ± 2.0 (61 ± 17%)	1.0 ± 0.2 (61 ± 12%)	5.4 ± 1.7 (55 ± 16%)	3.4 ± 0.5 (63 ± 6%)
SeO$_3$$^{2-}$ conc. (removal %)	mg Se/L	-	-	-	-	-	-	-	0.6 ± 0.2 (85 ± 6%)
NO$_3$$^-$ conc. (removal %)	mg NO$_3$$^-$-N/L	1.0 ± 0.2 (98 ± 3%)	0.4 ± 0.2 (99 ± 1%)	2.6 ± 2.2 (95 ± 5%)	5.4 ± 3.7 (88 ± 13%)	3.8 ± 1.8 (92 ± 4%)	13.4 ± 5.2 (74 ± 9%)	10.5 ± 3.4 (81 ± 6%)	13.7 ± 4.0 (70 ± 9%)
SO$_4$$^{2-}$ conc. (removal %)	g SO$_4$$^{2-}$/L	2.2 ± 0.2 (14 ± 9%)	1.1 ± 0.4 (47 ± 10%)	0.8 ± 0.1 (47 ± 16%)	1.0 ± 0.1 (33 ± 8%)	1.3 ± 0.1 (14 ± 9%)	1.3± 0.1 (13 ± 7%)	1.1 ± 0.2 (17 ± 12%)	1.1 ± 0.2 (32 ± 7%)
Total dissolved sulfide (TDS) conc.	mg TDS/L	70 ± 9	450 ± 22	425 ± 22	202 ± 6	83 ± 34	36 ± 11	15 ± 7	8 ± 6
COD conc. (removal %)	g COD/L	0.5 ± 0.3 (74 ±18%)	0.4 ± 0.1 (78 ± 7%)	0.5 ± 0.1 (72 ± 8%)	0.6 ± 0.2 (73 ± 4%)	0.9 ± 0.2 (58 ± 9%)	1.3 ± 0.2 (42 ± 10%)	1.2 ± 0.2 (44 ± 9%)	1.6 ± 0.1 (20 ± 9%)

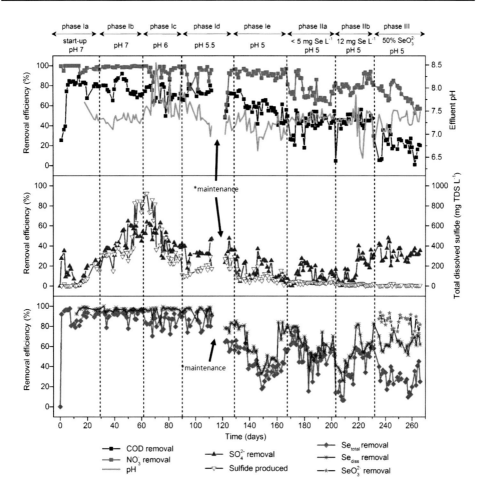

Fig. 5.2 Removal efficiencies (%) of the UASB reactor operated for 266 days: COD (black square), NO₃⁻ (green square), SO₄²⁻ (brown triangle), total Se (red diamond), dissolved Se (blue asterisk) and SeO₃²⁻ (yellow star). Effluent pH (orange line) was measured along with dissolved sulfide (blue & white inverted triangle). *Continuous operation was stopped to conduct maintenance (tubing and connectors replacement) but liquid re-circulation for the reactor was set-up to minimize biomass activity lost.

SO_4^{2-} removal, on the other hand, increased from 14 ($\pm$ 9)% in phase Ia to an average of 42 ($\pm$ 8)% during phases Ib-Id (influent pH 7.0 to 5.5, *P-value* $\leq$ 0.0001), but decreased again to 14 ($\pm$ 9)% in phase Ie (influent pH 5.0, *P-value* $\leq$ 0.0001). The sulfide concentrations in the effluent were initially at 70 ($\pm$ 9) mg TDS/L, increased to 359 ($\pm$ 137) mg TDS/L during phases Ib-Id and then exhibited a sudden decrease to 83 ($\pm$ 34) mg TDS/L.

Se_{tot} removal followed a similar trend from phase Ib until halfway through phase Id with high removal efficiencies of 91 ($\pm$ 6)%, 91 ($\pm$ 5)%, 84 ($\pm$ 6)%, 78 ($\pm$ 14)% Se_{tot}, for phases Ia-

137

d, respectively. Se_{tot} removal efficiencies were not statistically significantly different during phases Ia to Id ($0.45 \leq P\text{-}value \leq 1.0$) at an average of 88 ($\pm$ 7)% but decreased to 43 ($\pm$ 13)% for the remainder of phase Ie ($P\text{-}value \leq 0.003$). Se_{diss} removal efficiencies exhibited a similar trend to the Se_{tot} removal efficiencies, showing good removal at 95 ($\pm$ 3)% from start-up to halfway into phase Id. During phase Id (day 113-121), the reactor continuous operation had to be stopped for approximately one week due to maintenance needs such as replacement of tubing. Mixing was still maintained by providing an internal recirculation (from the biomass sampling port to the effluent port, **Fig. 5.1**) to maintain biomass activity. Due to the internal circulation arrangement, only a small portion of the sludge was not properly mixed during maintenance. This could have caused the temporary decrease in reactor performance during the end of phase Id which appeared to have recovered within a few days. During phase Ie (operated at pH 5.0), the removal efficiency of Se_{diss} further decreased to 61 ($\pm$ 17)%.

In summary, below pH 5.5, the reactor performance (measured by the removal efficiency) decreased by approximately 20-30% for all oxyanions. Increased Se_{tot} concentrations in the effluent and decreased COD, NO_3^- and Se_{diss} removal efficiencies could be an indication of decreased metabolic activity or stressed microbes due to the slightly acidic (pH 5.0) influent wastewater. However, despite the decline in removal efficiency, activity tests (**Appendix 3, Fig. S5.1**) indicated that the biomass still had the potential to be highly metabolically active. Averaged removal performance results for the control reactor over 92 days are shown in **Fig. S5.2 (Appendix 3)** and were found to not be statistically different from the low pH reactor (**Table S5.1**) for COD, NO_3^-, SO_4^{2-} and Se removal during phases Ia to Id.

5.3.2 Low pH reactor performance with changing Se concentration

The COD removal efficiency decreased from 58 ($\pm$ 9)% from phase Ie to an average of 43 ($\pm$ 5)% during phase II and finally to 20 ($\pm$ 9)% when SeO_3^{2-} was added to the low pH reactor (phase III, $P\text{-}value \leq 0.0001$). Similarly, the low pH reactor showed an overall 20% decrease in NO_3^- removal efficiency after phase Ie ($P\text{-}value \leq 0.01$), while the SO_4^{2-} removal performance was fairly consistent at an average of 13 ($\pm$ 4)%, similar to phase Ie ($0.52 \leq P\text{-}value \leq 1.0$). Sulfide concentrations in the effluent consistently decreased during each phase, reaching a concentration of 8 ($\pm$ 6) mg TDS/L at the end of the reactor run. Phase II and phase III attained a Se removal performance comparable to the one in phase Ie ($0.13 \leq P\text{-}value \leq 0.95$), calculating an average Se_{tot} and Se_{diss} removal efficiency of 40 ($\pm$18)% and 60 ($\pm$ 14)%, respectively. A SeO_3^{2-} removal efficiency of 85 ($\pm$ 6)% was achieved during phase III.

5.3.3 Granular sludge characterization

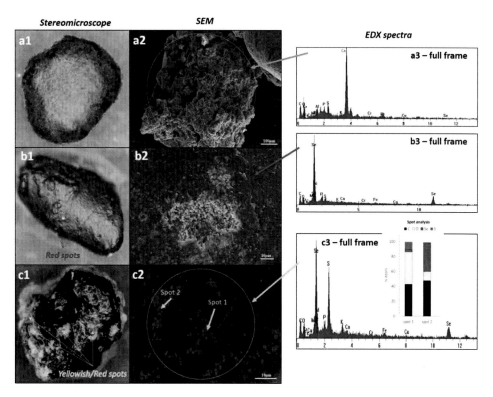

Fig. 5.3 Morphology of granular sludge samples taken from (a) inoculum, (b) control reactor and (c) low pH reactor after 266 days of continuous operation, assessed using a (1) stereomicroscope and (2) SEM with full frame and spot-based elemental analysis at 20 kV; (3) representative EDX spectra of selected region.

During the beginning of phase Ie, floating biomass was frequently observed in the low pH reactor as depicted in **Fig. S5.3a** (**Appendix 3**), while no floating biomass was observed in the control reactor (**Appendix 3**, **Fig. S5.3b**). However, starting with phase IIb until the end of the low pH reactor run, floating biomass was rarely observed. The sludge from the control reactor retained its granular shape with red particles deposited on the surface (**Fig. 5.3 b1**). In contrast, the biomass from the low pH reactor became fluffier with yellowish/reddish particles deposited on the surface (**Fig. 5.3 c1**).

The total Se concentration in the low pH reactor biomass at the end of phase Ie (day 168) was lower than in the control reactor biomass (**Table 5.3**). However, the Se concentrations increased by the end of phase III (day 266), indicating that the biomass was still capable of accumulating additional Se. SEM images (**Fig. 5.3 b2** and **c2**) and EDX spectra (**Fig. 5.3 b3**

139

and **c3**) indicated that the biomass from both reactors formed Se^0 within the granules. Spherical as well as a few rod-shaped Se^0 structures were observed within the low pH reactor biomass, similar to Se^0 spheres observed in other studies at neutral pH (Oremland et al. 2004; Jain et al. 2015).

Table 5.4 Protein, polysaccharide and Se concentrations in the inoculum, control reactor and low pH reactor granular sludge. Soluble EPS was extracted from the granular sludge prior to quantification of protein and polysaccharides while Se content in the biomass was determined after combined acid-heat digestion.

Granular sludge	Inoculum	Control reactor (pH 7.0)	Low pH reactor (pH 5.0)	
Conditions	Untreated (day 0)	$NO_3^- + SO_4^{2-} + SeO_4^{2-}$ (day 92)	$NO_3^- + SO_4^{2-} + SeO_4^{2-}$ (day 168, phase Ie)	Addition of 50% SeO_3^{2-} (day 266, phase III)
Protein (mg/g TS)	153 (±2)	80 (± 4)	432 (± 17)	128 (± 6)
Polysaccharide (mg/g TS)	31 (±2)	17 (± 2)	6 (± 2)	52 (± 19)
Polysaccharide/ Protein (PS/PN)	0.2 (±0.06)	0.2 (± 0.06)	0.01 (± 0.01)	0.4 (± 0.2)
Total Se in biomass (mg Se/g TS)	0	45 (± 6)	12 (± 1)	86 (± 12)

The soluble EPS protein content from the low pH reactor biomass at the end of phase Ie (**Table 5.3**) was significantly higher at 432 (± 17) mg protein/g TS (*P-value* ≤ 0.001) compared to the inoculum and control reactor at 153 (± 2) and 80 (± 4) mg protein/g TS, respectively. In contrast, the soluble EPS polysaccharide concentration of the low pH reactor biomass during phase Ie was the lowest at 6 (± 2) mg polysaccharide g/TS (*P-value* ≤ 0.009). At the end of phase III (day 266), the soluble EPS protein content of the low pH reactor biomass had decreased considerably (*P-value* ≤ 0.0001), while the polysaccharide concentration had increased (*P-value* ≤ 0.002) as shown in **Table 5.3**.

5.3.4 Microbial community analysis

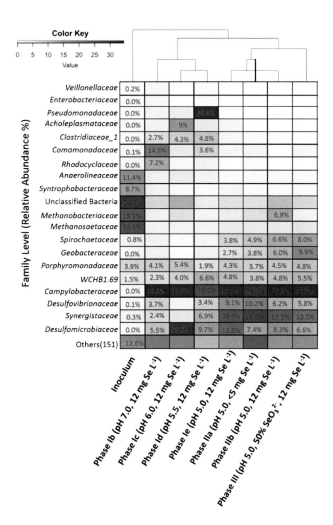

Fig. 5.4 Microbial community analysis of granular sludge taken from the UASB reactor at different sampling times shown as heat map at the Family level in % relative abundance of each phase of reactor operation. Samples were correlated and clustered according to relatedness as shown in the dendrogram. Taxa with < 5% relative abundance (only if relative abundance are < 5% for all samples) were grouped together as "others" for all samples.

The microbial composition of the inoculum and with each change in reactor operation are reported as prokaryotic relative abundance at the family-level (**Fig. 5.4**). The inoculum was dominated by archaea such as the *Methanosaetaceae* (19.1%) and *Methanobacteriaceae* (18.3%); the most abundant bacterial groups were *Anaerolineaceae* (11.4%) and

Syntrophobacteraceae (8.7%). A major change in microbial community was observed between the inoculum and the low pH reactor biomass samples due to a change of the relative archaeal abundance from 39% to 5%. A shift in bacterial diversity was noticed within the low pH reactor biomass samples, but to a lesser extent than between the inoculum and low pH reactor samples.

The proteobacterial families *Campylobacteraceae* and *Desulfomicrobiaceae* were enriched in the reactor relative to the inoculum and changed from non-detectable to relative abundances of 34.4% and 20.2%, respectively. The *Synergistaceae* and *Desulfovibrionaceae* increased from 6.9% and 3.4% (phase Id) to an average of 13 (± 4)% and 8 (± 2)%, respectively, from phase Ie to III. *Spirochaetaceae* and *Geobacteraceae* started appearing during phase Ie (at pH 5.0) and their abundance increased until the end of phase III (**Fig. 5.4**).

5.4 Discussion

5.4.1 Selenium bioremoval at low pH

This study, to our knowledge, is the first to investigate Se removal performance in a laboratory-scale UASB reactor treating a wastewater close to a real acid mine drainage: influent pH values as low as pH 5, the presence of co-electron acceptors (i.e. NO_3^- and SO_4^{2-}) and ambient temperature (20°C). The Se removal efficiencies obtained at an influent pH value of 5.0 showed a trend of approximately 20 to 30% decrease in reactor performance at 43 (± 13)% Se_{tot} and 61 (± 17)% Se_{diss} from influent concentrations of 10-12 mg Se L^{-1} as compared to operation at pH 5.5 to 7.0 with a standard error of 2.5%. The UASB reactor operation at an influent pH of 5.0 (phase Ie) showed unstable performance compared to an influent pH of 5.5 influent (phase Id). Therefore, future studies should investigate the reactor stability at pH 5.5 for Se removal with co-electron acceptors as the optimal low pH operation.

It should be noted that other studies pre-conditioned the system for optimal performance at neutral pH and mesophilic condition. Dessì et al. (2016) showed that the Se removal efficiency and Se retention within the biomass were greater at higher temperatures (55°C), presumably due to the increase in crystallinity size, shape and decrease in zeta-potential of the biogenic Se^0, allowing for better settleability and lowering the release of colloidal Se (Jain et al. 2017). Additionally, most of the previous studies did not consider the presence of co-electron acceptors, which can result in an added stress with regards to Se removal, e.g. competition or hindrance of bioreduction and toxicity. A study by Soda et al. (2011) used real Se wastewater taken from a Se refinery plant to evaluate biological Se reduction using a UASB reactor. The authors pre-conditioned the wastewater to increase the pH from < 1.0 to 6-9 and to decrease salinity from 7% to 1% before feeding the wastewater into the biological treatment system. With this pretreatment approach, Soda et al. (2011) achieved $> 90\%$ Se removal in the biological system regardless of the presence of various heavy metals (< 5 mg/L).

Espinosa-Ortiz et al. (2015) demonstrated 70% SeO_3^{2-} removal using pure culture fungal pellets at pH 4.5 under aerobic conditions and immobilization of biogenic Se^0 within the fungal pellets. However, compared to SeO_4^{2-}, SeO_3^{2-} is highly reactive and readily reduced to Se^0 by a number of microorganisms (Nancharaiah and Lens, 2015). Both low pH studies by Espinosa-Ortiz et al. (2015) and Kenward et al. (2006) employed pure cultures and as such, contamination becomes a concern when scaling-up. Concerns about contamination are largely eliminated when using mixed culture-based wastewater treatment systems, such as granular sludge as in this study. In any industrial operation, the stability of an anaerobic process is a key factor when scaling-up a system (Aslanzadeh et al. 2013). Long-term re-use and biosorption capacity of both fungal pellets and pure bacteria culture in a continuous bioreactor under low pH operation have yet to be established. In contrast to the previously mentioned studies, the UASB reactor used in this study was operated for a total of 176 days at pH 5.0 and managed to maintain a consistent Se removal efficiency (40% Se_{tot} and 60% Se_{diss}). Furthermore, the bioreactor was able to adapt quickly to operational changes and no irreversible toxicity to the granular sludge occurred due to the presence of SeO_3^{2-} at the low concentrations tested (4 ± 0.5 mg Se/L). In contrast, Jain et al. (2016) observed SeO_3^{2-} toxicity to the activated sludge under aerobic conditions leading to repeated reactor crashes and failure in continuous operation.

The specific Se removal mechanisms in the presence of both NO_3^- and SO_4^{2-} at low pH are largely unknown. Se oxyanion reduction to Se^0 can be facilitated by dissimilatory metal reduction or unspecific detoxification pathways of SO_4^{2-} reducers or NO_3^- reducers (Nancharaiah and Lens, 2015). Unspecific detoxification pathways pertain to the use of non-Se specific reductases in other microorganisms, i.e. SO_4^{2-}, NO_3^- or NO_2^- reductases for SeO_4^{2-} reduction (Lenz and Lens, 2009; Tucker et al. 1998; Müller et al. 1997). On the other hand, recent studies on the metal adsorption properties of EPS showed that under acidic conditions, bacteria-associated EPS might control anion and metal binding through reactive functional groups (D'Abzac et al. 2013). This was observed by Kenward et al. (2006), where *Shewanella putrefaciens* sorbed SeO_4^{2-} to its cell wall optimally at pH 3. Sorbed SeO_4^{2-} was then reduced to Se^0 from within the cell by an intracellular electron donor, presumably ubiquinone. Further studies are required related to the microbial sorption and role of EPS in SeO_4^{2-} reduction at different pH values in the presence of NO_3^- and SO_4^{2-} to fully decipher the Se removal mechanisms.

Though the influent pH was at 5.0, the system was always able to neutralize the incoming low pH to an effluent pH of > 7.0. This is most likely due to both the reactor re-circulation and the ongoing denitrification and sulfate reduction processes that generate alkalinity and, therefore, neutralize the pH during reactor operation (Koschorreck and Tittel 2007; Shen et al. 2013). Alkalinity produced during denitrification and SO_4^{2-} reduction processes were estimated to be approximately 188 ($\pm$ 30) mg $CaCO_3$/L (**Appendix 3, Fig. S5.4**).

NO_3^- removal was high and stable during phase I which lasted for 169 days. During this period, removal of NO_3^- was associated with nitrite concentrations < 1 mg NO_2^--N/L (**Appendix 3, Fig. S5.5**). However, NO_3^- removal became unstable in phases II and III coinciding with higher nitrite concentrations in the effluent reaching up to 3 mg NO_2^--N/L. Low pH values have been suggested to hinder or even inhibit denitrification because of nitrite accumulation (Shen et al. 2013; Nancharaiah et al. 2017). Nitrite at low pH can exist as nitrous acid and has been shown to inhibit denitrification activity at a concentration as low as 0.04 mg HNO_2^--N/L (Baeseman et al. 2006). Shen et al. (2013) observed an approximately 28% decrease in denitrification rates when the initial pH was gradually lowered to 4.5 from 7.2. Similarly, Nancharaiah et al. (2017) observed that it took longer for complete NO_3^- removal to occur in a sequential batch reactor when the initial pH was lowered from 7.0 to 5.0 and 4.0. Therefore, unstable and 20% decrease in NO_3^- removal can be due increase in nitrite levels from approximately 0.5 to 3 mg NO_2^--N L^{-1} caused by the feeding of lower influent pH at 5.0.

The SO_4^{2-} removal and sulfide production performance of the low influent pH reactor were fairly low (**Fig. 5.2**), which could be due to the presence of NO_3^- (He and Yao 2010) and/or the low temperature (20°C) operation (Costabile et al. 2011). At pH 5.0, the SO_4^{2-} reduction was even lower, presumably because SO_4^{2-} reducers present in the Eerbeek sludge are neutrophilic (Roest et al. 2005), though there are studies on acid tolerant and acidophilic SO_4^{2-} reducers (Sánchez-Andrea et al. 2014). In contrast to this study, Lopes et al. (2008) were able to achieve nearly complete SO_4^{2-} reduction at pH 4.0; however, the system was operated at thermophilic conditions, SO_4^{2-} was the sole electron acceptor and the organic loading rate was lower (0.8-1.9 g COD/L·d).

5.4.2   UASB reactor biomass characterization and formation

Unlike in other UASB reactor studies conducted at low pH (Lopes et al. 2008; Gonzalez-Gil et al. 2012), the granular shape of the low pH reactor biomass was not well maintained during low pH UASB reactor operation (**Fig. 5.3**). The biomass growing at pH 5.0 was fluffier with filamentous structures and sticky substances surrounding the biomass, the bulk of which is likely EPS. The gradual formation of light, fluffy flocs was one of the likely causes for the floating biomass (**Appendix 3**, **Fig. S5.3a**) observed during phase Ie, potentially caused by gas (N_2, H_2S or CH_4) entrapped within the biomass. Loss of minerals from the low pH biomass can also contribute to the change in granular structure, though both studies at low pH from Gonzalez-Gil et al. (2012) and Lopes et al. (2008) maintained a granular structure despite mineral leaching (Fe, Zn, Cu and Al).

Another possible factor explaining the floating biomass from the low pH reactor is the polysaccharide (PS) to protein (PN) ratio (PS/PN) of the EPS, which can affect biomass properties. Shin et al. (2000) were able to demonstrate that higher PS/PN ratios correlate with a lower negative surface charge of the biomass or higher hydrophobicity of the cells. This correlates well with the results from this study with the PS/PN ratio of the biomass in the low pH reactor being 15 times lower (*P-value* ≤ 0.02) than the PS/PN ratio of both the inoculum and the control reactor (**Table 5.3**, phase Ie). Furthermore, floating biomass was noticed less frequently during operation of the low pH reactor at pH 5.0, coinciding with an increase in PS/PN ratio reaching similar values to both the inoculum and the control reactor (phase III, *P-value* = 1.0). This could indicate either adaptation of the microbial community to the low pH conditions or a decrease in EPS production. Geyik and Çeçen (2014), for instance, observed that a lower PS/PN correlated with larger flocs of less compact biomass. These observations were attributed to the increase in hydrophobicity and cohesion between aggregates.

Another important aspect in microbial SeO_4^{2-} reduction is the immobilization of Se^0 within the biomass to prevent discharge into the environment. Red deposits observed in the low pH granular sludge (**Fig. 5.3 c1**) are strong indicators of elemental Se precipitation in the biomass. Further evidence of Se precipitation was obtained from the detection of Se, at 86 (± 12) mg Se/g TS (**Table 5.3**), in the biomass and the SEM-EDX imaging analysis (**Fig. 5.3**) of the low pH granular sludge. The Se immobilization capacity for long-term operation of the UASB reactor at low pH has yet to be fully defined, despite a clear indication of Se^0 entrapment within the biomass. Although there were no changes in Se removal efficiency between days 138-266 (during pH 5.0 operation), the eventual release of Se^0 can be expected due to biomass cell wall rupture/damage (Gonzalez-Gil et al. 2016). Therefore, further studies on biogenic Se^0 toxicity to biomass and Se^0 retention by granular sludge in a given time frame should be conducted in order to establish the necessary frequency of biomass removal and the addition of fresh biomass.

5.4.3 Microbial ecology of anaerobic granular sludge at low pH

The microbial community composition changed noticeably from the inoculum (day 0), where the majority of the community consisted of archaeal methanogens to a bacterially-dominated diverse community mainly composed of *Deltaproteobacteria* and *Epsilonproteobacteria*. In contrast, no major changes in bacterial diversity were observed among the biomass samples taken during different phases of reactor operation.

Microorganisms capable of reducing SeO$_4^{2-}$ are found in a wide range of phylotypes and appear to be largely related to both denitrifying and sulfate-reducing bacteria (Nancharaiah and Lens, 2015). Families of *Campylobacteraceae* and *Desulfomicrobiaceae* were dominant and consistently found in the Se removing granular sludge (**Fig. 5.4**). Additionally, analysis at the genus level (**Appendix 3, Fig. S5.6**) indicated that the population shifted from predominantly known methanogenic to denitrifying and sulfidogenic (SO$_4^{2-}$ reducing) community members. This was evident by the increase in relative abundance of *Sulfospirillum* gen. (belonging to the *Campylobacteraceae* family), followed by *Aminivibrio* gen. (belonging to the *Synergistaceae* family), *Desulfomicrobium* gen. and *Desulfovibrio* gen. (both belonging to the *Desulfomicrobiaceae* family). These genera are not only recognized as denitrifiers, e.g. *Sulfospirillum barnesii*, and SO$_4^{2-}$ reducers, e.g. *Desulfovibrio desulfuricans*, but have also been reported to reduce SeO$_4^{2-}$ along with NO$_3^-$ and SO$_4^{2-}$ (Lai et al. 2016; Lenz et al. 2009; Nancharaiah and Lens 2015; Ontiveros-Valencia et al. 2016; Tomei et al. 1995; Truong et al. 2013). This is most likely the reason for the high Se removal efficiencies noticed in phases Ia to Id before low pH 5.0 impacted the reactor performance. Moreover, the dominant phyla were *Proteobacteria* and *Synergistestes* which have many known denitrifying bacteria representatives that can explain the high NO$_3^-$ removal efficiencies (> 70%) attained throughout the low-pH reactor operation. In addition, low SO$_4^{2-}$ removal efficiencies (> 30%) noticed during the entire reactor operation could be due to the lower abundance of SO$_4^{2-}$ reducers in the granular sludge.

Interesting to note was the increase of *Spirochaetaceae* and *Geobacteraceae* when operating at pH 5.0. The various species of *Geobacteraceae* are well-known to be metal reducers, including SeO_3^{2-} (Nancharaiah and Lens 2015; Pearce et al. 2009), at circum-neutral pH. In this study, the enrichment of *Geobacteraceae* was only observed after the pH was decreased to approximately 5.0. Correspondingly, *Geobacteraceae*-related sequences have been identified abundantly in an acidic mine lake at pH 5.5-5.9 (Lu et al. 2013). Likewise, the family *Spirochaetaceae* was found to be one of the dominant phylotypes in an acidic (Sánchez-Andrea et al. 2014) as well as a NO_3^- and SO_4^{2-} rich (Liao et al. 2014) system. There are, however, thus far, no reports on Se oxyanion reduction by *Spirochaetaceae*.

Further studies regarding changes in microbial community structure at low pH are interesting to pursue, particularly during the start-up operation of the UASB reactor. In order to avoid a pH shock to the system, gradual lowering of the operational pH was employed in this study and therefore it took more than 100 days before the reactor was operated at pH 5.0. Using sludge bioaugmentation with *Geobacteraceae* and *Spirochaetaceae*, it might be possible to accelerate reactor start-up at pH 5.0 without major concerns for reactor failure due to high acidity.

5.4.4 Industrial implications

This study indicates the possibility of treating acidic Se-laden wastewater directly in a bioreactor without the additional cost of pre-treatment for wastewater neutralization. Assuming a wastewater treatment capacity of 15,000 m^3 per day and caustic soda cost of $0.44 per kg, the cost of increasing the pH of a wastewater from 4 to 5, using 30% caustic solution, would be approximately $35 per day, compared to $39 per day when increasing the pH to 7. It might be possible to further lower the pH adjustment cost, depending on the load of acidic Se-laden wastewater, by increasing the HRT (e.g. by increasing the reactor volume or decreasing the influent flow rate).

There are many aspects that need further investigation in order to further improve the understanding of reactor operations at low pH values. Exploration of possible changes in reactor design or operation should be considered to evaluate the optimal removal performance of Se operated at low pH with exposure to the co-electron acceptors NO_3^- and SO_4^{2-}. Changes in reactor operation that can be explored to increase the Se removal performance of a bioreactor are: (a) reactor configuration, e.g. promoting a biofilm-dominated system as described in Ontiveros-Valencia et al. (2016); (b) increasing reactor temperature (Dessì et al. 2016); or (c) varying NO_3^- and SO_4^{2-} concentration (**chapter 3**).

A techno-economical assessment of the treatment process should be conducted to assess whether Se removal at low pH values would be competitive with pH adjustment of acidic Se-wastewaters. Establishing the granular sludge Se retention capacity of a UASB reactor over time to prevent a sudden Se release potentially caused by biomass lysis (Gonzalez-Gil et al. 2016) would be an important aspect of this assessment. Another aspect to consider is the long-term Se toxicity to the biomass and microbial stress caused by low pH exposure. Working knowledge regarding the biomass stress response would be useful in order to establish a proper biomass removal cycle and the frequency of fresh biomass addition.

5.5 Conclusions

UASB bioreactor performance for the removal of SeO_4^{2-}, SO_4^{2-} and NO_3^- using lactate as the electron donor at 20°C and at pH values between 7.0 and 5.0 was investigated over 266 days. Upon operation of UASB reactor at influent pH 5.0, the reactor performance showed a trend of 20 to 30% decrease with a standard error of 2.5% achieving a Se_{tot} and Se_{diss} removal efficiency of 43 (± 13)% and 61 (± 17)%. A higher removal efficiency was achieved for NO_3^- (> 70%), while SO_4^{2-} removal efficiencies were lower than 50% for all pH values investigated. The biomass which developed at pH 5.0 consisted of large, fluffy and filamentous granules. Microbial community analysis showed a shift from a dominantly archaeal methanogenic population in the inoculum to a bacterially dominated community during reactor operation. The bacterial community in the reactor was mainly composed of known denitrifiers and SO_4^{2-} reducers, i.e. *Sulfospirillum* and *Desulfovibrio*. An enrichment of *Geobacteraceae* and *Spirochaetaceae* was observed once the reactor was operated at pH 5.0.

5.6 References

Albalasmeh, A.A., Berhe, A.A., Ghezzehei, T.A., 2013. A new method for rapid determination of polysaccharide and total carbon concentrations using UV spectrophotometry. *Carbohydr. Polym.* 97, 253-261.

Association, A.P.H., Association A.W.W., Federation, W.E., 2005. Standard methods for examination of water and wastewater, 5th ed. American Public Health Association, Washington, DC, USA.

Aslanzadeh, S., Rajendran, K., Jeihanipour, A., Taherzadeh, M., 2013. The effect of effluent recirculation in a semi-continuous two-stage anaerobic digestion system. *Energies* 6, 2966-2981.

Baeseman, J.L., Smith, R.L., Silverstein, J., 2006. Denitrification potential in stream sediments impacted by acid mine drainage: Effects of pH, various electron donors, and iron. *Microb. Ecol.* 51, 232-241.

Chung, J., Nerenberg, R., Rittmann, B.E., 2006. Bioreduction of selenate using a hydrogen-based membrane biofilm reactor. *Environ. Sci. Technol.* 40, 1664-1671.

Costabile, A.L., Canto, C.S., Ratusznei, S.M., Rodrigues, J.A., Zaiat, M., Foresti, E., 2011. Temperature and feed strategy effects on sulfate and organic matter removal in an AnSBB. *J. Environ. Manage.* 92, 1714-1723.

d'Abzac, P., Bordas, F., Joussein, E., van Hullebusch, E.D., Lens, P.N.L., Guibaud, G., 2013. Metal binding properties of extracellular polymeric substances extracted from anaerobic granular sludges. *Environ. Sci. Pollut. Res. Int.* 20, 4509-4519.

Dessì, P., Jain, R., Singh, S., Seder-Colomina, M., van Hullebusch, E.D., Rene, E.R., Ahammad, S.Z., Carucci, A., Lens, P.N.L., 2016. Effect of temperature on selenium removal from wastewater by UASB reactors. *Water Res.* 94, 146-154.

Erosa, D., Höll, W.H., Horst, J., 2009. Sorption of selenium species onto weakly basic anion exchangers: I. Equilibrium studies. *React. Funct. Polym.* 69, 576-585.

Espinosa-Ortiz, E.J., Rene, E.R., van Hullebusch, E.D., Lens, P.N.L., 2015b. Removal of selenite from wastewater in a *Phanerochaete chrysosporium* pellet based fungal bioreactor. *Int. Biodeterior. Biodegrad.* 102, 361-369.

Geyik, A.G., Çeçen, F., 2014. Production of protein- and polysaccharide-EPS in activated sludge reactors operated at different carbon to nitrogen ratios. *J. Chem. Technol. Biotechnol.* 91, 522-531.

Gonzalez-Gil, G., Lens, P.N.L., Saikaly, P.E., 2016. Selenite reduction by anaerobic microbial aggregates: microbial community structure, and proteins associated to the produced selenium spheres. *Front. Microbiol.* 7, 571.

Gonzalez-Gil, G., Lopes, S.I.C., Saikaly, P.E., Lens, P.N.L., 2012. Leaching and accumulation of trace elements in sulfate reducing granular sludge under concomitant thermophilic and low pH conditions. *Bioresour. Technol.* 126, 238-246.

He, Q., Yao, K., 2010. Microbial reduction of selenium oxyanions by *Anaeromyxobacter dehalogenans*. *Bioresour. Technol.* 101, 3760-3764.

Jain, R., Jordan, N., Tsushima, S., Hübner, R., Weiss, S., Lens, P.N.L., 2017. Shape change of biogenic elemental selenium nanoparticles from nanospheres to nanorods decreases their colloidal stability. *Environ. Sci. Nano* 4, 1054-1063.

Jain, R., Matassa, S., Singh, S., van Hullebusch, E.D., Esposito, G., Lens, P.N.L., 2016. Reduction of selenite to elemental selenium nanoparticles by activated sludge. *Environ. Sci. Pollut. Res.* 23, 1193-1202.

Jain, R., Seder-Colomina, M., Jordan, N., Dessì, P., Cosmidis, J., van Hullebusch, E.D.,Weiss, S.,

Farges, F., Lens, P.N.L., 2015b. Entrapped elemental selenium nanoparticles affect physicochemical properties of selenium fed activated sludge. *J. Hazard.Mater.* 295, 193-200.

Kenward, P.A., Fowle, D.A., Yee, N., 2006. Microbial selenate sorption and reduction in nutrient limited systems. *Environ. Sci. Technol.* 40, 3782-3786.

Koschorreck, M., Tittel, J., 2007. Natural alkalinity generation in neutral lakes affected by acid mine drainage. *J. Environ. Qual.* 36, 1163-1171.

Lai, C.-Y., Wen, L.-L, Shi, L.-D., Zhao, K.-K., Wang, Y.-Q., Yang, X., Rittman, B.E., Zhou, C., Tang, Y., Zheng, P., Zhao, H.-P., 2016. Selenate and nitrate bioreductions using methane as the electron donor in a membrane biofilm reactor. *Environ. Sci. Technol.* 50, 10179-10186.

Lemly, A.D., 2014. Teratogenic effects and monetary cost of selenium poisoning of fish in Lake Sutton, North Carolina. *Ecotoxicol. Environ. Saf.* 104, 160-167.

Lenz, M., Enright, A.M., O'Flaherty, V., van Aelst, A.C., Lens, P.N.L., 2009. Bioaugmentation of UASB reactors with immobilized *Sulfurospirillum barnesii* for simultaneous selenate and nitrate removal. *Appl. Microbiol. Biotechnol.* 83, 377-388.

Li, D.-B., Cheng, Y.-Y.,Wu, C., Li,W.-W., Li, N., Yang, Z.-C., Tong, Z.-H., Yu, H.-Q., 2014. Selenite reduction by *Shewanella oneidensis* MR-1 is mediated by fumarate reductase in periplasm. *Sci. Rep.* 4, 3735.

Liao, R., Li, Y., Yu, X., Shi, P., Wang, Z., Shen, K., Shi, Q., Miao, Y., Li, W., Li, A., 2014. Performance and microbial diversity of an expanded granular sludge bed reactor for high sulfate and nitrate waste brine treatment. *J. Environ. Sci.* 26, 717-725.

Lopes, S.I., Capela, M.I., Dar, S.A., Muyzer, G., Lens, P.N.L., 2008. Sulfate reduction at pH 4 during the thermophilic (55°C) acidification of sucrose in UASB reactors. *Biotechnol. Prog.* 24, 1278-1289.

Lu, S., Chourey, K., Reiche, M., Nietzsche, S., Shah, M.B., Neu, T.R., Hettich, R.L., Küsel, K., 2013. Insights into the structure and metabolic function of microbes that shape pelagic iron-rich aggregates ("Iron Snow"). *Appl. Environ. Microbiol.* 79, 4272-4281.

Lueders, T., Manefield, M., Friedrich, M.W., 2004. Enhanced sensitivity of DNA- and rRNA-based stable isotope probing by fractionation and quantitative analysis of isopycnic centrifugation gradients. *Environ. Microbiol.* 6, 73-78.

Nancharaiah, Y.V., Lens, P.N.L., 2015. Selenium biomineralization for biotechnological applications. *Trends Biotechnol.* 33, 323-330.

Nancharaiah, Y.V., Krishna Mohan, T. V., Satya Sai, P.M., Venugopalan, V.P., 2017. Denitrification of high strength nitrate bearing acidic waters in granular sludge sequencing batch reactors. *Int. Biodeterior. Biodegrad.* 119, 28-36.

Ontiveros-Valencia, A., Penton, C.R., Krajmalnik-Brown, R., Rittmann, B.E., 2016. Hydrogen-fed biofilm reactors reducing selenate and sulfate: Community structure and capture of elemental selenium within the biofilm. *Biotechnol. Bioeng.* 113, 1736-44.

Oremland, R.S., Herbel, M.J., Blum, J.S., Langley, S., Beveridge, T.J., Ajayan, P.M., Sutto, T., Ellis, A.V., Curran, S., 2004. Structural and spectral features of selenium nanospheres produced by Se-respiring bacteria. *Appl. Environ. Microbiol.* 70, 52-60.

Roest, K., Heilig, H.G.H.., Smidt, H., De Vos, W.M., Stams, A.J.M., Akkermans, A.D.L., 2005. Community analysis of a full-scale anaerobic bioreactor treating paper mill wastewater. *Syst. Appl. Microbiol.* 28, 175-85.

Pearce, C.I., Pattrick, R.A., Law, N., Charnock, J.M., Coker, V.S., Fellowes, J.W., Oremland, R.S., Lloyd, J.R., 2009. Investigating different mechanisms for biogenic selenite transformations: *Geobacter sulfurreducens*, *Shewanella oneidensis* and *Veillonella atypica*. *Environ. Technol.* 30, 1313-1326.

Sánchez-Andrea, I., Sanz, J.L., Bijmans, M.F., Stams, A.J., 2014. Sulfate reduction at low pH to remediate acid mine drainage. *J. Hazard. Mater.* 269, 98-109.

Schloss, P.D., Westcott, S.L., Ryabin, T., Hall, J.R., Hartmann, M., Hollister, E.B., Lesniewski, R.A., Oakley, B.B., Parks, D.H., Robinson, C.J., Sahl, J.W., Stres, B., Thallinger, G.G., Van Horn, D.J., Weber, C.F., 2009. Introducing mothur: Open-source, platform-independent, community-supported software for describing and comparing microbial communities. *Appl. Environ. Microbiol.* 75, 7537-7541.

Shen, Z., Zhou, Y., Hu, J., Wang, J., 2013. Denitrification performance and microbial diversity in a packed-bed bioreactor using biodegradable polymer as carbon source and biofilm support. *J. Hazard. Mater.* 250-251, 431-438.

Shin, H.S., Kang, S., Nam, S.T., 2000. Effect of polysaccharides to protein ratio in EPS on sludge settling characteristics. *Biotech. Bioprocess Eng.* 5, 460-464.

Soda, S., Kashiwa, M., Kagami, T., Kuroda, M., Yamashita, M., Ike, M., 2011. Laboratory-scale bioreactors for soluble selenium removal from selenium refinery wastewater using anaerobic sludge. *Desalination* 279, 433-438.

Takahashi, S., Tomita, J., Nishioka, K., Hisada, T., Nishijima, M., 2014. Development of a prokaryotic universal primer for simultaneous analysis of bacteria and archaea using next-generation sequencing. *PLOS One* 9, e105592.

Tan, L.C., Nancharaiah, Y.V., van Hullebusch, E.D., Lens, P.N.L., 2016. Selenium: Environmental significance, pollution, and biological treatment technologies. *Biotechnol. Adv.* 34, 886-907

Tomei, F.A., Barton, L.L., Lemanski, C.L., Zocco, T.G., Fink, N.H., Sillerud, L.O., 1995. Transformation of selenate and selenite to elemental selenium by *Desulfovibrio desulfuricans*. *J. Ind. Microbiol. Biotechnol.* 14, 329-336.

Truong, H.Y.T., Chen, Y.W., Belzile, N., 2013. Effect of sulfide, selenite and mercuric mercury on the growth and methylation capacity of the sulfate reducing bacterium *Desulfovibrio desulfuricans*. *Sci. Total Environ.* 449, 373-384.

CHAPTER 6

Comparative performance of anaerobic attached biofilm and granular sludge reactors for the treatment of model mine drainage wastewater containing selenate, sulfate and nickel

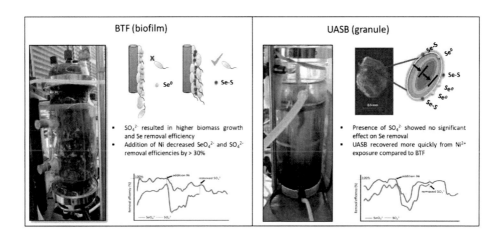

This chapter has been published in modified form:

Tan, L.C, Papirio, S., Luongo, V., Nancharaiah, Y.V., Cennamo, P., Esposito, G., van Hullebusch, E., Lens, P.N.L. 2018. Comparative performance of anaerobic attached biofilm and granular sludge reactors for the treatment of model mine drainage wastewater containing selenate, sulfate and nickel. *Chem. Eng. J.* 345, 545-555. doi:10.1016/j.cej.2018.03.177.

Abstract

Acid mine drainage and mining wastewaters contain, depending on the source, elevated concentrations of metals, e.g. nickel (Ni^{2+}), and oxyanions, e.g. selenate (SeO_4^{2-}) and sulfate (SO_4^{2-}). This study compared the performance of two reactor configurations, a biotrickling filter (BTF) and an upflow anaerobic sludge blanket (UASB) reactor, for the treatment of model mining wastewater contaminated with SeO_4^{2-}, SO_4^{2-} and Ni^{2+}. The Se removal efficiency of the BTF biofilm was improved by > 70% in the presence of SO_4^{2-}. In contrast, the Se removal performance of the UASB reactor was not affected by the presence of SO_4^{2-}. The addition of Ni^{2+} decreased the oxyanion (SO_4^{2-} and SeO_4^{2-}) removal efficiency of both the BTF and UASB reactor by > 30%. However, the UASB reactor recovered more quickly (~10 days after Ni^{2+} addition) from Ni^{2+} toxicity as compared to the BTF (~22 days after Ni^{2+} addition). A Ni^{2+} removal efficiency of more than 80% was achieved for both the BTF and UASB reactor. Ni^{2+} was mainly removed via sulfidic (HS^-) precipitation by forming nickel sulfide (Ni_3S_4). Scanning electron microscopy coupled with energy dispersive X-ray spectroscopy (SEM-EDX) and powder X-ray energy diffraction (P-XRD) revealed entrapped Se and selenium monosulfide (SeS) in the biofilm and granules of both reactor configurations, which could be potentially recovered for further reuse applications. This study demonstrated that the feed wastewater characteristics and the reactor configuration are key factors in achieving an efficient treatment of Se-laden mine drainage wastewater.

Keywords: selenate wastewater; sulfate; nickel; biofilm; granules; reactor configuration

6.1 Introduction

Selenium (Se) is a vital trace metalloid for all living organisms, but also a toxic element to natural ecosystems due to its high bioaccumulation potential (Lenz and Lens 2009). Industrial sources of Se contribute for about 40% of the Se emissions in the atmosphere and include irrigation drain water, fly-ash from coal-fired power plants, release from metallic ore mining and smelting as well as acid mine drainage (Wen and Carignan 2007). While there is a debate over the exact nature of Se toxicity, Se discharge clearly has adverse environmental impacts, i.e. excess Se in impounded areas and irrigation ditches resulted in the death of numerous waterfowls at the Kesterson National Wildlife Refuge (NWR) at San Joaquin Valley (California, USA) (Ellis and Salt 2003). Se in wastewater must thus be removed prior to discharge to prevent environmental impacts on surrounding wildlife and ecosystems.

The application of biological processes for the removal of the mobile and toxic Se species, i.e. selenate (SeO_4^{2-}), has shown potential as a green and cost-effective alternative to the traditional physicochemical SeO_4^{2-} removal methods (Lenz and Lens 2009; Tan et al. 2016). Moreover, bioremediation processes give an added advantage of converting SeO_4^{2-} into a less mobile and non-toxic biogenic elemental Se (Se^0), which can be recovered and reused for future applications (Hageman et al. 2017). However, Se-laden wastewaters also contain various co-contaminants at elevated concentrations, i.e. sulfate (SO_4^{2-}) and heavy metals that can affect the Se bioreduction process and speciation of bioreduced Se.

Se shares a close chemical similarity with sulfur (S) and is often associated as a by-product of metal sulfide ore mineral processing (Wang et al. 2016). Concentrations of SO_4^{2-} in Se-laden wastewater can reach up to > 1 g/L (Lenz et al. 2008; Tan et al. 2016), while heavy metals can range from 0.1 to > 100 mg/L depending on the wastewater stream (Mal et al. 2016). Heavy metals are non-biodegradable and toxic elements (Mal et al. 2016; Kiran et al. 2017). Similar to SeO_4^{2-}, heavy metals can also be transformed from the soluble and toxic form into immobile and less toxic forms through metal complexation and/or precipitation with other elements. Bioreactors operating under SO_4^{2-} reducing conditions have shown potential for metal (Kiran et al. 2017) and Se (**chapter 4**) bioremediation. Under anaerobic conditions, SO_4^{2-} is reduced to sulfide which can act as a precipitant for metals (Kiran et al. 2017) and possibly as an enhancement agent for the removal or reduction of Se oxyanions, i.e. SeO_4^{2-} or selenite (SeO_3^{2-}) (Hageman et al. 2013; Hockin and Gadd 2006; **chapter 4**).

The studies exploring the effect of SO_4^{2-}, metals or both on biological SeO_4^{2-} removal are limited. Previous studies have showed that the concomitant presence of SO_4^{2-} can have a

155

positive effect on SeO_4^{2-} removal (**chapters 3 and 4**) from mine drainage and mining wastewaters. Experiments in serum bottles using anaerobic granular sludge showed that a SO_4^{2-}/SeO_4^{2-} molar ratio of ≤ 100 enhanced SeO_4^{2-} removal by 20-30% when compared to only SeO_4^{2-} (**chapter 3**). In the case of a short-term (10 days) continuously operated biofilm system, the presence of SO_4^{2-} was a controlling parameter in promoting both biofilm growth and Se removal efficiencies, when compared to biofilm exposed to SeO_4^{2-} alone or to both SeO_4^{2-} and nitrate (NO_3^-) (**chapter 4**). Recently, Mal et al. (2016) investigated the bioreduction of SeO_3^{2-} in the presence of heavy metals (i.e. Cd, Pb and Zn) by anaerobic granular sludge in serum-bottles under fed-batch conditions. Cd was found to have an inhibitory effect on SeO_3^{2-} reduction only at > 150 mg/L, while SeO_3^{2-} reduction was not inhibited in the presence of Pb and Zn individually, even at concentrations of 400 mg/L (Mal et al. 2016).

To the best of our knowledge, there are no studies on the simultaneous removal of oxyanions, e.g. SeO_4^{2-} and SO_4^{2-}, and heavy metals (e.g. nickel), by mixed microbial communities. The aim of this study was, therefore, to evaluate the application of two reactor configurations, i.e. a biofilm in a biotrickling filter (BTF) and an upflow anaerobic sludge blanket (UASB) reactor, for the simultaneous removal of SeO_4^{2-} in the presence of SO_4^{2-} and nickel (Ni^{2+}). The two reactor configurations were chosen based on previous studies (**chapters 3 and 4**) as these allow two compare to microbial growth systems, i.e. biofilms in the BTF and granules in the UASB reactor. Nickel was chosen because it is one of the most commonly used heavy metals in the mining industry present in oxide ores and one of the main heavy metal co-contaminants found in Se-laden wastewaters (Wang et al. 2016). The Se removal performance of the growth systems was determined in the presence of the co-contaminants under four different operational conditions: (1) SeO_4^{2-} alone, (2) $SeO_4^{2-} + SO_4^{2-}$, (3) $SeO_4^{2-} + SO_4^{2-} + Ni^{2+}$ and (4) $SeO_4^{2-} + Ni^{2+}$. Scanning electron microscopy with dispersive X-ray spectroscopy (SEM-EDX) and powder X-ray energy diffraction (P-XRD) were used to characterize Se, S and Ni minerals present in the BTF biofilm and UASB granular sludge.

6.2 Materials and methods

6.2.1 Source of inoculum

Anaerobic granular sludge taken from a full-scale UASB reactor treating a paper-mill wastewater (Industriewater Eerbeek B.V., Eerbeek, The Netherlands) was used as the inoculum for both the BTF and UASB reactor. A detailed description of the inoculum can be found in

Roest et al. (2005). The UASB reactor was inoculated with 10% (w/v) anaerobic granular sludge. The BTF was initially seeded with 1% ((w/v) crushed anaerobic granular sludge in a 2 L mineral medium containing 1.2 g COD/L and incubated for two days in batch mode at 30°C for microbial attachment/biofilm development onto the silicon tubing prior to the continuous operation. After 2 days under batch incubation, the liquid solution was drained from the BTF before starting the continuous operation.

6.2.2   Synthetic mine drainage wastewater

The reactors were fed with a synthetic acid mine drainage wastewater containing a mineral medium, organic carbon source and the different contaminants of interest, i.e. SeO_4^{2-}, SO_4^{2-} and Ni^{2+}. The mineral medium was prepared as per Lenz et al. (Lenz et al. 2008) containing (in mg/L): 100 $CaCl_2$, 10 $MgCl_2$, 300 NH_4Cl, 40 $KHCO_3$ and 1 mL/L each of an acid and alkaline trace element solution. The detailed composition of the acid and alkaline trace element solution can be found in Lenz et al. (Lenz et al. 2008). A phosphate buffer (53 mg/L Na_2HPO_4 and 41 mg/L KH_2PO_4) was used to maintain the pH in the reactor at 7.5 ($\pm$ 0.2). Sodium lactate (60% purity, ACROS Organics, Belgium) was provided as the sole electron donor and carbon source. Sodium selenate (Na_2SeO_4; Sigma-Aldrich, Italy), potassium sulfate (K_2SO_4; Panreac AppliChem, Spain) and nickel (II) chloride hexahydrate ($NiCl_2 \cdot 6H_2O$; Sigma-Aldrich, Italy) were used as sources of SeO_4^{2-}, SO_4^{2-} and Ni^{2+}, respectively.

6.2.3   Reactor configurations and start-up

The BTF (**Fig. 1a**) was made of Pyrex glass with an internal diameter of 0.1 m and a height of 0.3 m. The biofilm carrier material consisted of silicone tubing (peroxide-cured silicon tubing, Algam, Italy) cut into 1 cm length having an internal diameter of 2 mm. An average biofilm carrier density of 0.40 ($\pm$ 0.03) g/L was used during each run. The total volume of the BTF was 2.4 L, while the empty filter bed volume (biofilm carrier) was 1.8 ($\pm$ 0.1) L. The empty bed contact time (EBCT) was calculated to be 5.8 ($\pm$ 0.4) h.

The UASB reactor (**Fig. 6.1b**) was made of Pyrex glass with an internal diameter of 0.13 m and a height of 0.26 m. The total working volume was 3.5 L and the calculated hydraulic retention time (HRT) was 11.3 h.

6.2.4 Reactor operating parameters

The chemical oxygen demand (COD), oxyanion and Ni^{2+} concentrations in the feed solution as well as the reactor operating conditions are given in **Table 6.1**. The same feed wastewater was supplied to both reactors at a flow rate of 7.3 (± 0.7) L/d using peristaltic pumps (520U, Watson Marlow, USA) during all the experimental phases. Due to the difference in reactor volume and HRT/EBCT between the two reactors, the organic loading rate for the BTF was approximately 2-fold higher than that used for the UASB reactor.

The reactor conditions were separated into two main operational phases, RC1 and RC2. During phase RC1, the reactors were operated only with SeO_4^{2-} for 45 days with a COD concentration of 1.1 g COD/L. Upon termination of phase RC1, both reactors were stopped and cleaned before using a fresh inoculum and new biofilm carriers for the BTF. Phase RC2 was conducted for 84 days and divided into 4 periods of approximately 20 days each: (period RC2a) $SeO_4^{2-} + SO_4^{2-}$ with a COD/SO_4^{2-} (g/g) ratio of 1.1; (period RC2b) $SeO_4^{2-} + SO_4^{2-}$ with a COD/SO_4^{2-} ratio of 2.8; (period RC2c) $SeO_4^{2-} + SO_4^{2-} + Ni^{2+}$ with a COD/SO_4^{2-} ratio of 2.8 and (period RC2d) $SeO_4^{2-} + Ni^{2+}$ with a COD concentration of 3.2 g COD/L.

6.2.5 Physico-chemical analysis

Total dissolved sulfide (HS⁻) and COD concentrations were measured using colorimetric methods (APHA/AWWA/WEF 2005) and analyzed using a UV-VIS spectrophotometer (Photolab 6600, UV-VIS series, WTW, USA). Se speciation in the treated effluent was assessed through SeO_4^{2-}, SeO_3^{2-}, total Se (Se_{tot}) and dissolved Se (Se_{diss}) measurements. For Se_{diss}, SeO_4^{2-} and SeO_3^{2-} measurement, effluent samples were centrifuged at 14,000 rpm for 15 min and filtered through a 0.45 μm membrane filter to remove suspended cells and colloidal Se^0 particles. Se_{tot} and Se_{diss} were analyzed after 1:1 acid digestion with 65% HNO_3 at 90°C for 30 min using hydride generation atomic fluorescence spectrometry (HG-AFS, AFS-8220, Beijing Titan Instruments, China) following instrument manual instructions. SeO_3^{2-} was analyzed without acid digestion using the HG-AFS.

SO_4^{2-} and SeO_4^{2-} concentrations were measured by ion chromatography (761 Compact IC, Metrohm, Switzerland) equipped with an IonPac AS12A 4 x 200 mm column (Dionex, USA) at a retention time of approximately 14 and 17 min, respectively. The IC mobile phase solution used as eluent was 0.1 M NaOH and 0.4 M Na_2CO_3 at a flow rate of 0.7 mL/min. An atomic absorption spectrophotometer (AAS, SpectraAA 50, Varian, Germany) with Lumina lamps Ni (PerkinElmer, USA) was used for analyzing the Ni^{2+} concentration.

Table 6.1 Feed composition and reactor conditions of the BTF and UASB reactor for the removal of SeO_4^{2-}, SO_4^{2-} and Ni using lactate as the electron donor and carbon source.

Phase/Period	Days	Lactate (g COD/L)	SeO_4^{2-} (mg Se/L)	SO_4^{2-} (mg S/L)	Ni^{2+} (mg/L)	COD/SO_4^{2-}
RC1[1]	0-45	1.1 (± 0.2)	10.9 (± 1.3)	-	-	-
RC2a[2]	0-24	1.2 (± 0.2)	9.4 (± 1.0)	384.2 (± 12.3)	-	1.1 (± 0.9)
RC2b	25-46	3.1 (± 0.3)	8.6 (± 0.8)	375.4 (± 16.6)	-	2.8 (± 0.3)
RC2c	47-66	3.6 (± 0.2)	9.6 (± 1.0)	385.1 (± 6.8)	49.4 (± 2.0)	2.9 (± 0.1)
RC2d	67-84	3.5 (± 0.2)	10.0 (± 0.9)	-	52.1 (± 3.2)	-

[1] Only operated for 45 days and biomass was completely removed from both reactors; [2] New operation provided with new biofilms and biofilm carriers were placed into the reactors

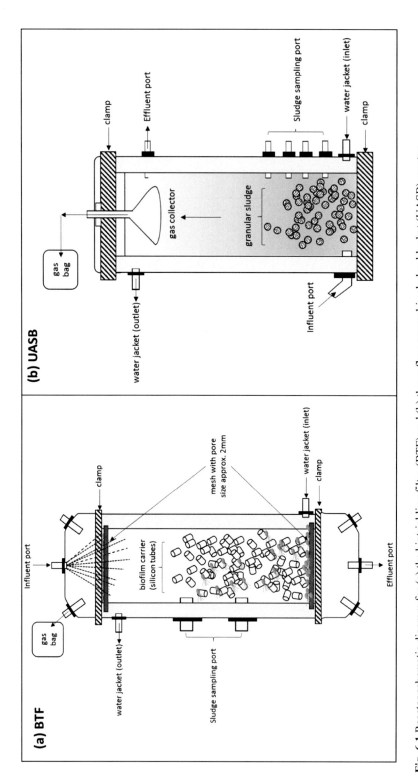

Fig. 6.1 Reactor schematic diagram for (a) the biotrickling filter (BTF) and (b) the upflow anaerobic sludge blanket (UASB) reactor.

The volatile fatty acids (VFAs) were quantified by a high-pressure liquid chromatograph (HPLC, LC 25 Chromatography Oven, Dionex, USA) equipped with a 250×4.60 mm column (Synergi 4u Hydro RP 80A, Phenomenex, USA) and a UV detector (AD25 Absorbance Detector, Dionex, USA) reading at 210 nm.

Biomass samples consisting of biofilm carriers and granules from both reactors were taken to analyze total solids (TS) and volatile solids (VS) according to the standard methods (APHA/AWWA/WEF 2005). The biomass activity was assessed as the consumption of lactate by bacteria, while the biomass growth was based on the TS and VS concentrations measured at the end of each reactor condition. Total Se and Ni content in the biofilm and granular sludge were determined after acid (65% HNO_3) digestion treatment with a microwave (5 pKo Temp Microwave MARS, CEM, Italy) oven using HG-AFS and AAS, respectively (Mal et al. 2016). A portion of the biofilm and granules was lyophilized overnight using a freeze dryer (Christ Alpha 2-4 LSC plus, Christ, Germany) and then stored prior to analysis.

Lyophilized biomass samples were examined for morphology and elemental semi-quantitative analysis using scanning electron microscopy (SEM, Vega 3, TESCAN, Czech Republic) with an energy dispersive X-ray spectrometry (EDX, QUANTAX, Bruker, USA) analyzer (Cennamo et al. 2012). The identification of relevant mineralogical phases present in the lyophilized biomass samples was carried out using a powder X-ray diffractometer (P-XRD, Miniflex Diffractometer, Rigaku, USA) at 30 kV and 3600 seconds count time with a 15 mA cobalt tube (Cennamo et al. 2016). The peak identification in the P-XRD was performed with a match on the Joint Committee on Powder Diffraction Standards (JCPDS) database.

6.2.6 Extracellular polymeric substances (EPS) analysis

EPS were extracted from the biofilms and biogranules taken during the different reactor phases of operation using a cation exchange resin (CER, Dowex Marathon C) method as described by Frølund et al. (1996). Briefly, a known amount (containing 2 g VS) of biomass sample was weighed and centrifuged at 2000 g for 15 min to remove the reactor liquid suspension. The biomass pellets were re-suspended in a phosphate buffer and mixed with CER (70 g CER/g VS) and stirred for 17 h at 4°C. After extraction, the supernatant was taken, centrifuged, filtered through a 0.45 µm membrane (Whatman) and stored at -20°C until analysis.

Total organic carbon (TOC) content of the extracted EPS was determined using a TOC analyzer (Shimadzu TOC-VCPN analyzer, Kyoto, Japan) to determine the total EPS concentration. 3D excitation (220-400 nm) and emission (300-500 nm) fluorescent spectroscopy of extracted EPS (total organic carbon concentration of 1.0 mg/L) were carried out using a FluoroMax-3 spectrofluorometric (HORIBA Jobin Yvon, Edison) instrument. The extracted EPS were analyzed for polysaccharides using the phenol-sulfuric acid method (Dubois et al. 1956), proteins and humic-like substances using a modified Lowry method (Frølund et al. 1995).

6.2.7 Calculations and statistical analysis

Specific removal rates (SRR) and removal efficiencies were calculated according to Eq. (6.1) and Eq. (6.2), respectively:

$$\text{SRR (mg/d} \cdot \text{gVS)} = \frac{(\text{concentration}_{in} - \text{concentration}_{out}) \left(\frac{mg}{L}\right)}{\text{biomass concentration} \left(\frac{g\,VS}{L}\right) \times \text{HRT or EBCT (d)}} \qquad \text{Eq. (6.1)}$$

$$\text{Removal efficiency (\%)} = \frac{(\text{concentration}_{in} - \text{concentration}_{out})}{\text{concentration}_{in}} \qquad \text{Eq. (6.2)}$$

Colloidal Se^0 and unaccounted Se_{diss} in the liquid effluent were determined according to Eq. (6.3) and Eq. (6.4), respectively:

$$\text{Colloidal Se}^0 = Se_{tot} - \text{total Se}_{diss} \qquad \text{Eq. (6.3)}$$

$$\text{Unaccounted Se}_{diss} = \text{total Se}_{diss} - SeO_4^{2-} - SeO_3^{2-} \qquad \text{Eq. (6.4)}$$

Average and standard deviations were calculated for SRR, removal efficiency and residual concentration of contaminants in the effluent. Statistical differences in SRR, removal efficiency and effluent concentrations between the two reactors and among the reactor conditions were evaluated using analysis of variance (ANOVA) with the Tukey tests (Minitab 17 statistical software, Minitab Inc.). Two comparing data sets were considered statistically different when a $p\text{-value} \leq 0.05$ was obtained.

6.3 Results

6.3.1 Reactor performance - SeO_4^{2-} and total Se removal and Se speciation

Fig. 6.2 shows the concentration profiles of SeO_4^{2-} and Se_{tot} in the effluent in the BTF and UASB reactor under different reactor conditions. Under SeO_4^{2-} reducing conditions (phase RC1), the effluent SeO_4^{2-} concentration remained at an average value of 5.4 ($\pm$ 0.7) mg Se/L from day 10 to 45 (**Fig. 6.2a**). In contrast, the effluent SeO_4^{2-} concentration in the UASB reactor

(**Fig. 6.2a**) gradually decreased starting from ~5 mg Se/L (days 3-15), to ~2 mg Se/L (days 16-33) and finally reached < 0.4 mg Se/L (day 34-45). In the case of the effluent Se_{tot} concentration (**Fig. 6.2c**), the BTF and UASB reactor showed an average effluent Se_{tot} concentration of 8.9 ($\pm$ 1.4) and 5.7 ($\pm$ 1.6) mg Se/L, respectively. Overall, the Se_{tot} removal efficiency in phase RC1 was < 20% for the BTF, while the UASB reactor attained ~50% (**Table 6.2**).

Fig. **6.2b** and **6.2d** report the effluent SeO_4^{2-} and Se_{tot} concentrations, respectively, during phase RC2. For the BTF, the effluent SeO_4^{2-} concentration of ~0.6 mg Se/L during periods RC2a to RC2b was significantly lower when compared to that obtained during phase RC1 (*p-value* < 0.0001) achieving a > 90% SeO_4^{2-} removal efficiency. For the UASB reactor, the effluent SeO_4^{2-} concentration profile during the periods RC2a and RC2b was statistically similar (*p-value* > 0.13) to that achieved in phase RC1. A step-wise decrease of concentration from ~4 mg Se/L (day 1-15) to ~2.2 mg Se/L (day 16-33) and finally reaching < 0.3 mg Se/L (day 34-46) was noticed. The effluent Se_{tot} concentration reached ~2 and ~4 mg Se/L in the BTF and UASB reactor, respectively, during periods RC2a to RC2b (**Fig. 6.2d**). Upon addition of Ni^{2+} in the feed solution under SO_4^{2-} reducing conditions (period RC2c), the BTF and UASB reactor showed an increase in the effluent SeO_4^{2-} concentration reaching ~3 and ~6 mg Se/L, respectively, from days 47 to 55 (**Fig. 6.2b**). Subsequently, both reactors recovered from the Ni^{2+} impact achieving a lower effluent SeO_4^{2-} concentration back to ~1 mg Se/L from day 56 to 67. Once SO_4^{2-} was removed from the feed solution (period RC2d), the BTF exhibited an immediate increase in effluent SeO_4^{2-} concentration to 3.4 ($\pm$ 0.7) mg Se/L from day 69 to 84. On the other hand, the UASB reactor showed the same effluent SeO_4^{2-} concentration from day 60 (period RC2c) until day 75 (period RC2d) before a rapid increase to ~4 mg Se/L was observed from day 76 until day 84. The overall SeO_4^{2-} removal during periods RC2c and RC2d for both reactors was statistically similar (*p-value* > 0.05) to that observed during periods RC2a and RC2b, with an average removal efficiency of 75 ($\pm$ 2)%. Comparing the two reactors, the effluent Se_{tot} concentration during periods RC2c and RC2d showed statistically similar values (*p-value* > 1.00), with an average Se_{tot} concentration of 3.4 ($\pm$ 1.2) mg Se/L (**Fig. 6.2d**). The Se_{tot} in the effluent mainly consisted of four Se species: (1) unreduced SeO_4^{2-}, (2) intermediate SeO_3^{2-}, (3) colloidal Se^0 and (4) unaccounted Se_{diss}. **Fig. 6.3** shows the Se speciation (%) of the Se_{tot} detected in the treated effluent during the five operating periods. During phase RC1, unaccounted Se_{diss} accounted for 30 ($\pm$ 11)% of the effluent Se_{tot} concentration for both reactors. A significant difference in Se speciation (*p-value* < 0.04) was observed during phase RC2, when unaccounted Se_{diss} was only < 6% of the effluent Se_{tot}

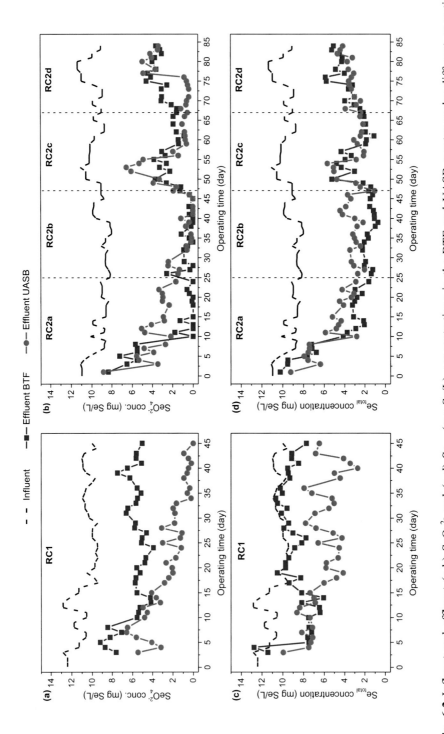

Fig. 6.2 Influent and effluent (a-b) SeO$_4^{2-}$ and (c-d) Se$_{tot}$ (mg Se/L) concentrations in the BTF and UASB reactor under different operating conditions: (a-c) period RC1 and (b-d) period RC2.

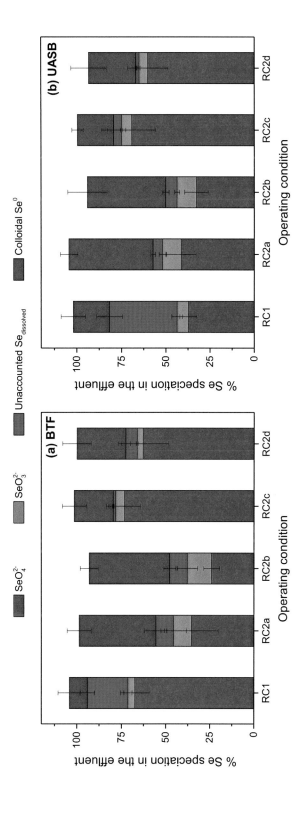

Fig. 6.3 Averaged Se speciation (%) of the effluent Se_{tot} concentration in (a) the BTF and (b) the UASB reactor under different operating conditions over entire period/phase.

concentration. Both phases RC1 and RC2 showed a similar SeO_3^{2-} production (*p-value* > 0.44), reaching an average SeO_3^{2-} concentration of 0.72 ($\pm$ 0.53) and 0.35 ($\pm$ 0.24) mg Se/L in the BTF and UASB reactor, respectively. A similar colloidal Se^0 release, i.e. 35 ($\pm$ 16)%, of the effluent Se_{tot} concentration (*p-value* > 0.70), was achieved during phase RC1 and periods RC2a and RC2b in both reactors. The percentage of colloidal Se^0 slightly decreased to 24 ($\pm$ 3)% upon addition of Ni^{2+} to both the reactors.

Table 6.3 shows the calculated SRR for Se_{tot} and SeO_4^{2-} for both reactors. For the BTF, despite the higher SeO_4^{2-} removal efficiency attained during the periods RC2a and RC2b compared to phase RC1, the SeO_4^{2-} SRR was statistically similar from phase RC1 to period RC2b at an average value of 1.8 ($\pm$ 0.5) mg Se/gVS·d (*p-value* > 0.63). Upon addition of Ni^{2+}, the SeO_4^{2-} SRR decreased to < 1.0 mg Se/gVS·d. The Se_{tot} SRR for the BTF was within the range of 0.6 to 1.4 mg Se/gVS·d for both phases RC1 and RC2 (*p-value* > 0.20). For the UASB reactor, the SRR for SeO_4^{2-} and Se_{tot} were similar (*p-value* > 0.20) for both phases RC1 and RC2 at an average value of 2.0 ($\pm$ 0.5) and 1.6 ($\pm$ 0.5) mg Se/gVS·d, respectively.

6.3.2 Reactor performance - SO_4^{2-} removal

Fig. 6.4 shows the SO_4^{2-} and dissolved sulfide (HS^-) concentration profiles in the BTF and UASB reactor under SO_4^{2-} reducing conditions in the presence and absence of Ni^{2+}. During period RC2a, both reactors were operated at a COD/SO_4^{2-} ratio of 1.1 from day 0 to 24. SO_4^{2-} reduction only reached ~22% (**Table 6.2**) and resulted in an effluent SO_4^{2-} concentration of 300.7 ($\pm$ 21.4) mg S/L (**Fig. 6.4a**) in the BTF during period RC2a. For the UASB reactor, the effluent SO_4^{2-} concentration was stable at 303.4 ($\pm$ 42.2) mg S/L until day 19, before decreasing to 203.9 ($\pm$ 5.9) mg S/L on day 24 (**Fig. 6.4b**). For both reactors, the effluent HS^- concentration increased from an initial value of ~1 up to ~75 mg S/L on day 24. During period RC2a, the HS^- production was statistically similar in both the BTF and UASB reactor (*p-value* > 0.95).

During period RC2b, the COD/SO_4^{2-} ratio was increased to 2.8 starting from day 25 until the end of the reactor run. **Fig. 6.4a** showed that the effluent SO_4^{2-} concentration was, on average, lower during period RC2b than period RC2a (*p-value* < 0.001). SO_4^{2-} varied from ~250 mg S/L (day 26) to < 120 mg S/L (day 29 to 46). Similarly, the effluent HS^- concentration increased from ~75 mg S/L (day 24) up to 150-210 mg S/L (day 28-46). For the UASB reactor, the effluent SO_4^{2-} concentration was 167.7 ($\pm$ 36.0) mg S/L from day 25-36 before decreasing by approximately 2.5 times from day 38 to 46 (**Fig. 6.4b**). Likewise, the effluent HS^- concentration increased from ~70 to ~170 mg S/L from day 25 to 46. For both reactors, the overall SO_4^{2-} removal efficiency, at a COD/SO_4^{2-} ratio of 2.8, was averagely 67 ($\pm$ 15)%, i.e. 3 times higher than that obtained with a lower COD/SO_4^{2-} ratio (*p-value* < 0.001).

A COD/SO_4^{2-} ratio of 2.8 was maintained when Ni^{2+} was added to the SO_4^{2-} reducing system starting from day 47 (period RC2c) (**Fig. 6.4**). The effluent SO_4^{2-} concentration rapidly increased up to 220-330 mg S/L and remained stable from day 49 to 57 for both reactors. From day 59 to 66, the effluent SO_4^{2-} concentration decreased and was steady at ~150 and ~60 mg S/L in the BTF and UASB reactor, respectively. The effluent HS^- concentration was lower than 10 mg S/L from day 47 to 56 before increasing to a range of 50-150 mg S/L from day 57 to 66 for both reactors. The overall SO_4^{2-} removal efficiency during period RC2c for the BTF was significantly lower (*p-value* < 0.001) than during period RC2b at an average value of 36 ($\pm$ 18)%. In contrast, during period RC2c, the overall SO_4^{2-} removal efficiency for the UASB reactor was statistically similar (*p-value* > 0.11) to period RC2b at an average value of 54 ($\pm$ 26)% (**Table 6.2**).

During period RC2b, the SO_4^{2-} SRR was 145% and 319% higher than that achieved during period RC2a for the BTF and UASB reactor, respectively. Upon addition of Ni^{2+}, the BTF was significantly impacted (*p-value* < 0.001) decreasing the SO_4^{2-} SRR from 42.1 ($\pm$ 8.5) to 15.4 ($\pm$ 7.6) mg S/gVS·d (**Table 6.2**). Opposite to this, the SO_4^{2-} SRR for the UASB reactor was only slightly affected by the presence of Ni^{2+} from ~71 to ~64 mg S/gVS·d (**Table 6.2**).

6.3.3   Reactor performance - Ni^{2+} removal

Ni^{2+} was added to both reactors during period RC2 from day 47 onwards. The profile of the effluent Ni^{2+} concentration is shown in **Fig. 6.4** for periods RC2c (SeO_4^{2-} and SO_4^{2-} reducing conditions) and RC2d (SeO_4^{2-} reducing conditions). Both reactors showed similar Ni^{2+} concentration profiles and removal performance. During period RC2c, Ni^{2+} removal exceeded 90%, with an effluent Ni^{2+} concentration < 4 mg Ni/L in both reactors (*p-value* > 0.87). The effluent Ni^{2+} concentration started to increase during period RC2d reaching a value of ~20 mg Ni/L on day 84 and achieving an average removal efficiency of 74 ($\pm$ 20)%. The Ni^{2+} SRR for the BTF was statistically similar (*p-value* > 0.37) between periods RC2c and RC2d at an average value of 5.9 ($\pm$ 0.9) mg Ni/gVS·d (**Table 6.2**). In contrast, the Ni^{2+} SRR of the UASB reactor significantly decreased from period RC2c to period RC2d (*p-value* < 0.0001) from ~16 to ~11 mg Ni/ gVS·d (**Table 6.2**).

Table 6.2 Specific removal rates (SRR) and removal efficiencies obtained for the BTF and UASB reactor under different operating conditions.

BTF

Condition	Se_{tot}		SeO_4^{2-}		SO_4^{2-}		Ni^{2+}	
	mg Se/gVS·d	%	mg Se/gVS·d	%	mg S/gVS·d	%	mg Ni/gVS·d	%
RC1	0.6 (± 0.3)	16 (± 14)	1.6 (± 0.5)	44 (± 11)	n/a	n/a	n/a	n/a
RC2a	1.5 (± 0.8)	50 (± 27)	2.3 (± 0.9)	74 (± 30)	29.1 (± 6.7)	22 (± 5)	n/a	n/a
RC2b	1.2 (± 0..1)	83 (± 6)	1.4 (± 0.2)	95 (± 8)	42.1 (± 8.5)	66 (± 13)	n/a	n/a
RC2c	0.8 (± 0.1)	72 (± 12)	0.8 (± 0.1)	78 (± 8)	15.4 (± 7.6)	36 (± 18)	5.2 (± 0.3)	92 (± 6)
RC2d	0.9 (± 0.2)	59 (± 10)	1.1 (± 0.1)	68 (± 8)	n/a	n/a	6.5 (± 1.5)	77 (± 17)

UASB

Condition	Se_{tot}		SeO_4^{2-}		SO_4^{2-}		Ni^{2+}	
	mg Se/gVS·d	%	mg Se/gVS·d	%	mg S/gVS·d	%	mg Ni/gVS·d	%
RC1	1.2 (± 0.3)	47 (± 11)	2.0 (± 0.3)	77 (± 14)	n/a	n/a	n/a	n/a
RC2a	0.9 (± 0.4)	43 (± 19)	1.2 (± 0.3)	60 (± 15)	22.3 (± 12.2)	27 (± 15)	n/a	n/a
RC2b	1.6 (± 0.1)	66 (± 10)	2.2 (± 0.3)	92 (± 9)	71.1 (± 16.8)	69 (± 16)	n/a	n/a
RC2c	2.1 (± 0.3)	69 (± 11)	2.2 (± 0.5)	74 (± 20)	63.6 (± 29.8)	54 (± 26)	16.3 (± 1.1)	97 (± 3)
RC2d	2.0 (± 0.3)	65 (± 7)	2.4 (± 0.5)	79 (± 17)	n/a	n/a	10.8 (± 0.8)	72 (± 23)

n/a – not applicable

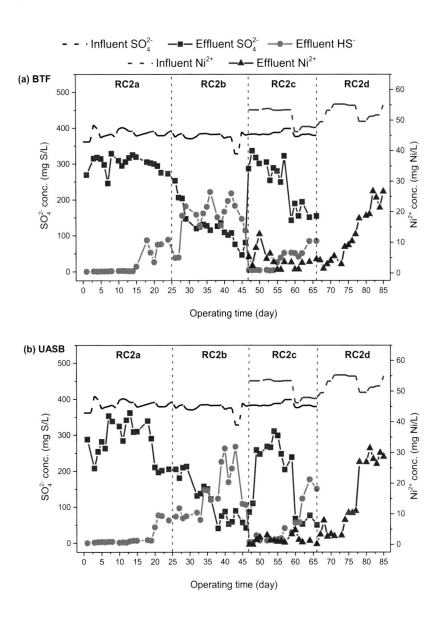

Fig. 6.4 Feed and effluent SO_4^{2-}, HS^- and Ni^{2+} concentrations in (a) the BTF and (b) the UASB reactor under different operating conditions (phase RC2).

6.3.4   Reactor performance - COD removal

The total COD removal efficiency was stable at < 30% during phases RC1 and RC2 in both reactors (**Appendix 4, Fig. S6.1 a-c**). The lactate removal efficiency was within a range of 70-90% for both reactors in phases RC1 and RC2 (**Appendix 4, Fig. S6.2**). Acetate and propionate were the only detected VFAs produced during lactate oxidation in both the BTF and UASB reactor. The acetate and propionate concentrations in the effluent ranged from 100-1000 and 200-1500 mg COD/L, respectively, in both reactors throughout the operation (**Appendix 4, Fig. S6.1 d-f**).

6.3.5 Biomass activity and growth

Table 6.3 shows the TS and VS concentration of the biomass samples withdrawn from both reactors at the end of each experimental period. The BTF fed with SeO_4^{2-} alone showed the lowest TS and VS concentrations of 15.1 ($\pm$ 0.6) g TS/L and 11.7 ($\pm$ 1.5) g VS/L, respectively. During the addition of SO_4^{2-}, the TS and VS concentrations were statistically similar to those obtained during phase RC1 (*p-value* > 1.00) till day 25, before doubling to 29.3 ($\pm$ 2.8) g TS/L and 24.1 ($\pm$ 0.4) g VS/L on day 45. Upon addition of Ni^{2+}, the biofilm weight doubled during period RC2c, but slightly decreased during period RC2d when SO_4^{2-} was no longer fed. The volatile organic fraction during phase RC1 and periods RC2a to RC2b was ~80%, which decreased to ~60% when Ni^{2+} was added. In contrast to the biofilm, the granular weight remained statistically similar (*p-value* > 0.90) during phases RC1-RC2 and to the inoculum with an average TS and VS concentration of 12.0 ($\pm$ 1.7) g TS/L and 8.3 ($\pm$ 1.4) g VS/L, respectively (**Table 6.3**). The volatile organic fraction for the granules during phases RC1-RC2 was in the range of 65 to 75% (**Table 6.3**).

6.3.6 Characterization of mineral immobilized in the biofilm/granules

Se and Ni were detected in the air dried biomass samples collected at the end of each reactor phase/period (**Table 6.4**). The entrapped Se in the BTF biofilm during phase RC2 was 200% higher than in phase RC1, where a maximum Se concentration of ~70 mg Se/g VS was detected during periods RC2b and RC2c. In contrast, the entrapped Se in the UASB granules was statistically similar at a value of 27.4 ($\pm$ 5.9) mg Se/g TS (*p-value* > 0.8) during the whole UASB reactor run. The entrapped Ni content within the BTF biofilm ranged between 150-470 mg Ni/g VS, while the UASB granules contained an average of 170.0 ($\pm$ 48.4) mg Ni/g VS during the periods RC2c and RC2d.

The mineral precipitates detected in the biofilm and granules were different during each reactor condition, but similar in both reactors. As such, only the SEM-EDX and P-XRD results obtained with the BTF biofilm under each operating condition are shown as a representative characterization analysis for both reactors (**Fig. 6.5** and **Fig. 6.6**).

Table 6.3 Total solids (TS), volatile solids (VS) and the volatile organic fraction for the biofilm and granules taken from the BTF and UASB reactor, respectively, under different operating conditions. The volatile organic fraction was calculated as VS/TS.

	BTF			UASB		
	Total solids (g TS/L)	Volatile solids (g VS/L)	Volatile organic fraction (%)	Total solids (g TS/L)	Volatile solids (g VS/L)	Volatile organic fraction (%)
Inoculum	-	-	-	13.9 (± 0.6)	9.5 (± 0.8)	69 (± 2)
RC1	15.1 (± 0.6)	11.7 (± 1.5)	78 (± 3)	13.4 (± 0.9)	8.8 (± 1.0)	66 (± 1)
RC2a	15.3 (± 0.7)	12.2 (± 1.0)	80 (± 3)	13.3 (± 1.2)	10.0 (± 0.9)	75 (± 5)
RC2b	29.3 (± 2.8)	24.1 (± 0.4)	82 (± 1)	10.8 (± 0.8)	7.7 (± 0.9)	71 (± 1)
RC2c	57.3 (± 3.2)	37.1 (± 2.2)	65 (± 1)	10.8 (± 0.7)	6.9 (± 0.8)	64 (± 5)
RC2d	40.8 (± 1.7)	25.7 (± 0.9)	63 (± 6)	9.9 (± 0.5)	6.8 (± 1.2)	69 (± 4)

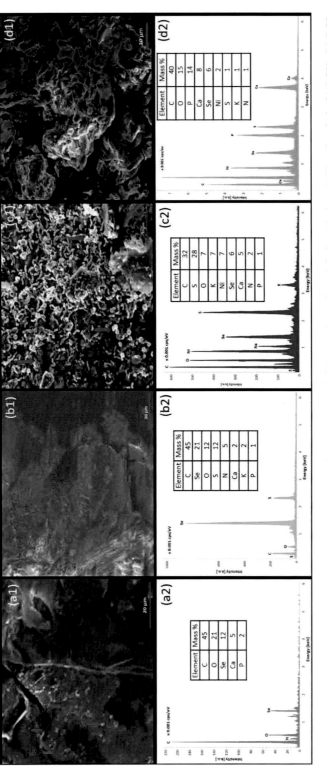

Fig. 6.5 (1) SEM imaging and (2) EDX semi-quantitative analysis for the BTF biofilm samples collected during (a) phase RC1 and periods (b) RC2b, (c) RC2c and (d) RC2d. SEM images were acquired in secondary electron mode at a magnification 2500× with EDX conducted in a back-scattered mode using 20 kV. Red marks in SEM images indicate the ratio of elements (Se, S and Ni) detected as mass percentage.

During phase RC1, a light reddish coloration was observed both on the BTF biofilm and in the UASB reactor liquid suspension (**Appendix 4, Fig. S6.3**). The SEM images (**Fig. 6.5 a1**) showed spherical shaped nanoparticles and the EDX analysis confirmed the presence of the element Se (**Fig. 6.5 a2**). P-XRD spectra of the biofilm during phase RC1 showed peak broadening possibly due to a highly amorphous nature of the Se^0 precipitates (**Fig. 6.6a**).

Table 6.4 Amount of entrapped Se and Ni (mg/g VS) in the BTF biofilm and UASB granules.

	Selenium (mg Se/g VS)	Nickel (mg Ni/g VS)
BTF biofilm		
RC1	33.5 ($\pm$ 4.8)	n.d.
RC2b	69.0 ($\pm$ 3.2)	n.d.
RC2c	72.7 ($\pm$ 4.7)	479.3 ($\pm$ 12.2)
RC2d	56.5 ($\pm$ 6.4)	164.1 ($\pm$ 4.5)
UASB granules		
RC1	27.6 ($\pm$ 7.2)	n.d.
RC2b	19.1 ($\pm$ 3.2)	n.d.
RC2c	31.2 ($\pm$ 4.5)	135.8 ($\pm$ 3.3)
RC2d	31.8 ($\pm$ 4.8)	204.2 ($\pm$ 6.8)

n.d. - below detection limit

When SO_4^{2-} was added to the feed solution, the BTF showed a more intense red coloration onto the biofilm, while the UASB reactor showed a yellowish-red coloration in the liquid suspension (**Appendix 4, Fig. S6.3**). EDX analysis (**Fig. 6.5 b2**) of the SEM image (**Fig. 6.5 b1**) revealed the presence of both S and Se elements in the biomass. The P-XRD pattern identified mainly SeS (JCPDS card no. 00-002-0320) mineral deposits in the BTF biofilm and UASB granules (**Fig. 6.6b**).

Upon addition of Ni^{2+} to the reactor, both the BTF biofilm and UASB reactor liquid suspension turned black (**Appendix 4, Fig. S6.3**). During SO_4^{2-} reducing conditions with Ni^{2+} addition (period RC2c), SEM images of the biofilm showed the presence of rod-shaped bacteria, which were not observed during the previous reactor conditions (**Fig. 6.5 c1**). The EDX analysis revealed deposits of Se, S, and Ni at 6, 28, and 7% on a mass basis, respectively (**Fig. 6.5 c2**). The P-XRD spectrum confirmed a mixture of Ni_3S_4 (JCPDS card no. 00-043-

1469) and SeS mineral deposits on the biomass (**Fig. 6.6c**). When SO_4^{2-} was excluded from the feed solution (period RC2d), the EDX spectra of the biofilm showed a low Se and Ni elemental mass percentage at 6 and 2%, respectively, with a higher elemental P mass percentage of 14% (**Fig. 6.5 d2**). Also in this case, a peak broadening was observed in the P-XRD spectrum due to the mainly amorphous nature of the precipitates (**Fig. 6.6d**).

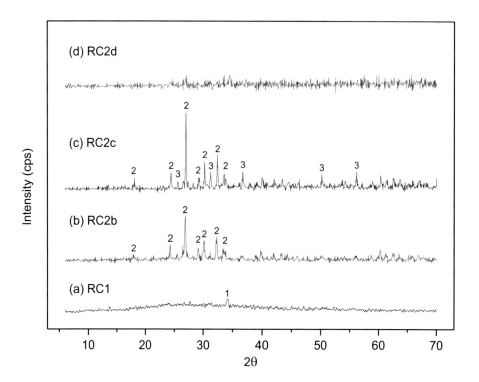

Fig. 6.6 P-XRD spectra of the BTF biofilm samples taken during (a) phase RC1 and periods (b) RC2b, (c) RC2c and (d) RC2d. The identified minerals are marked with the following peak signatures according to the standard database JCPDS: 1 - Se (03-065-3404), 2 - SeS (00-002-0320) and 3 - Ni_3S_4 (00-043-1469).

6.3.7 Biochemical and EEM fluorescence analysis of extracted EPS

Table 6.5 summarizes the biochemical characteristics and composition of the EPS extracted from the inoculum and biomass taken from the BTF and UASB reactor under different reactor conditions. The EPS was mainly made up of polysaccharides (PS), proteins (PN) and humic-like substances (HL-S).

The total EPS concentration for the BTF biofilm was 45 to 63% higher than the inoculum with the highest EPS concentration of 126.0 ($\pm$ 1.4) mg TOC/g VS for biofilms grown during the SO_4^{2-} reducing conditions (period RC2b). The EPS of the BTF biofilm showed the increasing content of PS with the addition of SO_4^{2-} and Ni^{2+} by 40% and 70%, respectively, when compared to biofilms fed with SeO_4^{2-} only. Similarly, the HL-S content increased by > 80% when both SeO_4^{2-} and SO_4^{2-} were supplemented. Opposite to this, the PN was 28 and 46% lower upon addition of SO_4^{2-} and Ni^{2+}, respectively, when compared to 86.0 ($\pm$ 4.2) mg PN/g VS achieved during RC1. The total Se content in the extracted EPS of the BTF biofilm was ~ 1.4 mg/g VS at phase RC1 and period RC2b and increased to ~11.8 mg/g VS when Ni^{2+} was fed into the BTF reactor (period RC2c-d). The total Ni content in the extracted EPS of the BTF biofilm was in the range 4 to 7 mg Ni/g VS.

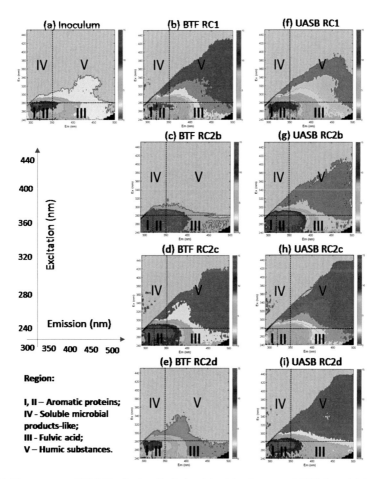

Fig. 6.7 3D Fluorescence EEM (Excitation-Emission Matrix) of extracted EPS from (a) inoculum and reactor samples of the (b-e) BTF and (f-i) UASB operated under different operating conditions.

Fluorescence EEM are divided in 5 regions at excitation/emission wavelengths (Ex/Em): I and II - 280/350 nm derived from proteins (tyrosine and tryptophan) corresponding to aromatic proteins; III - 280/500 nm for fulvic-like acids; IV - <280/350 nm for soluble microbial by-product like (SMP) substances and; V - <280/<350 nm for humic-like substances. The TOC concentration for all samples was adjusted to 0.5 mg/L prior to analysis.

The total EPS concentration of the UASB granules was 25-50% higher than that observed in the inoculum, with the exception of the period RC2c period when Ni^{2+} was initially added. The biochemical composition of the EPS for the UASB granules showed a similar pattern as the BTF biofilm. The PS and HL-S content increased up to ~25 and ~8 mg/g VS, respectively, in the presence of both SeO_4^{2-} and SO_4^{2-}. On the other hand, PN decreased from ~64 (phase RC1) to ~26 mg/g VS when the UASB granules were fed with also SO_4^{2-}. The total Se content in the extracted EPS of the UASB granules was ~0.7 mg/g VS from phase RC1 to period RC2c, but significantly increased to 15.8 (± 1.0) mg/g VS during period RC2d. The total Ni content similarly increased from 0.5 (± 0.1) to 4.5 (± 0.3) mg/g VS from periods period RC2c to period RC2d.

3D fluorescence spectra of excitation/emission wavelengths (FEEM) were also determined identify possible protein-like and humic-like substances present in the extracted EPS (**Fig. 6.7**). The 3D spectra were divided into five different regions, corresponding to different compounds such as aromatic proteins (region I-II), fulvic acid (region III), soluble microbial products (region IV), and humic substances (region V) (Chen et al. 2003). The changes in biochemical composition of extracted EPS from the BTF biofilms and UASB biogranules were reflected in the 3D FEEM spectra as shown in **Fig. 6.7 b-e** and **f-i**, respectively. Reactor biomass fed with SeO_4^{2-} alone (**Fig. 6.7 b** and **f**) showed similar spectra as with the inoculum (**Fig. 6.7 a**). An increase in fluorescence intensity was observed for region I-II for both the reactor biomass at the end of the period RC2b (**Fig. 6.7 c** and **g**). Upon addition and further operation of the BTF reactor with Ni^{2+} (**Fig. 6.7 d** and **e**), fluorescence intensity for region I-II decreased. In contrast, the fluorescence intensity of region III for UASB biogranules initially fed with Ni^{2+} decreased dramatically (**Fig. 6.7 h**), but increased again by the end of the reactor run (**Fig. 6.7 i**). The fluorescence intensity in region III (fulvic acid) remained constant for all reactor conditions, while the fluorescence intensity for HL-S (region V) spread out during the whole phase RC2.

Table 6.5 Characterization of the EPS extracted from the BTF and UASB biofilms under different reactor conditions. The extracted EPS samples were analyzed for total organic carbon (TOC), polysaccharide (PS), total protein (PN), humic-like substances (HL-S), and metal contents (Se, Ni). The TOC concentration indicates the total EPS concentration in the extracted sample.

	Inoculum	BTF				UASB			
		RC1	RC2b	RC2c	RC2d	RC1	RC2b	RC2c	RC2d
Biochemical composition of EPS (mg/g VS)									
TOC	47.2 (± 0.4)	89.2 (± 1.4)	126.0 (± 1.4)	113.0 (± 13.8)	88.6 (± 3.4)	62.1 (± 0.8)	74.2 (± 2.5)	44.8 (± 0.1)	97.1 (± 1.0)
PS	10.8 (± 4.1)	8.6 (± 1.2)	14.3 (± 0.9)	35.4 (± 1.9)	31.4 (± 1.8)	14.9 (± 4.8)	21.5 (± 1.5)	8.3 (± 0.7)	25.0 (± 4.4)
PN	24.2 (± 7.7)	86.0 (± 4.2)	61.9 (± 5.5)	52.3 (± 2.9)	40.5 (± 1.7)	63.8 (± 7.1)	45.5 (± 2.1)	26.4 (± 0.1)	30.9 (± 4.3)
HL-S	1.4 (± 1.2)	1.4 (± 0.7)	9.2 (± 0.4)	14.6 (± 0.1)	8.0 (± 1.3)	1.0 (± 0.6)	3.8 (± 0.4)	1.8 (± 0.2)	7.8 (± 0.4)
PN/PS	2.2	9.9	4.3	1.5	1.3	4.3	2.1	3.2	1.2
PN/HL-S	17.5	59.9	6.7	3.6	5.0	67.0	11.9	14.8	3.9
Metals in EPS (mg/g VS)									
Se	n/a	1.9 (± 0.1)	0.8 (± 0.2)	11.9 (± 1.5)	11.7 (± 0.3)	0.9 (± 0.2)	0.3 (± 0.2)	1.0 (± 0.1)	15.8 (± 1.0)
Ni	n/a	n/a	n/a	7.1 (± 0.3)	4.6 (± 0.1)	n/a	n/a	0.5 (± 0.1)	4.5 (± 0.3)

6.4   Discussion

6.4.1   Influence of SO_4^{2-} on Se removal efficiency (phase RC1 vs. periods RC2a to RC2b)

This study showed that addition of SO_4^{2-} (periods RC2a to RC2b) to the feed solution had a significant effect on the Se removal performance of (**Fig. 6.2**) and biofilm growth in (**Table 6.3**) the BTF. Differences in the BTF removal performance between phase RC1 and periods RC2a to RC2b could be related to i) the increase in biofilm growth by 200% and ii) the different bioreduced Se species and mineral precipitates observed when SO_4^{2-} was present in the feed solution. Hockin and Gadd (2006) similarly observed that the net SeO_4^{2-} removal was more rapid with excess SO_4^{2-} (898 mg S/L) and resulted in a higher protein production using *Desulfomicrobium* sp. biofilms in serum bottles. A previous study (**chapter 4**) similarly showed that a biofilm grown with SeO_4^{2-} alone only led to < 30% SeO_4^{2-} removal and a thin biofilm layer (~0.2 mm) in comparison to a biofilm grown with both SO_4^{2-} and SeO_4^{2-}, which had three times higher biofilm thickness and SeO_4^{2-} removal efficiency.

During phase RC1, high signals of Se, likely Se^0, were mainly observed in the BTF biofilm (**Fig. 6.5 a1**), with about 30% of the Se_{tot} discharged (~2 mg Se/L) as unaccounted Se_{diss} at (**Fig. 6.3a**). In contrast to phase RC1, high signals of both S and Se (**Fig. 6.5 b2**), identified as SeS complexes (**Fig. 6.6b**), were observed during periods RC2a to RC2b along with a lower unaccounted Se_{diss} concentration (~0.3 mg Se/L). The unaccounted Se_{diss} could be related to soluble organo-Se species such as selenocystamine and selenocyanatoacetic acid molecules (Tan et al. 2016). Different studies have shown that both organo-Se and Se^0 nanoparticles have a strong antimicrobial effect that prevents bacterial colonization and biofilm formation. Shakibaie et al. (2015) and Zonaro et al. (2015) observed that biologically synthesized Se^0 nanoparticles have the ability to both completely inhibit the biofilm formation as well as eradicate an established biofilm at a concentration of 16 to 150 mg Se/L. Vercellino et al. (Vercellino et al. 2013a, 2013b) reported a reduced biofouling in reverse osmosis membranes by covalently attaching two organo-Se compounds (1% *w/w* concentration) to the membrane and feed spacer. The antibacterial ability of Se species has been linked to the formation of the superoxide radicals that can enter the bacterial membrane and disrupt the cell wall, causing eradication of the bacterial cells and preventing bacterial attachment and growth (Tran et al. 2017). It is, therefore, feasible that the presence of SO_4^{2-} and subsequent bioreduced S forms alter the metabolic reduction of SeO_4^{2-}, decreasing the formation of organo-Se and allowing for

SeS complex formation. While Se^0 nanoparticles possess antimicrobial activity (Shakibaie et al. 2015; Zonaro et al. 2015; Tran et al. 2017), no reports are available on the effect of biogenic Se-S minerals on microbial metabolism and biofilms.

In contrast to the BTF, a gradual increase of the Se removal performance was observed in the UASB reactor (**Fig. 6.2**), which can indicate the adaptation of the granular microbial community to Se. Lenz et al. (2008) observed an abrupt improvement of the SeO_4^{2-} removal efficiency after 58 days of UASB reactor operation, suggesting the gradual development of a specialized Se-converting microbial population. Additionally, the authors noted that a SO_4^{2-}/SeO_4^{2-} ratio of > 87 was optimal to achieve the highest SeO_4^{2-} removal efficiency. The SO_4^{2-}/SeO_4^{2-} ratio used in this study was 59, which was the optimal ratio observed in previous batch experiments using the same inoculum as this study (**chapter 3**).

Although biogenic Se^0 was also produced in the UASB reactor (**data not shown**), the Se removal performance increased over time showing that the produced Se^0 was not a significant factor for the inhibition of the granules. Zonaro et al. (2015) observed that a concentration > 150 mg Se/L of biogenic Se^0 nanoparticles would be required to eradicate an already formed *E. coli* and *S. aureus* biofilm in a 96-well microtiter plate. Since the UASB reactor was seeded with pre-formed anaerobic granular sludge, it is likely that the amount of biogenic Se^0 at the initial stage of UASB reactor operation was not enough to prevent further bacterial and granule growth. Furthermore, the location of the Se^0 deposits in the granules is a critical factor. Gonzalez-Gil et al. (2016) observed that Se oxyanion reduction occurred and Se^0 deposited in the distinct outer layer (~200 μm from the surface) of the anaerobic granular sludge. Additionally, though smaller Se^0 nanoparticles showed higher antimicrobial and eradication activity (Zonaro et al. 2015), they can also behave as seeds for further nanoparticles size increase through maturing. Therefore, it can be assumed that in this study the Se^0 nanoparticles likely deposited in the outer layer of the UASB granules, which over time led to the formation of bigger Se particles, thus resulting in a lower Se antimicrobial effect on the UASB granules compared to the BTF biofilm.

6.4.2 Reactor removal performance with increased COD supply (periods RC2a to RC2b)

An increase of COD/SO_4^{2-} ratio from 1.1 (period RC2a) to 2.8 (period RC2b) resulted in a higher SO_4^{2-} removal (**Fig. 6.4**). Velasco et al. (2008) and Najib et al. (2017) reported that a COD/SO_4^{2-} ratio of 2.5 was optimal in achieving the highest SO_4^{2-} removal efficiency and sulfide concentration. Similarly, Papirio et al. (2013) observed that a COD/SO$_4$$^{2-}$ ratio > 3.0 resulted in a 97% SO$_4$$^{2-}$ removal efficiency, as lactate was not completely oxidized to acetate, which can be inhibitory for SO$_4$$^{2-}$ reduction. In contrast, Lenz et al. (2008) did not observe an improvement on the SO$_4$$^{2-}$ removal despite the increase of the COD/SO$_4$$^{2-}$ ratio from 0.5 to 10, attributing this to the inhibition of SO$_4$$^{2-}$ reduction by SeO$_4$$^{2-}$.

The higher feed COD concentration resulted in a COD/SeO$_4$$^{2-}$ ratio increasing from 56 to 167, which had no influence on the Se removal efficiency (**Fig. 6.2**). This was likely because the complete reduction of SeO$_4$$^{2-}$ requires only a small amount of lactate, i.e. 0.5 moles of lactate per mole of SeO$_4$$^{2-}$ (Tan et al. 2016). Hageman et al. (2013) observed that an oversupply of lactate at 70 mM led to the inhibition of SeO$_4$$^{2-}$ reduction due to the accumulation of high propionate and acetate concentrations (1188 and 440 mg/L, respectively). However, this was not observed in this study despite propionate and acetate concentrations increased up to approximately 1500 and 1000 mg COD/L, respectively, in both the reactors (**Appendix 4, Fig. S6.1**).

6.4.3   Impact of Ni^{2+} addition on the oxyanion removal performance (periods RC2c to RC2d)

The main impact of Ni^{2+} on both reactor configurations was the sudden decrease of the oxyanion removal efficiency by > 50% for SO$_4$$^{2-}$ (**Fig. 6.4**) and 20-30% for Se (**Fig. 6.2**). The SEM image of the BTF biofilm samples during period RC2c shows the presence of short rod shaped bacteria (**Fig. 6.5 c1**), which could be the result of Ni^{2+} stress on the biofilm/granules. Stress response caused by metals can induce adaptive mechanisms and changes in the metabolic activity and cell morphology of the microorganisms (Prabhakaran et al. 2016) or the emergence of persistent cells of slow-growing or non-growing sub-populations (Koechler et al. 2015). Similarly, Zou et al. (2015) evaluated the negative effects of Ni^{2+} in the presence of SO$_4$$^{2-}$ and NO$_3$$^-$ as concomitant oxyanions. Denitrification was alleviated when Ni^{2+} precipitated with the biologically produced HS$^-$ by SO$_4$$^{2-}$ reduction.

High Ni^{2+} removal efficiencies (> 90%) were achieved for the entire period RC2c (**Fig. 6.4**) and the formation of Ni-S precipitates was observed in the biomass by SEM-EDX (**Fig. 6.5**) and P-XRD (**Fig. 6.6**), confirming that Ni^{2+} was removed via HS$^-$ precipitation. A similar Ni^{2+} removal mechanism was observed by Sierra-Alvarez et al. (2006) and Najib et al. (2017) using anaerobic SO$_4$$^{2-}$ reducing sludge for the treatment of acid mine wastewater.

Upon removal of SO_4^{2-} from the feed solution, a 10% decrease of the Ni^{2+} removal efficiency was observed before reaching a final removal efficiency of 50% by the end of period RC2d in both reactors (**Fig, 6.4**). The use of anaerobic granular sludge for Se oxyanion reduction showed evidence of selenide formation, which could be used to form metal selenide when the aqueous heavy metal concentration equals or exceeds the available selenide (Mal et al. 2016). However, no Ni-Se minerals were observed in both reactor configurations: EDX showed only a low signal of detected Se and Ni, i.e. < 6% in terms of mass (**Fig. 6.5 d2**). It is likely that the low SeO_4^{2-} concentrations fed to the reactor resulted in lower selenide concentrations, which was not enough to form a Ni-Se precipitates or the Ni-Se concentration was below the detection limit of the P-XRD and SEM-EDX analysis.

A simulation by the Visual MINTEQ software (**data not shown**) confirmed that precipitation of Ni-Se minerals at the applied initial conditions was unlikely, but that Ni was more likely to precipitate with phosphate in the absence of HS^-. The increase in mass percentage of element P detected in the BTF biofilm during period RC2d compared to the other operational conditions (**Fig. 6.5 d2**) could be an indication of nickel phosphate precipitation. Pümpel et al. (2003) observed that nickel-phosphorus interactions highly contribute to microbially mediated Ni^{2+} sequestration in the treatment of Ni^{2+} plating wastewater. Biosorption of Ni^{2+} onto bacteria is another possible removal mechanism in the absence of biogenic HS^-, although Ni^{2+} has a low affinity for biosorption (Pümpel et al. 2003).

6.4.4 Complexation and mineral formation in the biofilm/granules

The BTF biofilm and UASB granules had similar mineral complexation identified by P-XRD as Se-S and Ni-S precipitates (**Fig, 6.6**). The formation of these minerals is an added value of using biological treatment for the remediation of mine drainage wastewater, since these minerals can be recovered and reused for further applications.

Other characterization studies on Se-S indicated that cyclic Se-S eight rings can have similar spectra (Geoffroy and Demopoulos 2011). P-XRD results observed from Geoffroy and Demopoulos (2011) showed similar spectra (identified as SeS_7) to those found in this work (identified to be SeS). While SeS_2 has been used for decades in the pharmaceutical industry, selenium monosulfide (SeS) has been classified as carcinogenic. Luo et al. (2014) demonstrated that the SeS_x composite with carbonized polyacrylonitrile is a promising cathode material for lithium ion batteries providing longer life and higher power density. It is known that SeO_3^{2-} can abiotically react with HS^- (Hageman et al. 2017). Elucidation of the formation mechanism of the Se-S minerals requires nevertheless further research. It is difficult to differentiate the percentage of abiotic reaction between the biologically produced SeO_3^{2-} and/or HS^- and biotic reduction occurring within the reactors under the applied operating conditions and time frame. The understanding of the Se-S bond would provide useful information during the reactor operation for the production and recovery of Se-S minerals. Mineral recovery from the biomass was not done in this study and is recommended to be explored in the future along with identifying the location of precipitates within the biofilm/granules for the development of improved recovery process applications.

For metal sulfide, Villa-Gomez et al. (2011) demonstrated that control of the biogenic sulfide concentration can steer the location of the metal precipitates for an easier recovery. The authors reported that at the highest HS^- concentration (i.e. 600 mg/L), the precipitated metals were fine particles located in the bulk liquid, while the lowest HS^- concentration (i.e. 0.26 mg/L) resulted in entrapment of the precipitated metals in the biofilm. Ni-S minerals are an important class of the metal sulfide family and can be reused in a wide variety of applications as supercapacitors and dye-sensitive solar cells. A study by Molla et al. (2016) demonstrated the use of synthesized Ni-S particles for rapid catalytic decomposition of organic dyes through the generation of reactive oxygen species (ROS). The Ni-S catalyst was reusable, scalable and capable of functioning for long periods of time without depletion.

6.4.5 EPS composition under different feed conditions

Biofilm EPS play a crucial role in the complexation and sequestration of metal ions (Koechler et al. 2015). EPS are composed by hydrated biomolecules where polysaccharides, proteins, and humic-like substances are the major components and have a significant influence on the physicochemical properties of sludge in biological wastewater treatment systems (Prabhakaran et al. 2016). Depending on the EPS components in the biofilm matrix, architectural changes and stability can be affected and functional efficiency such as adhesive properties, molecular exchange processes, matrix water content, charge and sorption properties can be influenced (Flemming and Wingender 2010). This research was able to investigate the composition changes of the biofilm EPS matrix under varying influent feed conditions. 3D FEEM spectra of the EPS extracted during phases RC1 and RC2 from the biofilms and biogranules showed a change in organic structure and components in the EPS due to the change in reactor operation. It is evident that the fluorescence intensity of the aromatic PN-like associated to tyrosine and/or tryptophan significantly decreased (region I-II, **Fig. 6.7**). Aromatic

PN has been reported to be important in maintaining the stable structure of biofilms, particularly in the case of granular sludge (Flemming and Wingender 2010). Additionally, the PN ratio with PS and HL-S for the extracted EPS from biofilms and biogranules decreased at increasing addition of contaminants.

The decrease in PN content, particularly when SeO_4^{2-} was present with co-contaminants, suggests the prominent role of PN in the metabolic reduction of SeO_4^{2-} and protection of the biofilm structure, integrity, properties, and function. An EPS study related to Se oxyanion reduction showed that PN was a major component involved in the capping of biogenic produced Se^0 nanoparticles (Jain et al. 2015; Gonzalez-Gil et al. 2016). Gonzalez-Gil et al. (2016) observed that smaller particles of Se^0 nanoparticles were observed at higher PN concentrations. The higher PN content found in the EPS of period RC1 biofilms/biogranules, particularly in the BTF biofilm, can imply the formation of smaller size Se^0 nanoparticles, which likely induced the antimicrobial activity observed during period RC1 and resulted in a lower Se removal (**Table 6.5**). Lampis et al. (2017) also observed that PNs can bind to Se^0 nanoparticles depending on their spatial configuration and physiochemical properties of the amino acid composition.

The EPS concentration of PS and HL-S components was higher when biomass was exposed to SO_4^{2-} and Ni^{2+} as compared to biomass fed with SeO_4^{2-} only, particularly for BTF biofilms (**Table 6.5**). Koechler et al. (2015) observed that the increase of the PS content in the EPS matrix occurred in the complex 3D architecture structure of the biofilm upon exposure of arsenic (III). Furthermore, the authors noticed that changing the PS content in the biofilm exposed to the metal ions probably led to an improvement of the metal trapping process. An increase in the Se content in the EPS was observed with increasing PS concentration along with the detected Ni^{2+} concentration in the extracted EPS (**Table 6.5**).

The exact reason for the change in EPS composition largely remains unknown, especially when a variety of contaminants is present. While EPS matrix of biofilm/biogranules includes moieties that are capable of binding various metal cations, interactions of the EPS composition and changes in the presence of oxyanions, bioreduced products, and metal are not clearly understood. It is therefore suggested that additional studies on the changes in biochemical composition and fingerprint of EPS upon feeding varying types of influent wastewater to the biofilms/biogranules and understanding EPS-mineral complex interactions need to be further conducted. Such information would be useful to develop bioreactor applications for wastewater treatment as well as give a better understanding on the engineering of nanoparticle biosynthesis.

6.5 Conclusions

This study demonstrated that two reactor configurations employing different biomass retention systems (biofilms in BTF and granules in UASB) result in a diverse removal performance when treating a synthetic mine drainage wastewater containing SeO_4^{2-}, SO_4^{2-} and Ni^{2+}. Supplementation of SO_4^{2-} resulted in increased bacterial colonization, biofilm formation and improved Se removal efficiency by > 200% in the BTF. In contrast, the UASB reactor showed a similar behavior in terms of Se removal efficiency in both the presence and absence of SO_4^{2-}. An increase of the COD/SO_4^{2-} ratio from 1.1 to 2.8 led to an almost complete SO_4^{2-} removal. When Ni^{2+} was supplemented, the elevated HS^- concentration (~170 mg S/L) resulted in a > 80% Ni^{2+} removal efficiency through Ni-S precipitation. Ni^{2+} initially affected SeO_4^{2-} and SO_4^{2-} removal by approximately 20-30% and > 50%, respectively, but then the processes recuperated, with the UASB reactor showing a faster recovery compared to the BTF. Formation of Se-S and Ni-S mineral deposits onto the biofilm and granules was observed, which gives an added value in using biological treatment techniques for recovery and reuse of biosynthesized minerals.

6.6 References

Association, A.P.H., Association A.W.W., Federation, W.E., 2005. Standard methods for examination of water and wastewater, 5th ed. American Public Health Association, Washington, DC, USA.

Cennamo, P., Marzano, C., Ciniglia, C., Pinto, G., Cappelletti, P., Caputo, P., Pollio, A., 2012. A survey of the algal flora of anthropogenic caves of campi flegrei (Naples, Italy) archeological district. *J. Cave Karst Stud.* 74, 243-250.

Cennamo, P., Montuori, N., Trojsi, G., Fatigati, G., Moretti, A., 2016. Biofilms in churches built in grottoes. *Sci. Total Environ.* 543, 727-738.

Chen, W., Westerhoff, P., Leenheer, J.A., Booksh, K., 2003. Fluorescence excitation - emission matrix regional integration to quantify spectra for dissolved organic matter. *Environ. Sci. Technol.* 37, 5701-5710.

Dubois, M., Gilles, K.A., Hamilton, J.K., Rebers, P.A., Smith, F., 1956. Colorimetric method for determination of sugars and related substances. Anal. Chem. 28, 350-356.

Ellis, D.R., Salt, D.E., 2003. Plants, selenium and human health. *Curr. Opin. Plant Biol.* 6, 273-279.

Flemming, H.-C., Wingender, J., 2010. The biofilm matrix. *Nat. Rev. Microbiol.* 8, 623-633.

Frølund, B., Griebe, T., Nielsen, P.H., 1995. Enzymatic activity in the activated sludge floc matrix. *Appl. Microbiol. Biotechnol.* 43, 755-761.

Frølund, B., Palmgren, R., Keiding, K., Nielsen, P.H., 1996. Extraction of extracellular polymers from activated sludge using cation exchange resin. *Water Res.* 30, 1749-1758.

Geoffroy, N., Demopoulos, G.P., 2011. The elimination of selenium (IV) from aqueous solution by precipitation with sodium sulfide. *J. Hazard. Mater.* 185, 148-154.

Gonzalez-Gil, G., Lens, P.N.L., Saikaly, P.E., 2016. Selenite reduction by anaerobic microbial aggregates: Microbial community structure, and proteins associated to the produced selenium spheres. *Front. Microbiol.* 7, 571.

Hageman, S.P.W., van der Weijden, R.D., Stams, A.J., van Cappellen, P., Buisman, C.J.N., 2017. Microbial selenium sulfide reduction for selenium recovery from wastewater. *J. Hazard. Mater.* 329, 110-119.

Hageman, S.P.W., van der Weijden, R.D., Weijma, J., Buisman, C.J.N., 2013. Microbiological selenate to selenite conversion for selenium removal. *Water Res.* 47, 2118-2128.

Hockin, S., Gadd, G.M., 2006. Removal of selenate from sulfate-containing media by sulfate-reducing bacterial biofilms. *Environ. Microbiol.* 8, 816-826.

Jain, R., Jordan, N., Weiss, S., Foerstendorf, H., Heim, K., Kacker, R., Hübner, R., Kramer, H., van Hullebusch, E.D., Farges, F., Lens, P.N.L., 2015. Extracellular polymeric substances govern the surface charge of biogenic elemental selenium nanoparticles. *Environ. Sci. Technol.* 49, 1713-1720.

Kiran, M.G., Pakshirajan, K., Das, G., 2017. An overview of sulfidogenic biological reactors for the simultaneous treatment of sulfate and heavy metal rich wastewater. *Chem. Eng. Sci.* 158, 606-620.

Koechler, S., Farasin, J., Cleiss-Arnold, J., Arsène-Ploetze, F., 2015. Toxic metal resistance in biofilms: Diversity of microbial responses and their evolution. *Res. Microbiol.* 166, 764-773.

Lampis, S., Zonaro, E., Bertolini, C., Cecconi, D., Monti, F., Micaroni, M., Turner, R.J., Butler, C.S., Vallini, G., 2017. Selenite biotransformation and detoxification by *Stenotrophomonas maltophilia* SeITE02: Novel clues on the route to bacterial biogenesis of selenium nanoparticles. *J. Hazard. Mater.* 324, 3-14.

Lenz, M., Lens, P.N.L., 2009. The essential toxin: the changing perception of selenium in environmental sciences. *Sci. Total Environ.* 407, 3620-3633.

Lenz, M., van Hullebusch, E.D., Hommes, G., Corvini, P.F.X., Lens, P.N.L., 2008. Selenate removal in methanogenic and sulfate-reducing upflow anaerobic sludge bed reactors. *Water Res.* 42, 2184-2194.

Luo, C., Zhu, Y., Wen, Y., Wang, J., Wang, C., 2014. Carbonized polyacrylonitrile-stabilized SeS$_x$ cathodes for long cycle life and high power density lithium ion batteries. *Adv. Funct. Mater.* 24, 4082-4089.

Mal, J., Nancharaiah, Y.V., van Hullebusch, E.D., Lens, P.N.L., 2016. Effect of heavy metal co-contaminants on selenite bioreduction by anaerobic granular sludge. *Bioresour. Technol.* 206, 1-8.

Molla, A., Sahu, M., Hussain, S., 2016. Synthesis of tunable band gap semiconductor nickel sulphide nanoparticles: Rapid and round the clock degradation of organic dyes. *Sci. Rep.* 6, 26034.

Najib, T., Solgi, M., Farazmand, A., Heydarian, S.M., Nasernejad, B., 2017. Optimization of sulfate removal by sulfate reducing bacteria using response surface methodology and heavy metal removal in a sulfidogenic UASB reactor. *J. Environ. Chem. Eng.* 5, 3256-3265.

Papirio, G., Esposito, G., Pirozzi, F., 2013. Biological inverse fluidized-bed reactors for the treatment of low pH- and sulphate-containing wastewaters under different COD/SO$_4^{2-}$ conditions. *Environ. Technol.* 34, 1141-1149.

Prabhakaran, P., Ashraf, M.A., Agma, W.S., 2016. Microbial stress response to heavy metal in the environment. *RSC Adv.* doi: 10.1039/C6RA10966G.

Pümpel, T., Macaskie, L.E., Finlay, J.A., Diels, L., Tsezos, M., 2003. Nickel removal from nickel plating waste water using a biologically active moving-bed sand filter. *BioMetals* 16, 567-581.

Roest, K., Heilig, H.G.H.., Smidt, H., De Vos, W.M., Stams, A.J.M., Akkermans, A.D.L., 2005. Community analysis of a full-scale anaerobic bioreactor treating paper mill wastewater. *Syst. Appl. Microbiol.* 28, 175-185.

Shakibaie, M., Forootanfar, H., Golkari, Y., Mohammadi-khorsand, T., Shakibaie, M.R., 2015. Anti-biofilm activity of biogenic selenium nanoparticles and selenium dioxide against clinical isolates of *Staphylococcus aureus*, *Pseudomonas aeruginosa*, and *Proteus mirabilis*. *J. Trace Elem. Med. Biol.* 29, 235-241.

Sierra-Alvarez, R., Karri, S., Freeman, S., Field, J.A., 2006. Biological treatment of heavy metals in acid mine drainage using sulfate reducing bioreactors. *Water Sci. Technol.* 54, 179-185.

Tan, L.C., Nancharaiah, Y.V., van Hullebusch, E.D., Lens, P.N.L., 2016. Selenium: Environmental significance, pollution, and biological treatment technologies. *Biotechnol. Adv.* 34, 886-907.

Tran, P.L., Huynh, E., Pham, P., Lacky, B., Jarvis, C., Mosley, T., Hamood, A.N., Hanes, R., Reid, T., 2017. Organoselenium polymer inhibits biofilm formation in polypropylene contact lens case material. *Eye Contact Lens* 43, 1-6.

186

Velasco, A., Ramírez, M., Volke-Sepúlveda, T., González-Sánchez, A., Revah, S., 2008. Evaluation of feed COD/sulfate ratio as a control criterion for the biological hydrogen sulfide production and lead precipitation. *J. Hazard. Mater.* 151, 407-413.

Vercellino, T., Morse, A., Tran, P., Hamood, A., Reid, T., Song, L., Moseley, T., 2013a. The use of covalently attached organo-selenium to inhibit *S. aureus* and *E. coli* biofilms on RO membranes and feed spacers. *Desalination* 317, 142-151.

Vercellino, T., Morse, A., Tran, P., Song, L., Hamood, A., Reid, T., Moseley, T., 2013b. Attachment of organo-selenium to polyamide composite reverse osmosis membranes to inhibit biofilm formation of *S. aureus* and *E. coli*. *Desalination* 309, 291-295.

Villa-Gomez, D., Ababneh, H., Papirio, S., Rousseau, D.P.L., Lens, P.N.L., 2011. Effect of sulfide concentration on the location of the metal precipitates in inversed fluidized bed reactors. *J. Hazard. Mater.* 192, 200-207.

Wang, C., Li, S., Wang, H., Fu, J., 2016. Selenium minerals and the recovery of selenium from copper refinery anode slimes. *J. South. African Inst. Min. Metall.* 116, 593-600.

Wen, H., Carignan, J., 2007. Reviews on atmospheric selenium: Emissions, speciation and fate. *Atmos. Environ.* 41, 7151-7165.

Zonaro, E., Lampis, S., Turner, R.J., Junaid, S., Vallini, G., 2015. Biogenic selenium and tellurium nanoparticles synthesized by environmental microbial isolates efficaciously inhibit bacterial planktonic cultures and biofilms. *Front. Microbiol.* 6, 658.

Zou, G., Papirio, S., van Hullebusch, E.D., Puhakka, J.A., 2015. Fluidized-bed denitrification of mining water tolerates high nickel concentrations. *Bioresour. Technol.* 179, 284-290.

CHAPTER 7

Amberlite® IRA-900 ion exchange resin for the sorption of selenate and sulfate: Equilibrium, kinetic and regeneration studies

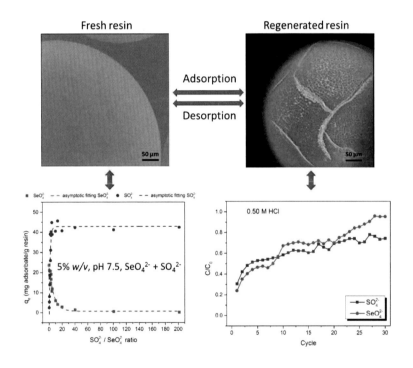

This chapter has been published in modified form:

Tan, L.C., Calix, E.M., Rene, E.R., Nancharaiah, Y.V., van Hullebusch, E.D., Lens, P.N.L. 2018. Adsorption and desorption of selenate and sulfate using Amberlite® IRA-900 ion exchange resin. *J. Environ. Eng.* doi:10.1061/(ASCE)EE.1943-7870.0001453.

Abstract

A synthetic strong base anion exchange resin Amberlite® IRA-900 (IRA-900) was investigated for its adsorption capacity to remove selenate (SeO_4^{2-}) and sulfate (SO_4^{2-}) from single and binary anion solutions at pH 7.5 and 20°C. At an initial SeO_4^{2-} and SO_4^{2-} concentration of 428.9 and 1441.1 mg/L, respectively, IRA-900 adsorbed 85 ($\pm$ 1)% SeO_4^{2-} and 75 ($\pm$ 5)% SO_4^{2-} from single anion solutions at 2% (w/v) resin dosage. In binary anion solutions, a 20% decrease for both the SeO_4^{2-} and SO_4^{2-} adsorption efficiency was observed. Batch kinetic experimental data indicated that the adsorption rate of IRA-900 for both SeO_4^{2-} and SO_4^{2-} in the single and binary anion solutions fitted well to the pseudo-second order kinetic model (R^2 = 0.99). For the single anion solutions, both SeO_4^{2-} and SO_4^{2-} adsorbed as monolayers onto the adsorbent, as evidenced from the Langmuir model fit (R^2 = 0.99). For the binary anion solutions, the modified Langmuir isotherm complete competition model fitted the experimental data well, showing 24% and 10% average relative error from theoretical calculations for SeO_4^{2-} and SO_4^{2-}, respectively. Exhausted resin was regenerated for the next cycle of adsorption using either 0.25 or 0.5 M HCl for 20 min. The regenerated resin was tested for 30 adsorption-desorption cycles. Results showed that resin was suitable for reuse at an optimal adsorption-desorption of 6 cycles without compromising the removal efficiencies. Efficient and fast adsorption of SeO_4^{2-} and SO_4^{2-} is promising for developing an IRA-900 based ion exchange process for remediation of Se-contaminated (waste)waters.

Keywords: selenate; sulfate; adsorption; binary solution; resin regeneration; Amberlite® IRA-900 resin

7.1 Introduction

Selenium (Se) is both an essential trace element needed for all the living organisms and a toxic element (Wiramanaden et al. 2010; Latorre et al. 2013). Se is typically found in lower concentrations (< 10 mg Se/L) in wastewaters as compared to other co-existing contaminants. But Se has the propensity to accumulate in the environment over time and reach 1.5 to 6-fold increase in concentration (Lemly 2004). A recent case of Se contamination caused by wastewater discharge from a coal-fired power plant into Lake Sutton (North Carolina, US) reported teratogenic effects and an increase in the mortality rate in fishes living in the lake (Lemly 2014). This in turn lead to an annual monetary loss and livelihood damages amounting to $8.6 million. As such, Se in wastewater must be removed before discharge to prevent environmental impacts on the surrounding wildlife, accumulation of Se entering the food-chain and ultimately humans. A strict freshwater aquatic discharge limit of 5 µg Se/L was imposed by the United States Environmental Protection Agency (USEPA) as a chronic aquatic life criterion (Santos et al. 2015).

Se has different forms and among them, the water-soluble oxyanion selenate (SeO_4^{2-}) is the most difficult to remove (Awual et al. 2014; Santos et al. 2015) and is often found in effluents from coal mining (Johansson et al. 2015) and flue-gas desulfurization (Staicu et al. 2017). Over the last few decades, various technologies and approaches have been developed to treat Se-laden wastewater such as membrane filtration, ferrihydrite adsorption and biological reduction (Tan et al. 2016; Staicu et al. 2017). Adsorption through ion exchange (IX) resins is a relatively low-cost method, simple and able to achieve rapid removal of the pollutants using simple operational conditions (Mac Namara et al. 2015). However, IX processes have seen little success for SeO_4^{2-} removal due to competition with sulfate (SO_4^{2-}) anions, which are present at one to two orders of magnitude higher than the SeO_4^{2-} concentration, quickly saturating the resins and causing low SeO_4^{2-} removal efficiencies (NSMP 2007; Staicu et al. 2017).

IX processes reversibly exchange the adsorbate (ion contaminant of interest) for a similarly charged ion that is attached to an immobilized resin on a solid surface material resin, e.g. synthetic or natural zeolites. Different kinds of both natural adsorbents and synthetic resins are available for removing Se oxyanions (Santos et al. 2015). The use of commercially available organic synthetic resins is a well-understood technology, where strong basic anion (SBA), weak basic anion (WBA) or chelating resins are employed for Se removal from polluted water. Synthetic resins can be anionic or cationic resins and can be regenerated by simple backwashing with an acid or alkaline solution (Erosa et al. 2009). SBA type materials were identified as the best candidate for anion removal due to their fixed positively-charged groups, can act as anion exchangers in wider pH ranges, and can withstand to at least 100°C (Mac Namara et al. 2015). SBA functionalization is generally based on fixed quaternary amines, i.e. ammonium cations, which promote an electrical attraction between the adsorbate and adsorbent surface following an IX mechanism (Clifford and Weber Jr. 1983).

Studies on the use of synthetic resins for SeO_4^{2-} removal have shown to be inhibited by SO_4^{2-}. As such, pre-treatment of Se-laden wastewater with high concentrations of SO_4^{2-} is always recommended (Staicu et al. 2017). This pre-treatment for removing SO_4^{2-} can be done through precipitation via barium chloride or lead (II) nitrate addition (Kijjanapanich et al. 2014). Furthermore, few studies have attempted to model the adsorption behavior of synthetic resins in binary anion solutions. Therefore, the main aim of this study was to optimize the batch operational parameters for the simultaneous removal of SeO_4^{2-} and SO_4^{2-} using Amberlite® IRA-900. A series of batch experiments were performed in single and binary anion solutions to i) find optimum parameters for removing SeO_4^{2-} and SO_4^{2-} and ii) understand the sorption mechanism(s) responsible for SeO_4^{2-} and SO_4^{2-} removal. Furthermore, a regeneration procedure was adopted for desorbing SeO_4^{2-} and SO_4^{2-} from exhausted resin and evaluated the resin reusability during multiple adsorption-desorption cycles.

7.2   Material and methods

7.2.1   IX resin Amberlite® IRA-900

The commercial IX resin Amberlite® IRA-900 Cl, hereafter referred to as IRA-900, was used as the main adsorbent in this study. It was procured from Sigma-Aldrich (Milan, Italy) and used without any pre-treatment. The main characteristics of this IRA-900 resin are shown in **Table 7.1**.

Table 7.1 Characteristics of IRA-900 (source: www.sigmaaldrich.com).

Parameters	Characteristics
Type	Macroreticular polystyrene type 1 strong base exchange resin
Physical form	Tan spherical beads
Matrix	Styrene divinylbenzene copolymer
Functional group	Trimethyl ammonium
Total exchange capacity	1.0 meq/mL by wetted bed volume
Moisture holding capacity	58 to 64%
Particle size	650 to 820 µm
Operating pH	0 to 13

7.2.2   Adsorption experiments

A preliminary screening of different commercially available resins was done to compare the adsorption capacity of IRA-900. Three SBA resins (Dowex 21K XLT, Dowex Marathon A2, IRA-900), one WBA resin (Amberlite® IRA-96) and granular activated carbon were screened. Batch adsorption studies were carried out in 100 mL flasks, 2% (w/v) resin dosage, at pH 7.5, and agitated at 100 rpm on an orbital shaker (INNOVA 2100, New Brunswick Scientific, New Jersey, USA). All experiments were carried out in duplicates. SeO_4^{2-} and SO_4^{2-} were either present alone (single anion) or as a mixture (binary anion) in the solutions, supplied as sodium selenate (Sigma-Aldrich) and potassium sulfate (Ensure), respectively.

Adsorption experiments comprised of 7 different process tests in ultra-pure water (Milli-Q water, 18MΩ cm) to investigate the effect of different process parameters: (i) temperature; (ii) adsorbent dose; (iii) contact time; (iv) initial concentration of SeO_4^{2-} and SO_4^{2-} in the single anion solutions; and (v) initial concentrations of SeO_4^{2-} and SO_4^{2-} in the binary anion solutions.

The effect of temperature on the adsorption capacity was tested at 20 and 30°C, respectively, while varying masses of the adsorbent dose between 0.5 and 10% (w/v). The effect of contact time and oxyanion competition were tested for a period of 24 h using 2% (w/v) resin dosage in a 100 mL solution of (i) SeO_4^{2-} alone, (ii) SO_4^{2-} alone, and (iii) $SeO_4^{2-} + SO_4^{2-}$ at 20°C. Samples were withdrawn at regular time intervals from 0-24 h and analyzed for residual SeO_4^{2-} and SO_4^{2-} concentrations. The effect of the initial SeO_4^{2-} or SO_4^{2-} concentration in single and binary anion solutions was tested by varying their initial concentrations from low to high values. The initial SeO_4^{2-} concentration was varied as follows: 7.1, 14.4, 35.7, 71.5, 107.2, 143.0, 286.9, 428.9 and 714.8 mg/L with SeO_4^{2}. The initial SO_4^{2-} concentration was varied as follows: 96.1, 192.1, 288.2, 480.4, 672.5, 960.7, 1152.8, 1441.1 and 1921.4 mg/L with SO_4^{2-}. In binary anion solutions, a set of experiments varied the initial SO_4^{2-} concentration from 96.1-1921.4 mg/L with co-exposure of SeO_4^{2-} at 428.9 mg/L and another set of experiments varied initial SeO_4^{2-} concentration from 7.1-714.8 mg/L with co-exposure of SO_4^{2-} at 1441.1 mg/L.

7.2.3   Resin regeneration

Resin regeneration tests were performed using 0.25 and 0.50 M HCl as the regenerants. Regeneration was performed after adsorption was carried out for 7 h, at resin dosage 5% (w/v), pH 7.5, temperature 20°C, and 100 rpm. After the adsorption step, the spent solution was removed and the resin was washed with ultra-pure water twice before adding either 0.25 or 0.50 M HCl for the elution of adsorbates. Regeneration with HCl was conducted for 60 min at sampling intervals of 0, 5, 10, 20, 30, 40, 50 and 60 min to determine the desorption profile. After determining the optimum desorption time, the reusability of the resin was investigated by conducting 30 adsorption-desorption batch cycles, i.e. 4 h of adsorption phase followed by 20 min regeneration, in single and binary anion solutions using 0.25 or 0.50 M HCl as the regenerant solution.

7.2.4 Calculations

The adsorption uptake capacity, i.e. the amount of oxyanion adsorbed onto the adsorbent, and the efficiency under equilibrium conditions were calculated using Eqs. (7.1) and (7.2), respectively:

$$q_e = \frac{(C_0 - C_e) \times V}{m} \qquad\qquad \text{Eq. (7.1)}$$

$$Adsorption\ efficiency\ (\%) = \frac{C_0 - C_e}{C_0} \qquad\qquad \text{Eq. (7.2)}$$

Where: q_e = amount adsorbed on the adsorbent at equilibrium (mg/g), C_0 = initial concentration of oxyanion (mg/L), C_e = equilibrium concentration of oxyanion in the solution (mg/L), V = volume of the solution (L) and m = mass of the adsorbent (g).

In order to determine the adsorption kinetics of SeO_4^{2-} and SO_4^{2-} onto IRA-900, the experimental data were fitted to the pseudo-first order, pseudo-second order and Elovich kinetic models. The transformed linear forms for each kinetic model are shown in Eqs. (7.3)-(7.5), respectively (Espinosa-Ortiz et al. 2016; Orakwue et al. 2016):

$$\ln(q_e - q_t) = \ln(q_e) - k_1 t \qquad\qquad \text{Eq. (7.3)}$$

$$\frac{t}{q_t} = \frac{1}{k_2 q_e^2} + \frac{t}{q_e} \qquad\qquad \text{Eq. (7.4)}$$

$$q_t = \frac{1}{\beta}\ln(\alpha\beta) + \frac{1}{\beta}\ln(t) \qquad\qquad \text{Eq. (7.5)}$$

Where: k_1 = rate constant of pseudo-first order (1/h), k_2 = rate constant of pseudo-second order (g/mg/h), t = time (h), q_e = amount adsorbed in the adsorbent at equilibrium (mg/g), q_t = adsorption at any given point of time (mg/g), β_E = Elovich desorption constant (g/mg) and α_E = Elovich initial adsorption rate (mg/g/h).

194

The transport of adsorbate from a liquid medium to the surface of the adsorbent through diffusion was also determined by fitting the experimental data to the intra-particle (Eq. 7.6) and liquid film (Eq. 7.7) diffusion models (Sheha and El-Shazly 2010):

$$q_t = K_{id}(t)^{0.5} + C \qquad\qquad \text{Eq. (7.6)}$$

$$\log\left(1 - \frac{q_t}{q_e}\right) = -\frac{K_{fd}}{2.303}t \qquad\qquad \text{Eq. (7.7)}$$

Where: K_{id} = intra-particle diffusion rate constant (mg/g/h$^{0.5}$), C empirical constant for boundary layer effect (mg/g) and K_{fd} = liquid film diffusion rate constant (1/h).

The experimental data from single anion solutions at different initial SeO_4^{2-} and SO_4^{2-} concentrations were fitted to four of the commonly used isotherm models, namely the Langmuir (Eq. 7.8), Freundlich (Eq. 7.9), Redlich-Peterson (Eq. 7.10) and Dubinin-Radushkevich (Eq. 7.11) isotherms to estimate the adsorption behavior of SeO_4^{2-} and SO_4^{2-} to IRA-900 (Allen et al. 2004; Foo and Hameed 2010). An essential characteristic of the Langmuir isotherm is the dimensionless constant R_L known as the separation factor and can be calculated as $R_L = 1/(1 + (K_L \times C_0))$. The nature of the adsorption process can be assumed favorable when $0 < R_L < 1$, while $R_L > 1$ indicates an unfavorable process (Foo and Hameed 2010). Non-linear regression analysis was carried out using OriginPro 9.0 software (OriginLab, USA).

$$q_e = \frac{q_m K_L C_e}{1 + K_L C_e} \qquad\qquad \text{Eq. (7.8)}$$

$$q_e = K_F (C_e)^{\frac{1}{n}} \qquad\qquad \text{Eq. (7.9)}$$

$$q_e = \frac{K_R C_e}{1 + \alpha_R (C_e)^{\beta_R}} \qquad\qquad \text{Eq. (7.10)}$$

$$q_e = (q_s) \exp\left(-K_{DR}\left(RT\left(\ln\left(1 + \frac{1}{C_e}\right)\right)\right)^2\right) \qquad\qquad \text{Eq. (7.11)}$$

Where q_m = Langmuir maximum monolayer coverage capacities (mg/g), K_L = Langmuir isotherm constant (L/mg), K_F = Freundlich isotherm constant (mg/g(mg/L)$^{-1/n}$) related to the adsorption capacity, n = adsorption intensity or surface heterogeneity ($1/n > 1$ indicates cooperative adsorption), K_R = Redlich-Peterson isotherm constant (L/mg), α_R = Redlich-Peterson isotherm constant (1/mg), β_R = Redlich-Peterson isotherm exponent, q_s = theoretical isotherm saturation capacity (mg/g), K_{DR} =

Dubinin-Radushkevich isotherm constant (mol^2/kJ^2), R = universal gas constant (8.314 J/mol/K) and, T = temperature (K).

The experimental data was also fitted to the Modified Langmuir isotherm complete competition model (CC-Modified Langmuir) and Ideal Adsorption Solution Theory Freundlich isotherm (IAST-Freundlich) model to observe the effect of adsorption in the binary anion solutions using equations Eqs. (7.12) and (7.13), respectively (Allen et al. 2004; Baig et al. 2009). The parameters calculated from the Langmuir and Freundlich single anion solutions (Eqs. 7.8 and 7.9) for each adsorbate and $q_{e,\ actual}$ or $C_{e,\ actual}$ values obtained from the experiments were used to predict the $q_{e,\ cal}$ or $C_{e,\ cal}$ for the binary system isotherms using Eqs. (7.12) and (7.13). The calculated values were compared to the actual $q_{e,\ actual}$ or $C_{e,\ actual}$ obtained from the experiments.

$$q_{e,i} = \frac{q_{m,i}K_{L,i}C_{e,i}}{1+\sum_{j=1}^{n}K_{L,j}C_{e,j}}$$
Eq. (7.12)

$$C_{e,i} = \frac{q_{e,i}}{\sum_{j=1}^{n}q_{e,j}}\left(\frac{\sum_{i=1}^{n}n_jq_{e,j}}{n_iK_{F,i}}\right)$$
Eq. (7.13)

An averaged relative error (ARE) was used to evaluate the goodness of fit between theoretical calculated data and the experimental data obtained using Eq. (7.14) (Foo and Hameed 2010).

$$Average\ relative\ error\ (\%) = \frac{100}{n}\sum_{i=1}^{n}\left|\frac{q_{e,actual}-q_{e,cal}}{q_{e,actual}}\right|$$
Eq. (7.14)

Desorption efficiencies of the regenerant solutions for SeO_4^{2-} and SO_4^{2-} desorption were calculated using Eq. (7.15):

$$Desorption\ efficiency\ (\%) = 1 - \frac{amount\ of\ anion\ left\ bound\ onto\ the\ resin}{amount\ anion\ initial\ bound\ onto\ resin}$$
Eq. (7.15)

7.2.5 Analytical methods

SO_4^{2-} and SeO_4^{2-} concentrations were measured by ICS-1100 (Dionex, Thermo Fisher Scientific Inc., USA) using an IonPac® AS4A-SC (solvent compatible) carbonate-selective anion exchange analytical column. The mobile phase solution was 1.7 mM $NaHCO_3$ and 1.8 mM Na_2CO_3 at a flow rate of 0.5 mL/min. Prior to the analysis, liquid samples were filtered through 0.45 µm cellulose acetate syringe filters (Sigma-Aldrich, USA) to remove any particulate matter. Surface morphology of fresh resin and after regeneration with HCl was analyzed using a scanning electron microscope (SEM, JEOL JSM6010LA) equipped with energy dispersive X-ray spectroscopy (EDS). Resin samples were dried at room temperature before analysis.

7.3 Results

7.3.1 Preliminary screening of IX resins

Fig. 7.1A shows the oxyanion adsorption efficiencies of each resin after 7 h of contact time at 20°C, pH 7.5, in binary anion solutions, and 2% (w/v) resin dosage. All the tested SBA resins, i.e. IRA-900, 21K XLT, Marathon A2 and IRA-400, showed similar results with an average adsorption efficiency of 69 (± 3)% for SeO_4^{2-} and 62 (± 2)% for SO_4^{2-}. On the other hand, Amberlite® IRA-96 (WBA resin) and GAC showed low adsorption capability for both oxyanions. Although all the tested SBA resins showed a similar adsorption performance, the IRA-900 resin is cheaper at €0.17/g as compared to the other SBAs (€0.45/g for 21K XLT and €0.22/g for both Marathon A2 and IRA-400). Therefore, IRA-900 was chosen as the resin for further batch adsorption tests for determining optimum conditions, develop a regeneration procedure and evaluate reusability.

The effect of the incubation temperature (20°C or 30°C) on SeO_4^{2-} and SO_4^{2-} adsorption by IRA-900 is shown in **Fig. 7.1B**. IRA-900 showed no marked change in the adsorption performance with an increase in temperature, attaining a q_e value after 7 h adsorption of 10.1 mg SeO_4^{2-}/g resin (64% adsorption efficiency) and 39.4 mg SO_4^{2-}/g resin (62% adsorption efficiency). Thus, a temperature of 20°C was used for the subsequent experiments. In order to select the optimal resin to solution ratio, the initial amount of resin was varied from 0.5% to 10.0% (w/v). **Fig. 7.1C** shows the adsorption efficiency with varying resin weight %. An increase in the resin dose from 0.5% to 5% (w/v) resulted in an exponential increase in the adsorption efficiency for SeO_4^{2-} (29 ± 1% to 75 ± 1%) and SO_4^{2-} (20 ± 1% to 68 ± 1%). The increase in adsorption percentage with increased resin weight was attributed to the availability of more sorption sites. However, an increase of resin dose beyond 5% (w/v) showed a minimal increase in the SeO_4^{2-} and SO_4^{2-} adsorption efficiencies.

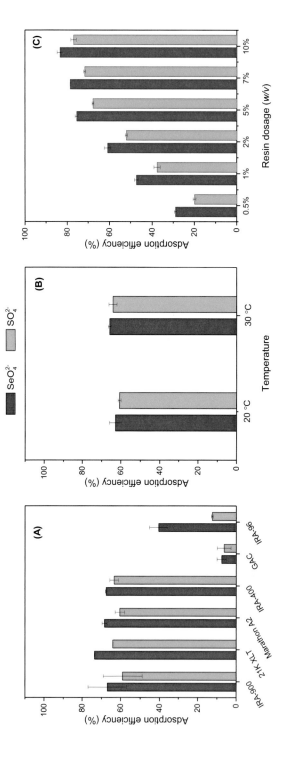

Fig. 7.1 (A) Comparison of IRA-900 adsorption efficiency with other adsorbents at 2% (w/v) resin dosage, (B) effect of temperature on IRA-900 adsorption performance in ultra-pure water containing SeO_4^{2-} + SO_4^{2-} at 2% (w/v) resin dosage, and (C) adsorption efficiency of SeO_4^{2-} and SO_4^{2-} with varying resin dosage. All experiments were conducted in binary anion solutions.

7.3.2 Effect of contact time and adsorption kinetics

The effect of contact time on the adsorption of SeO_4^{2-} and SO_4^{2-} by IRA-900 was studied for a period of 24 h to determine the time required to attain equilibrium. Additional changes in SeO_4^{2-} and SO_4^{2-} adsorption in the single and binary anion solutions were also determined. **Fig. 7.2A** shows the uptake of SeO_4^{2-} and SO_4^{2-} (q_t) at different contact times.

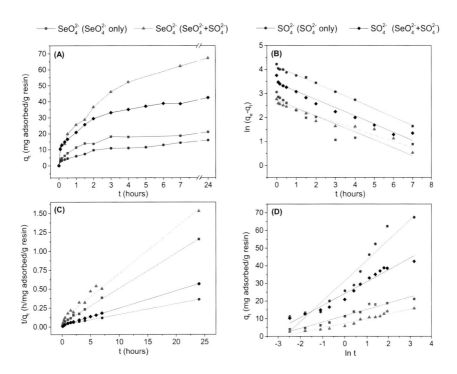

Fig. 7.2 Adsorption kinetics of SeO_4^{2-} and SO_4^{2-} on IRA-900 in different solutions (SeO_4^{2-} only, SO_4^{2-} only, and $SeO_4^{2-}+SO_4^{2-}$) showing (A) q_t as a function of contact time, (B) pseudo-first order, (C) pseudo-second order, and (D) Elovich kinetic model fit at an initial concentration of 428.9 mg/L SeO_4^{2-} and 1441.1 mg/L SO_4^{2-} at pH 7.5, 20°C and an adsorbent dose of 2% (w/v). Data points represent the experimental data while the dashed lines are fitted lines for the kinetic models.

There was a steady increase in the q_t values until ~7 h, after that the adsorption uptake increased by only 5% in 24 h, showing that equilibrium was reached after 7 h for both the single and binary anion solutions. At t = 7 h, the equilibrium adsorption amounts ($q_{e, actual}$) of 18.5 mg/L SeO_4^{2-} and 62.1 mg/L SO_4^{2-} from single anion solutions were 24% and 38% higher than those of SeO_4^{2-} and SO_4^{2-} $q_{e, actual}$ in the binary anion solutions at 14.0 mg/L SeO_4^{2-} and 38.6 mg/L SO_4^{2-}, respectively. After a contact time of 7 h, 87% SeO_4^{2-} and 80% SO_4^{2-} adsorption

199

efficiencies were reached in single anion solutions, while 67% SeO_4^{2-} and 59% SO_4^{2-} adsorption efficiencies were achieved in binary anion solutions. Based on these results, an equilibrium contact time of 7 h was selected for the subsequent experiments.

Fig. 7.2 B-D show the adsorption kinetic profiles for the pseudo-first order (B), pseudo-second order (C) and Elovich (D) models. Among the adsorption kinetic models, the pseudo-second order model was able to describe the adsorption process more precisely with high R^2 values (0.99). The linear plot of t/q_t versus t (Fig. 7.2C) was used to calculate the pseudo-second order kinetic constants, namely the q_e and k_2 as shown in Table 7.2. The pseudo-second order rate constant, k_2, was found to be higher for SeO_4^{2-} compared to SO_4^{2-} in both the single and binary anion solutions. A comparison between the $q_{e, cal}$ from the pseudo-second order kinetic model and $q_{e, actual}$ matches well for both SeO_4^{2-} and SO_4^{2-} in single and binary anion solutions. On the other hand, both the pseudo-first order and Elovich kinetic models showed lower correlation R^2 values and poor fitting between $q_{e, cal}$ and $q_{e, actual}$ (Appendix 5, Table S7.1). The $q_{e, actual}$ value for both SeO_4^{2-} and SO_4^{2-} in the binary anion solutions were ~0.7 times lower when compared to the values observed in single anion solutions.

The adsorption data for single and binary anion solutions were also fitted to the intra-particle (Eq. 7.6) and liquid film (Eq. 7.7) diffusion models to understand the adsorbate transport mechanism from the liquid media to the surface of the resin. Fig. S7.1 (Appendix 5) shows the plot for the two diffusion model, while Table 7.2 shows the estimated constant parameters. The experimental data for SeO_4^{2-} in both the single and binary anion solutions showed good correlation to both the intra-particle and liquid film diffusion models, while the SO_4^{2-} experimental data fitted only to the intra-particle diffusion model (Table 7.2). The two separate straight line fits of the intra-particle diffusion model indicate that mass transport (second straight line), apart from intra-particle diffusion (first straight line), also influenced the adsorption process (Appendix 5, Fig. S7.1A) for both oxyanions. The measure of boundary layer effects during intra-particle transport is represented by the intercept C_{id} (Table 7.2). Single anion solutions showed higher C_{id} values (~1.5 higher) compared to the binary anion solutions, indicating that the boundary layer effect is higher when SeO_4^{2-} and SO_4^{2-} are present individually in the solution. Similarly, the intra-particle diffusion rate constant (K_{id}) is significantly higher in the single anion solutions for SeO_4^{2-} and SO_4^{2-} by a factor of 1.7 and 2.7 times, respectively. The liquid film diffusion rate constant (K_{fd}) for SeO_4^{2-} showed minimal change between single and binary anion solutions (0.7-1.0/h). On the other hand, SO_4^{2-} showed higher (factor of 3.9) K_{fd} values in the single anion solutions.

200

Table 7.2 Pseudo-second order, intra-particle, and liquid film diffusion model constants at an initial concentration of 428.9 mg/L SeO_4^{2-} and 1441.1 mg/L SO_4^{2-} at 2% (w/v) resin dosage, 20°C, and pH 7.5.

Anion solutions	Oxyanion	Experimental (24 h)	Pseudo-second order kinetics			Diffusion model parameters					
						Intra-particle diffusion			Liquid film diffusion		
		$q_{e, actual}$	k_2	$q_{e, cal}$	R^2	K_{id}	C_{id} (intercept)	R^2	K_{fd}	C_{fd} (intercept)	R^2
		mg/g	mg/g/h	mg/g		mg/g/h$^{0.5}$	mg/g		1/h		
Single	SeO_4^{2-}	20.8	0.056	21.5	0.99	9.3	0.8	0.98	1	15.9	0.73
	SO_4^{2-}	67.1	0.029	70.9	0.99	25.1	1.2	0.99	4.4	46.3	0.98
Binary	SeO_4^{2-}	15.7	0.04	16.4	0.99	5.6	0.5	0.98	0.7	9.4	0.80
	SO_4^{2-}	42.4	0.01	43.6	0.99	9.3	0.8	0.98	1.2	33.3	0.89

7.3.3 Effect of varying SeO$_4^{2-}$ or SO$_4^{2-}$ initial concentration on adsorption performance

The effect of varying initial concentrations of SeO$_4^{2-}$ and SO$_4^{2-}$ in the single anion solutions were investigated in a range of 7.1 to 714.8 mg/L SeO$_4^{2-}$ and 96.1 to 1921.4 mg/L SO$_4^{2-}$, respectively (**Appendix 5, Fig. S7.2**). The SeO$_4^{2-}$ adsorption efficiency varied from 81 to 94% with increasing initial concentrations of SeO$_4^{2-}$ (**Appendix 5, Fig. S7.2A**). SO$_4^{2-}$ showed a stable adsorption efficiency (~98%) when the initial concentration was within the range of 96.1 to 672.5 mg/L SO$_4^{2-}$ before decreasing to 73 (± 1)% at an initial concentration of 1921.4 mg/L SO$_4^{2-}$. However, an increase in adsorption capacity (q_e) was noticed with an increase in initial concentration was applied for both oxyanions (**Appendix 5, Fig. S7.3**). The adsorption capacity of SeO$_4^{2-}$ (**Appendix 5, Fig. S7.3A**) ranged from 0.3 mg SeO$_4^{2-}$/g resin (at 7.1 mg/L SeO$_4^{2-}$) to 23.3 mg SeO$_4^{2-}$/g resin (at 714.8 mg/L SeO$_4^{2-}$). On the contrary, SO$_4^{2-}$ showed a gradual increase of q_e from 96.1 until 1152.8 mg/L SO$_4^{2-}$, from 3.6 to 41.6 mg SO$_4^{2-}$/g resin. Above this initial concentration, however, the q_e showed a minimal increase reaching a plateau of ~47 mg SO$_4^{2-}$/g resin at an initial concentration of 1921.4 mg/L SO$_4^{2-}$ (**Appendix 5, Fig. S7.3B**).

Table 7.3 Isotherm constants for SeO$_4^{2-}$ and SO$_4^{2-}$ adsorption onto IRA-900 in single anion solutions

Isotherm model	Parameters	Units	SeO$_4^{2-}$	SO$_4^{2-}$
Langmuir	K_L	L/mg	0.0038	0.0066
	q_m	mg/g	70.22	56.49
	R_L range		0.24-0.97	0.08-0.72
	R^2		0.99	0.99
Freundlich	K_F	mg/g(mg/L)$^{-1/n}$	0.44	3.14
	n		1.22	2.41
	R^2		0.99	0.90
Redlich-Peterson	K_R	L/mg	0.27	0.38
	α_R	L/mg	0.0038	0.0066
	β_R		1	1
	R^2		0.99	0.98
Dubinin-Radushkevich	K_{DR}	mol^2/kJ^2	n/a	0.00055
	q_s	mg/g	n/a	41.68
	E	kJ/mol	n/a	30.11
	R^2		n/a	0.90

Note: *n/a - not applicable*

Table 7.4 Summary of average relative error (ARE) in % for different isotherm models

Average relative error (%)	SeO_4^{2-}	SO_4^{2-}
Single anion solutions:		
Langmuir	18%	14%
Freundlich	15%	40%
Redlich-Peterson	18%	15%
Dubinin-Radushkevich	[1]n/a	45%
Binary anion solutions:		
[2]CC-Modified Langmuir	24%	10%
[3]IAST-Freundlich	22%	23%

[1]*not applicable*; [2] *Complete Competition*; [3] *Ideal Adsorption Solution Theory*

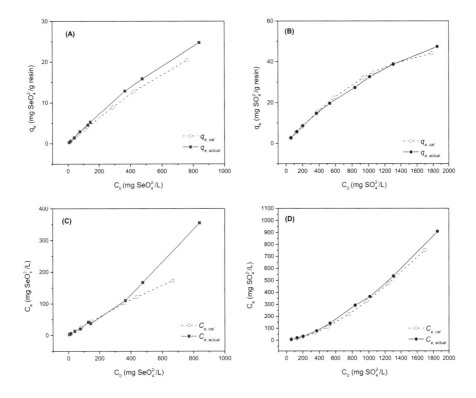

Fig. 7.3 Comparison of predicted values calculated from binary anion solution isotherm model for SeO_4^{2-} and SO_4^{2-} using (A-B) CC-Modified Langmuir and (C-D) IAST-Freundlich isotherm models. (Fig. 4 A and C) Initial concentration of SeO_4^{2-} was varied from 7.1 to 714.8 mg/L and co-expose to an initial SO_4^{2-} concentration of 1441.1 mg/L. (Fig. 4 B and D) Initial concentration of SO_4^{2-} was varied from 96.1 to 1921.4 mg/L and co-expose to an initial SeO_4^{2-} concentration of 428.9 mg/L.

203

Fig. S7.3 (**Appendix 5**) plotted q_e vs C_e with a non-linear fit of SeO_4^{2-} (**Appendix 5, Fig. S7.3A**) and SO_4^{2-} (**Appendix 5, Fig. S7.3B**) adsorption using the four different adsorption isotherms. **Table 7.3** presents the isotherm parameters and regression coefficients for all the isotherm plots. For the SeO_4^{2-} experimental data, the Langmuir, Freundlich and Redlich-Peterson isotherm models showed high R^2 values (**Table 7.3**), while the Dubinin-Radushkevich isotherm model was not able to provide a good fit due to the low concentration range used for SeO_4^{2-}. However, for SO_4^{2-}, only the Langmuir and Redlich-Peterson isotherm model showed a good correlation ($R^2 \sim 0.98$), which was also confirmed by the % ARE values (**Table 7.4**). The Langmuir maximum monolayer adsorption capacity (q_m) value for SeO_4^{2-} (70.2 mg SeO_4^{2-}/g resin) was higher than the q_m for SO_4^{2-} (54.5 mg SO_4^{2-}/g resin). The R_L value for SeO_4^{2-} ranges from 0.24-0.97, while it ranged from 0.08-0.72 for SO_4^{2-} (**Table 7.3**) showing that adsorption of both oxyanions was favorable. Additionally, the lower value of R_L implies that the adsorption was more favorable for SO_4^{2-} than it was for SeO_4^{2-} in the single anion solutions. **Table 7.3** shows $n > 1$ for both SeO_4^{2-} and SO_4^{2-}, which indicates that adsorption site heterogeneity is favored.

Fig. 7.3 compares predicted model values with the experimental data of SeO_4^{2-} and SO_4^{2-} in binary anion solutions, while the % ARE values are given in **Table 7.4.** As seen from the results, similar to the single anion solutions, both the CC-Modified Langmuir and IAST-Freundlich isotherm model were able to predict the behaviour of SeO_4^{2-} when co-existing with SO_4^{2-}, showing 20% ARE values. For the case of SO_4^{2-} in the presence of SeO_4^{2-}, a low ARE % (10%) was obtained using the Modified Langmuir isotherm complete competition model. However, both models showed that there was an underestimation of predicted values, particularly with SeO_4^{2-} (**Fig. 7.3**) at a higher initial concentration.

Fig. 7.4 shows the q_e values and adsorption efficiencies as a function of the initial concentration SO_4^{2-}/SeO_4^{2-} ratio. An opposite trend was observed for SeO_4^{2-} and SO_4^{2-}, where the uptake of SO_4^{2-} gives a steep increase at < 20 SO_4^{2-}/SeO_4^{2-} ratio before reaching a plateau (**Fig. 7.4A**). In contrast, q_e of SeO_4^{2-} showed a more gradual decrease with increasing SO_4^{2-}/SeO_4^{2-} ratio before reaching a plateau at a ratio of 30 (**Fig. 7.4A**). The regression line as a function of the initial SO_4^{2-}/SeO_4^{2-} ratio and q_e was an asymptotic formula (Eq. 7.16 and 7.17) with R^2 values of 0.95 and 0.92 for SeO_4^{2-} and SO_4^{2-}, respectively. Adsorption efficiencies with varying initial concentration for both SeO_4^{2-} and SO_4^{2-} did not show a clear trend and ranged between 60-80% for SO_4^{2-}/SeO_4^{2-} ratios ranging from 0.25 to 179 (**Fig. 7.4B**).

$$q_{e,\,SO_4^{2-}} = 46.4 - 48.2 \times 0.5^{\frac{SO_4^{2-}}{SeO_4^{2-}}} \qquad\qquad \text{Eq. (7.16)}$$

$$q_{e,\,SeO_4^{2-}} = 0.9 + 23.0 \times 0.8^{\frac{SO_4^{2-}}{SeO_4^{2-}}} \qquad\qquad \text{Eq. (7.17)}$$

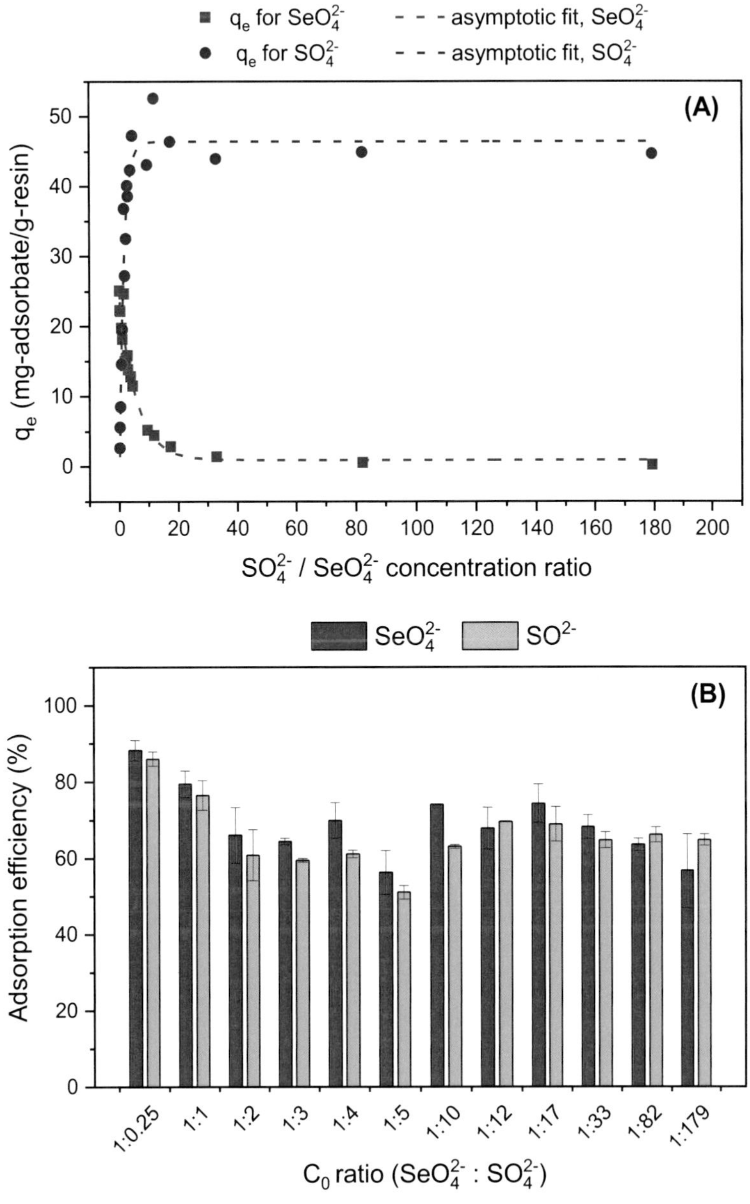

Fig. 7.4 Plotted (A) q_e and (B) removal efficiency at different initial concentration ratios of SO_4^{2-}/SeO_4^{2-}

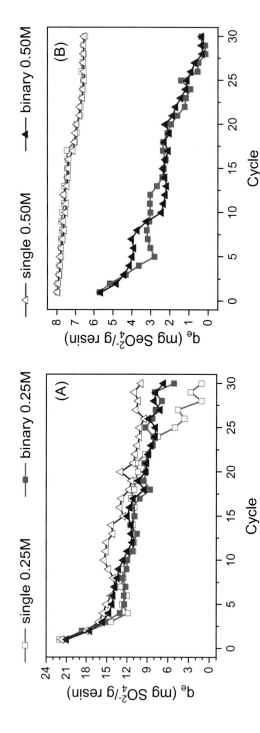

Fig. 7.5 Profile of amount oxyanion adsorbed (q_e) in the resin as a function of the number of cycles for (A) SO_4^{2-} and (B) SeO_4^{2-} in single and binary anion solutions. HCl concentrations used were 0.25 and 0.50 M as the regenerant for 20 min. Initial concentrations of SO_4^{2-} and SeO_4^{2-} were 1441 mg/L and 428 mg Se/L, respectively.

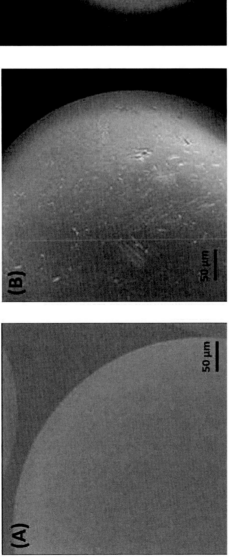

Fig. 7.6 Surface morphology of IRA-900 using SEM (SEI of 10 kV, width 10 mm and a magnification of 130×): (A) fresh resin and after 30 adsorption-desorption cycles for resin regenerated with (B) 0.25 and (C) 0.50 M HCl as the regenerant solutions.

7.3.4   Resin regeneration

Regeneration using HCl solution

Fig. S7.4 (**Appendix 5**) shows resin regeneration time profiles using 0.25 and 0.5 M HCl as the regenerant solution. It can be clearly seen that after 20 min, the adsorbed SO_4^{2-} and SeO_4^{2-} was completely desorbed into the solution. The desorption efficiencies using 0.25 M HCl for 20-60 min were 93 ($\pm$ 1)% and 95 ($\pm$ 1)% for, respectively, SO_4^{2-} and SeO_4^{2-}. For the case of 0.50 M HCl, the desorption efficiencies were 99 ($\pm$ 1)% and 92 ($\pm$ 2)% for SO_4^{2-} and SeO_4^{2-}, respectively. The desorption data was plotted and fitted to the pseudo-second order kinetic model to determine the kinetics of this process. The desorption data fitted well to the linearized pseudo-second order kinetic equation with $R^2 > 0.98$ (**Table 7.5** and **Appendix 5, Fig. S7.5**). Concerning the desorption rate constant (k_2) values, the use of 0.5 M HCl as the regenerant solution showed higher values for SeO_4^{2-} desorption than those obtained for SO_4^{2-} (**Table 7.5**).

Table 7.5 Kinetic parameters of pseudo-second order kinetic model for desorption tests using 0.25 M and 0.50 M HCl as the regenerant solution in binary anion solutions. Negative k_2 value indicates desorption process (release of adsorbed oxyanion from the resin).

Regenerant	Oxyanion	$q_{e, actual}$ (mg/g)	k_2 (mg/g/h)	$q_{e, cal}$ (mg/g)	R^2
0.25 M HCl	SO_4^{2-}	0.45	-1.70	0.43	0.99
	SeO_4^{2-}	0.18	-5.70	0.17	0.99
0.50 M HCl	SO_4^{2-}	0.03	-2.44	0.03	0.99
	SeO_4^{2-}	0.30	-6.50	0.28	0.99

Adsorption-desorption batch cycles

A total of 30 adsorption and desorption cycles was conducted for the single and binary anion solutions using both 0.25 M and 0.50 M HCl (**Fig. 7.5**). Overall, the q_e values reduced with repeated resin use and regeneration, reaching an > 70% decrease in the overall performance (cycle 30) for both oxyanions. Furthermore, irrespective of the strength of the regenerant (0.25 or 0.50 M HCl), the q_e profiles remained similar. The q_e of SO_4^{2-} did not vary between single or binary anion solutions. The q_e dropped by ~50% from cycle 5 to 15, after which the q_e dropped to > 60% in cycle 30 (**Fig. 7.5A**). Unlike SO_4^{2-}, with respect to each adsorption-desorption cycle, there was a clear distinction in the q_e values when SeO_4^{2-} was used in single or binary anion solutions (**Fig. 7.5B**). In single anion solutions, there was a minimal decrease in the q_e values for SeO_4^{2-} despite multiple regeneration steps. In contrast, in binary

208

anion solutions, there was a continuous drop in the q_e values with an increase in the number of cycles, > 80% decrease in the adsorption performance by cycle 30.

Surface morphology of resin after reuse

The resin regenerated with 0.25 M HCl (**Fig. 7.6B**) showed minor damages with small cracks and scratches on the surface of the regenerated resin. Resin regenerated with 0.5 M HCl (**Fig. 7.6C**) showed more surface damages compared to fresh resin (**Fig. 7.6A**). EDS was performed on the internal and external surfaces of the resin after its 30th cycle to determine if there was diffusion of SeO_4^{2-} and SO_4^{2-} oxyanions into the IX beads. No traces of S and Se were found on the fresh resin surface and the average percentages of the elements are: C (75%), N (9%), Cl (15%) and O (2%) (**Appendix 5, Table S7.2**). On the other hand, the resin exposed to 0.50 M HCl during the regeneration phase contained a small mass percentage of S (< 1%), while the resin treated with 0.25 M HCl contained a higher mass percentage of S (< 8%) and minimal presence of Se (< 1%) (**Appendix 5, Table S7.2**).

7.4   Discussion

7.4.1   Performance of IRA-900 for SeO_4^{2-} and SO_4^{2-} removal

This study showed that IRA-900 can be used to effectively adsorb both SO_4^{2-} and SeO_4^{2-} in binary anion solutions, allowing the direct use of an IRA-900 based IX process to treat Se-contaminated waters without any prior SO_4^{2-} pre-treatment (**Fig. 7.1** and **Fig. 7.4B**). An average 84 (± 1.4)% SO_4^{2-} and 88 (± 2)% SeO_4^{2-} removal efficiency was obtained at an initial concentration of 1441.1 mg SO_4^{2-}/L and 428.9 mg SeO_4^{2-}/L, pH 7.5 and 20°C using 5 g/L resin dosage (**Fig. 7.1C**). Irrespective of the increase in the initial concentrations of the oxyanions, i.e. 7.1-714.8 mg/L SeO_4^{2-} and 96.1-1921.4 mg/L SO_4^{2-} in the binary anion solutions, the adsorption efficiency remained within 60-80% for both SeO_4^{2-} and SO_4^{2-} (**Fig. 7.4B**).

Adsorption has been widely accepted as one of the simplest and effective treatment processes for removing oxyanion contaminants from wastewater (Santos et al. 2015). However, one of the major drawbacks of the adsorption process is the competition between adsorbates present in the solution, limiting the adsorption efficiencies of the targeted contaminants. In the case of Se-containing wastewater, the most common co-solute is SO_4^{2-}, which closely resembles SeO_4^{2-} in structure and ionic charge, therefore acting as a strong competitor during adsorption (Mac Namara et al. 2015). Pre-treatment of the wastewater for competing ions has been recommended prior to the IX process in order to avoid inhibition of Se oxyanion adsorption (Jegadeesan et al. 2005; Nishimura et al. 2007; Staicu et al. 2017).

SeO_4^{2-} is the most challenging Se form to be removed by most of the conventional treatment processes (Tan et al. 2016). Compared to other anions, SeO_4^{2-} is a weakly adsorbing anion, which is strongly affected by changes in ionic strength (Fernández-Martínez and Charlet 2009). It forms weak outer-sphere ionic surface complexes with adsorbents such as oxides and minerals (Jegadeesan et al. 2005). So far, there are very limited comparative studies investigating the adsorption of multi-component solutions containing both SeO_4^{2-} and other oxyanions commonly present in wastewater, e.g. SO_4^{2-} and NO_3^-.

One approach to mitigate the issue of competing ions is to increase the available active sites by increasing the resin load. However, this increases the cost of the process, regeneration and adsorbent post-treatment. Another possible solution is to selectively remove SO_4^{2-} prior to conducting the adsorption process in order to increase the SeO_4^{2-} removal efficiency. Jegadeesan et al. (2005) used a two-stage process where SO_4^{2-} (4000 mg/L) was first precipitated using $BaCl_2$ before using bimetallic NiFe particles to completely remove the remaining SeO_4^{2-} (1 mg/L). Staicu et al. (2017) pre-treated flue-gas desulfurization wastewater for the precipitation of 1115 mg SO_4^{2-}/L using $BaCl_2$ in order to improve the SeO_4^{2-} removal efficiency of from 3% (with SO_4^{2-}) to 80% (when SO_4^{2-} was pre-precipitated) using a commercially available synthetic resin FeerIX™ A33E. Johansson et al. (2015) investigated the effectiveness of waste biomass (*Gracilaria* extract) in treating real multi-element coal mine effluent and also concluded that SO_4^{2-} (640 mg/L) should be removed first to achieve any considerable biosorption of SeO_4^{2-} (0.1 mg/L). Similarly, although IRA-900 showed to be effective in removing both SeO_4^{2-} and SO_4^{2-} without preferring one oxyanion to another (**Fig. 7.4B**), this was done in pure water and as such, other contaminants present in real wastewater were not considered. Further studies should be conducted to investigate the applicability of IRA-900 in real wastewater components.

7.4.2 IRA-900 adsorption mechanism

Adsorption kinetics

To compare the adsorption of SeO_4^{2-} and SO_4^{2-} onto IRA-900, predict the adsorption rate and determine the rate-limiting steps, the experimental data were modelled using kinetic and diffusion model equations (Eq. 7.3 to 7.7). The adsorption data fitted well to the pseudo-second model, indicating that the mechanism largely depended on the adsorbate and adsorbent and that chemical adsorption or chemisorption is the rate-limiting step for the overall adsorption of the oxyanions under investigation onto the surface of the IRA-900 resin. Chemical adsorption is usually governed by co-valent forces and occurs through sharing or exchange of electrons between the adsorbent and adsorbate (Sheha and El-Shazly 2010; Sun et al. 2015).

The pseudo-second order k_2 value of SeO_4^{2-} (0.040 mg/g/h) and SO_4^{2-} (0.010 mg/g/h) in binary anion solutions showed a factor of 1.5 and 2.9 lower value, respectively, compared to the single anion solutions, which suggests that the adsorption rate decreased faster with time when two oxyanions were present and was more likely proportional to the number of unoccupied sites (Sun et al. 2015). Adsorption was perhaps attributed to the initial surface site available compared to the oxyanion concentration, but with multiple oxyanions present, the fraction of available sites rapidly diminished causing competition between ions for the adsorption sites.

For the conditions in the present study, SeO_4^{2-} showed a higher constant rate (k_2) compared to SO_4^{2-}, which indicated that SeO_4^{2-} was adsorbed faster compared to SO_4^{2-} (**Table 7.2**). Compared to SO_4^{2-}, SeO_4^{2-} has a higher charge density on its oxygen atoms and could therefore have a higher adsorption interaction (Mac Namara et al. 2015) with IRA-900. It is also possible that the higher k_2 value of SeO_4^{2-} could be due to the lower concentration of SeO_4^{2-} (428.9 mg/L) compared to SO_4^{2-} (1441.1 mg/L). Resin selectivity towards a particular oxyanion was also investigated in this study by normalizing the concentration gradient factor by performing experiments at equimolar concentrations, i.e. 3 mM, of SeO_4^{2-} and SO_4^{2-}. By performing experiments at equimolar concentrations, the ionic charges present in the solution were equal for SeO_4^{2-} and SO_4^{2-}. The equimolar concentration experiment also indicated that SeO_4^{2-} was adsorbed faster than SO_4^{2-} with a pseudo-second order kinetic rate constant (k_2) of 0.045 mg/g/h compared to SO_4^{2-} (0.039 mg/g/h) (**Appendix 5, Table S7.3**).

It is noteworthy to mention that most of the commercially available synthetic resins contain macroporous polymeric materials with the functional groups located within the pores and on the surface of the resin. Hence, it is important to understand the mechanism of adsorbate diffusion from the bulk liquid to the resin. The rate-limiting step in a liquid/solid adsorption process is generally controlled by either the liquid or intra-particle diffusion or a combination of these two. Based on the fitted data for both single and binary anion solutions for diffusion models (**Table 7.2**), a combination of liquid film and intra-particle diffusion fitted best (R^2 ~0.99) the SeO_4^{2-} transport from the solution phase to the resin, while SO_4^{2-} was more controlled by intra-particle diffusion (R^2 ~0.98). This clearly indicates that, for SO_4^{2-}, intra-particle diffusion is the rate-limiting step, while both liquid and intra-particle diffusion limits the adsorption of SeO_4^{2-} onto the resin (**Table 7.2**).

The boundary layer effect (C_{id} and C_{fd}) was higher by a factor of > 2 in single anion solutions compared to binary anion solutions (**Table 7.2**). This indicates that the concentration gradient is a driving force for the adsorption of both oxyanions since higher ionic concentrations (binary anion solutions) create a greater concentration gradient that accelerates the adsorption process (Ociński et al. 2016). Overall, the interaction adsorption mechanisms of IRA-900 for SeO_4^{2-} and SO_4^{2-} is more likely controlled by several sequential steps: diffusion of ions in the liquid film layer surrounding the resin, followed by diffusion through the polymeric matrix of the resin and finally chemical adsorption to the functional groups attached to the resin (Ociński et al. 2016; Soliman et al. 2016).

Adsorption isotherms

Among the tested isotherm models, the Langmuir model described the adsorption data for both SeO_4^{2-} and SO_4^{2-} in single anion solutions. This indicates that the IRA-900 resin has an equal adsorption potential for the oxyanions and provides finite and specific homogenous binding sites in a monolayer with no interaction between each other (Tuzen and Sari 2010). The Langmuir constant, K_L, describes the adsorption affinity or the strength of the adsorption reaction bond energy between adsorbate and adsorbent. The higher the K_L value, the higher the bond energy is between adsorbate and adsorbent (Sun et al. 2015). The K_L value of SO_4^{2-} was found to be twice higher than SeO_4^{2-} (**Table 7.3**), suggesting a higher adsorption affinity of the resin for SO_4^{2-}. The Modified Langmuir isotherm complete competition model was able to closely predict the behavior of SeO_4^{2-} and SO_4^{2-} in the binary anion solutions. This indicates that the adsorption of SeO_4^{2-} and SO_4^{2-} was a complete competitive process, where the two oxyanions inhibited each other's adsorption sites.

7.4.3 IRA-900 regeneration and reusability

The concentration of the regenerant solution (0.25 and 0.50 M HCl) used in this study was relatively low compared to other studies on resin regeneration. Adsorption of phosphate by La(III)-chelex resin was done using 6 M HCl for complete elution (Wu et al. 2007), while 12 M HCl was used to regenerate Dowex Marathon C after adsorption of olive mill wastewater (Víctor-Ortega et al. 2017). From a practical viewpoint, it is possible to increase the concentration of the regeneration solution to ensure 100% resin regeneration, particularly when operating a large-scale IX column. However, there is a possible drawback of resin damage using these highly concentrated regeneration solutions. Indeed, as seen from the SEM images (**Fig. 7.6**), when 0.5 M HCl was used, major damage was observed in the resin surface (rougher surface with visible cracks), which can lower the reuse capability of the resin and the number of regeneration cycles. Cracks on the resin indicate damage to the IX resin polymeric matrix that holds the functional group and possibily lessening the anion retention capability of the IRA-900 resin.

Based on the reusability assessment done on IRA-900, this resin showed good resistance to HCl even after 30 cycles, but its performance decreased after each regeneration step (**Fig. 7.5**). The reusability of the resin can be established according to the designated breakthrough point established for each process which is typically set at 50%. In this study, ~6 cycles were observed to be optimal before reaching the breakthrough point for both oxyanions investigated in the binary anion solution. Martins et al. (2012) used an IX pilot scale column to treat wastewater from coal mines and reported that the reusability of the resin was only for 6 cycles because of the diversity of compounds present in the influent. A study on molybdenum and vanadium adsorption in binary-component solutions using Amberlite IRA-400 and IRA-743 chelating resins was able to reuse the resin for only 3 cycles using 6 M H_2SO_4 or 1 M Na_2CO_3 as the regenerant solution (Polowczyk et al. 2017).

7.4.4 Practical implications

In dealing with complex wastewater streams, IX treatment can be coupled with a biological process as a pre-treatment step (CH2M HILL 2010; Staicu et al. 2017) or as process integration in one set-up for a simultaneous adsorption and bioreduction. Biological treatment offers a cheaper and green alternative to the conventional physical and chemical treatments by reducing Se oxyanions to less toxic and insoluble elemental Se (Se^0) which can later be recovered and re-used. However, despite the advances in biotechnology for Se reduction, biological methods are still limited in treating Se-laden wastewater to reach the low-discharge regulatory limit of 5 µg Se/L (Tan et al. 2016). As such, a post-treatment is generally required after the biological treatment in order to lower the concentrations to permissible levels before final discharge. This study showed that the IRA-900 resin simultaneously adsorbs SO_4^{2-} and SeO_4^{2-} without complete mutual inhibition. One example of process coupling is the combination of the mobile pilot-scale (2.8 m^3/day) IX reactor with electrochemical reduction (Selen-IX™) developed by BioteQ and established in Elk Valley, British Colombia (Mohammandi et al. 2014). The technology achieved Se reduction to levels as low as 1 µg Se/L with stabilized Se in the form of inorganic iron-Se during the electrochemical step.

IX can also be integrated with membrane bioreactors to treat inorganic pollutants based on the Donnan dialysis principle (Jaroszek and Dydo 2016). This process uses the IX material as a membrane barrier between the wastewater compartment and the bio-compartment where mixed microbial cultures degrade the pollutants. The technology is mainly used to avoid the additional treatment of the brine waste and has been applied to NO_3^-, perchlorate and bromate contaminated wastewater (Velizarov et al. 2008; Matos et al. 2009). This integrated process could also be applied to treat Se-laden wastewater containing high concentrations of NO_3^- and SO_4^{2-}. However, the selection of a suitable membrane is challenging for an IX membrane bioreactor. Membrane selectivity towards Se is not readily available in the market and hence membrane fabrication would be needed.

Based on the results obtained in this study, it is possible to fabricate membranes using IRA-900 resins as it shows good selectivity for the removal of both SeO_4^{2-} and SO_4^{2-}. Another potential synthetic resin Smopex® (-103 and -269) was recently developed by Johnson Matthey Water Technologies consisting of functionalized polymeric chains attached to polypropylene fiber backbones that is selective towards SeO_4^{2-} removal and does not compete with SO_4^{2-} (Mac Namara et al. 2015). Compared to other synthetic resins, Smopex® has a rod shape structure allowing for the functionality to be readily exposed to the solution because they are placed on the surface, and not contained within restricted pore spaces. However, there are still drawbacks to this approach, such as biomass coating the outer layer of the resin, general issues with biofouling, and long-term column operation. Further studies should be conducted for evaluating IRA-900 performance in continuous experiments for prolonged operation times as well as a possible application for process integration with bioreactors.

7.5 Conclusions

Amberlite® IRA-900 resin was able to successfully remove both SeO_4^{2-} and SO_4^{2-} from single and binary anion solutions and achieved an adsorption efficiency of more than 70% at pH 7.5 and 20°C using a resin dosage of 5 g/L. Varying the initial concentration from 7.1-714.8 mg/L SeO_4^{2-} and 96.1-1921.4 mg/L SO_4^{2-} showed no major changes in adsorption efficiencies reaching 60-80% for both oxyanions in binary anion solutions. The adsorption process of SeO_4^{2-} and SO_4^{2-} onto the resin surface was well described by the Langmuir isotherm model in single anion solution, suggesting a monolayer adsorption to finite and energetically equivalent homogenous active sites. Adsorption from binary anion solutions was best described by the Modified Langmuir isotherm complete competition model indicating that both oxyanions fully compete with each other for the active sites. The adsorption of both oxyanions by IRA-900 occurred in a two-step process by steady intra-particle diffusion into the polymer matrix followed by adsorption to the functionality of the resin. The optimal adsorption-desorption cycle was 6 and 0.50 M HCl was able to completely regenerate the IRA-900 resin within a short time (20 min). This study showcased the potential usage of IRA-900 resin in simultaneously removing SeO_4^{2-} and SO_4^{2-} for remediation applications of industrial wastewater containing high concentrations of both SeO_4^{2-} and SO_4^{2-}.

7.6 References

Allen, S.J., Mckay, G., Porter, J.F., 2004. Adsorption isotherm models for basic dye adsorption by peat in single and binary component systems. *J. Colloid Interface Sci.* 280, 322-333.

Awual, M.R., Hasan, M.M., Ihara, T., Yaita, T., 2014. Mesoporous silica based novel conjugate adsorbent for efficient selenium(IV) detection and removal from water. *Microporous Mesoporous Mater.* 197, 331-338.

Baig, K.S., Doan, H.D., Wu, J., 2009. Multicomponent isotherms for biosorption of Ni^{2+} and Zn^{2+}. *Desalination* 249, 429-439.

CH2M HILL, 2010. Review of available technologies for the removal of selenium from water - Final report prepared for North American Metal Council. (Available at) http://www.namc.org/docs/00062756.PDF.

Clifford, D., Weber Jr., W.J., 1983. The determinants of divalent/monovalent selectivity in anion exchangers. *React. Polym.* 1, 77-89.

Erosa, M.S.D., Höll, W.H., Horst, J., 2009. Sorption of selenium species onto weakly basic anion exchangers: I. Equilibrium studies. *React. Funct. Polym.* 69, 576-585.

Espinosa-Ortiz, E.J., Shakya, M., Jain, R., Rene, E.R., van Hullebusch, E.D., Lens, P.N.L., 2016. Sorption of zinc onto elemental selenium nanoparticles immobilized in *Phanerochaete chrysosporium pellets. Environ. Sci. Pollut. Res.* 23, 21619-21630.

Fernández-Martínez, A., Charlet, L., 2009. Selenium environmental cycling and bioavailability: A structural chemist point of view. *Rev. Environ. Sci. Biotechnol.* 8, 81-110.

Foo, K.Y., Hameed, B.H., 2010. Insights into the modeling of adsorption isotherm systems. *Chem. Eng. J.* 156, 2-10.

Jaroszek, H., Dydo, P., 2016. Ion-exchange membranes in chemical synthesis-a review. *Open Chem.* 14, 1-19.

Jegadeesan, G., Mondal, K., Lalvani, S.B., 2005. Selenate removal from sulfate containing aqueous solutions. *Environ. Technol.* 26, 1181-1188.

Johansson, C.L., Paul, N.A., de Nys, R., Roberts, D.A., 2015. The complexity of biosorption treatments for oxyanions in a multi-element mine effluent. *J. Environ. Manage.* 151, 386-392.

Kijjanapanich, P., Annachhatre, A. P., Esposito, G., Lens, P. N. L., 2014. Chemical sulphate removal for the treatment of construction and demolition debris leachate. *Environ. Technol.* 35, 1989–1996.

Latorre, C.H., García, J.B., Martín, S.G., Pe, R.M., 2013. Solid phase extraction for the speciation and preconcentration of inorganic selenium in water samples: A review. *Anal. Chim. Acta* 804, 37-49.

Lemly, A.D., 2014. Teratogenic effects and monetary cost of selenium poisoning of fish in Lake Sutton, North Carolina. *Ecotoxicol. Environ. Saf.* 104, 160-167.

Lemly, A.D., 2004. Aquatic selenium pollution is a global environmental safety issue. *Ecotoxicol. Environ. Saf.* 59, 44-56.

Mac Namara, C., Torroba, J., Deacon, A., 2015. New Smopex® ion exchange materials for the removal of selenium from industrial effluent streams: Material characterization, modeling and process implementation. *Johnson Matthey Technol. Rev.* 59, 334-352.

Martins, K., Johnson, J., Leber, K., Srinivasan, R., Heller, B., 2012. Bench- and pilot-scale testing of ion exchange and zero valent iron technologies for selenium removal from a surface coal mine run-off water, in: WEFTEC. pp. 318-338.

Matos, C.T., Sequeira, A.M., Velizarov, S., Crespo, J.G., Reis, M.A.M., 2009. Nitrate removal in a closed marine system through the ion exchange membrane bioreactor. *J. Hazard. Mater.* 166, 428-434.

Mohammadi, F., Littlejohn, P., West, A., Hall, A., 2014. Selen-IX™: Selenium removal from mining affected runoff using ion exchange based technology, in: *Hydrometallurgy 2014*, Victoria, Canada. pp. 1-13.

Nishimura, T., Hashimoto, H., Nakayama, M., 2007. Removal of selenium(VI) from aqueous solution with polyamine-type weakly basic ion exchange resin. *Sep. Sci. Technol.* 42, 3155-3167.

NSMP, 2007. Identification and assessment of selenium and nitrogen treatment technologies and best management practices. (Available at) http://www.ocnsmp.com/library.asp.

Ociński, D., Jacukowicz-Sobala, I., Mazur, P., Raczyk, J., Kociołek-Balawejder, E., 2016. Water treatment residuals containing iron and manganese oxides for arsenic removal from water -

Characterization of physicochemical properties and adsorption studies. *Chem. Eng. J.* 294, 210-221.

Orakwue, E.O., Asokbunyarat, V., Rene, E.R., Lens, P.N.L., Annachhatre, A., 2016. Adsorption of iron(II) from acid mine drainage contaminated groundwater using coal fly ash, coal bottom ash, and bentonite clay. *Water Air. Soil Pollut.* 227, 74-86.

Polowczyk, I., Cyganowski, P., Urbano, B.F., Rivas, B.L., Bryjak, M., Kabay, N., 2017. Amberlite IRA-400 and IRA-743 chelating resins for the sorption and recovery of molybdenum(VI) and vanadium(V): Equilibrium and kinetic studies. *Hydrometallurgy* 169, 496-507.

Santos, S., Ungureanu, G., Boaventura, R., Botelho, C., 2015. Selenium contaminated waters: An overview of analytical methods, treatment options and recent advances in sorption methods. *Sci. Total Environ.* 521-522, 246-260.

Sheha, R.R., El-Shazly, E. a., 2010. Kinetics and equilibrium modeling of Se(IV) removal from aqueous solutions using metal oxides. *Chem. Eng. J.* 160, 63-71.

Soliman, M.A., Mahmoud, M.R., Ali, A.H., Othman, S.H., 2016. The sorption mechanism of Selenium-75 on Amberlite MB9L. *J. Radioanal. Nucl. Chem.* 307, 567-575.

Staicu, L.C., Morin-Crini, N., Crini, G., 2017. Desulfurization: Critical step towards enhanced selenium removal from industrial effluents. *Chemosphere* 172, 111-119.

Sun, W., Pan, W., Wang, F., Xu, N., 2015. Removal of Se(IV) and Se(VI) by MFe_2O_4 nanoparticles from aqueous solution. *Chem. Eng. J.* 273, 353-362.

Tan, L.C., Nancharaiah, Y.V., van Hullebusch, E.D., Lens, P.N.L., 2016. Selenium: Environmental significance, pollution, and biological treatment technologies. *Biotechnol. Adv.* 34, 886-907.

Tuzen, M., Sari, A., 2010. Biosorption of selenium from aqueous solution by green algae (*Cladophora hutchinsiae*) biomass: Equilibrium, thermodynamic and kinetic studies. *Chem. Eng. J.* 158, 200-206.

Velizarov, S., Matos, C., Oehmen, A., Serra, S., Reis, M., Crespo, J., 2008. Removal of inorganic charged micropollutants from drinking water supplies by hybrid ion exchange membrane processes. *Desalination* 223, 85-90.

Víctor-Ortega, M.D., Ochando-Pulido, J.M., Martínez-Ferez, A., 2017. Impacts of main parameters on the regeneration process efficiency of several ion exchange resins after final purification of olive mill effluent. *Sep. Purif. Technol.* 173, 1-8.

Wiramanaden, C.I.E., Liber, K., Pickering, I.J., 2010. Selenium speciation in whole sediment using X-ray absorption spectroscopy and micro X-ray fluorescence imaging. *Environ. Sci. Technol.* 44, 5389-94.

Wu, R.S.S., Lam, K.H., Lee, J.M.N., Lau, T.C., 2007. Removal of phosphate from water by a highly selective La(III)-chelex resin. *Chemosphere* 69, 289-294.

CHAPTER 8

Simultaneous removal of sulfate and selenate from wastewater by process integration of an ion exchange column and upflow anaerobic sludge blanket bioreactor

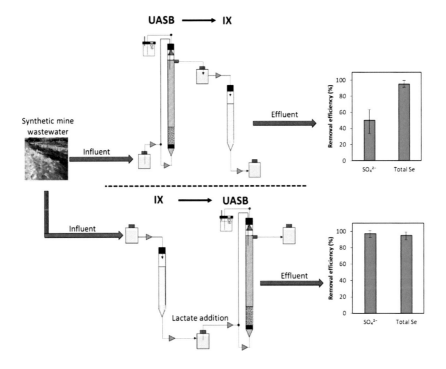

This chapter has been submitted in modified form:

Calix, E.M., Tan, L.C., Rene, E.R., Nancharaiah, Y.V., van Hullebusch, E.D., Lens, P.N.L. Simultaneous removal of sulfate and selenate from wastewater by process integration of an ion exchange column and upflow anaerobic sludge blanket bioreactor. *Sep. Sci. Technol.* (*under review*)

Abstract

A two-stage integrated treatment system comprising of an Amberlite® IRA-900 resin ion exchange (IX) column and an upflow anaerobic sludge blanket (UASB) reactor was evaluated for the treatment of synthetic mine wastewater containing 1441 mg/L SO_4^{2-} and 7.8 mg Se/L as SeO_4^{2-}. The operation of the IX column was optimized at a flow rate of 3 mL/min, empty bed contact time (EBCT) of 17 min and an adsorption time of 9 h to reach 50% breakthrough capacity. The IX column was packed with 60 g resin with a regeneration process of 20 min using 0.5 M HCl. The UASB reactor was operated at 20°C, 16 h hydraulic retention time, 2.2 mL/min influent flow rate and 1780 mg/L lactate at an organic loading rate of 7.7 g_{COD}/L·day, using 10% (*w/v*) Eerbeek granular sludge as inoculum. The removal efficiencies of the UASB reactor in the integrated system were 10% and 50% higher for total Se (Se_{tot}) and SO_4^{2-}, respectively, when the IX was used as the pre-treatment in comparison to its use as a post-treatment step. During the 42 days of continuous reactor operation, the integrated IX→UASB treatment system achieved an overall SO_4^{2-} and Se_{tot} removal efficiency of, respectively, 99% and 97% with the final treated water containing < 100 mg/L SO_4^{2-} and < 0.3 mg Se/L Se_{tot}. Bioregeneration of the loaded resin using the anaerobic granular sludge to biologically reduce the resin adsorbed SO_4^{2-} and SeO_4^{2-} was possible, allowing for reuse of the bioregenerated resin for 4 h IX operation before the breakthrough point was reached.

Keywords: ion exchange column; UASB reactor; selenate removal; adsorption; bioregeneration; integrated treatment system

8.1 Introduction

Selenium (Se) is a metalloid that is required in trace amounts by animals and humans. The Se distribution in the environment occurs through natural and anthropogenic sources. Anthropogenic sources include agricultural runoffs and wastewaters from coal combustion and mining (Butterman et al. 2004). Se, even in low concentrations (< 0.1 g/L), represents a risk to living organisms, particularly to aquatic animals because Se tends to bioaccumulate in different trophic levels (Lemly 2002, 2004). Se bioaccumulation in fishes causes acute effects in the reproductive system which reduces fish fertility and leads to a high loss in fishery livelihood (Lemly 2002, 2004; Spallholz et al. 2002). Additionally, due to expanding manufacturing industries, state and federal government regulations on effluent discharge have become very stringent in order to maintain the water quality of the receiving water bodies. The European Union (Council Directive 98/83/EC) has set a Se threshold concentration in drinking water as 10 μg/L, while the US Environmental Protection Agency (USEPA) has set a maximum value of 50 μg/L for drinking water and 5 μg Se/L for industrial effluent discharge to surface water bodies (USEPA 2001).

Se has two soluble oxyanions: selenate (SeO_4^{2-}) and selenite (SeO_3^{2-}), which are both toxic. The high reactivity of the SeO_3^{2-} to form bonds with organic matter allows for its easy and fast removal from waste streams, while SeO_4^{2-} is less reactive and therefore harder to remove from the wastewater. Various physical, chemical and biological techniques have been developed for removing Se oxyanions from wastewaters (Staicu et al. 2017; Tan et al. 2016). Physical (e.g. membrane separation or adsorption) and chemical (e.g. chemical reduction or coagulation) methods are relatively successful in removing Se oxyanions, but they are usually energy intensive and expensive (Staicu et al. 2017). Biological treatment systems, i.e. wetlands, phytoremediation and bioreactors, are more sustainable compared to other technologies and can be easily adapted to the local conditions (Tan et al. 2016). Although biological processes (i.e. upflow anaerobic sludge blanket (UASB) reactors) are able to simultaneously reduce both sulfate (SO_4^{2-}) and SeO_4^{2-}, it is not possible to meet the stringent discharge limits for Se stipulated by the regulatory bodies (Tan et al. 2016). Ion exchange (IX) processes, on the other hand, allow for the efficient removal and recovery of SeO_4^{2-} from wastewaters (Mac Namara et al. 2015). A major disadvantage of IX processes for SeO_4^{2-} removal is the presence of other oxyanions in real wastewater, i.e. SO_4^{2-}, which can compete with SeO_4^{2-} for the sorption sites (**chapter 7**). Typically, the SO_4^{2-} concentrations are several orders of magnitude higher than the SeO_4^{2-} concentrations in real wastewaters causing faster resin exhaustion due to SO_4^{2-}

oversaturation and inhibition of SeO_4^{2-} removal. Thus, stand-alone IX processes have limited applicability in treating Se containing wastewater (Mac Namara et al. 2015, **chapter 7**).

In this study, the integration of an IX column and a UASB bioreactor was studied for with an overall aim to improve the treatment of wastewater containing different oxyanions, i.e. SO_4^{2-} and SeO_4^{2-}. In the previous work (**chapter 7**) using the Amberlite® IRA-900 resin, batch adsorption parameters were optimized and modelled for the simultaneous removal of SO_4^{2-} and SeO_4^{2-} from binary solutions prepared in pure water. The IRA-900 resin is capable of adsorbing both SeO_4^{2-} and SO_4^{2-} at ~70% capacity in binary solutions using batch operation. The aim of this study was i) to investigate the adsorption capacity of the IRA-900 resin for SeO_4^{2-} and SO_4^{2-} removal from a modelled synthetic mine wastewater solution, ii) optimize the IX column operation for continuous operation and iii) evaluate the suitability of an IX column as a pre- or post-treatment step for a UASB bioreactor. In addition, bioregeneration of IRA-900 resin loaded with oxyanions was investigated using anaerobic granular sludge as a potential alternative for chemical regeneration.

8.2 Materials and methods

8.2.1 Resin selection

The synthetic resin Amberlite® IRA-900 (hereafter referred to as IRA-900) was purchased from Sigma-Aldrich (Italy) and used as received without any pre-treatment. IRA-900 has a good adsorption capacity for simultaneously removing SeO_4^{2-} and SO_4^{2-} (Mac Namara et al. 2015; **chapter 7**). IRA-900 (Sigma-Aldrich, Germany) is a type 1 polymeric strong base anionic exchange resin which contains quaternary amines as functional groups and styrene-divinylbenzene as the copolymer matrix (**chapter 7**). This resin has a chloride ion as the ion exchanger.

8.2.2 Source of biomass

Anaerobic granular sludge was collected from a full-scale UASB reactor treating effluents from four paper mills (Eerbeek, The Netherlands). It had a total solid (TS) and volatile solid (VS) content of 20% and 10% (*w/w*), respectively. A more detailed description of the microbial community of this UASB sludge can be found in Roest et al. (2005). Eerbeek anaerobic granular sludge (10% *w/v*) was used as microbial inoculum for the UASB bioreactor.

8.2.3 Composition of the synthetic mine wastewater

The synthetic mine wastewater contained 7.8 mg Se/L (0.1 mM) Se and 1441 mg/L (15 mM) SO_4^{2-} and mineral medium. The mineral medium has the following chemical composition (mg/L) (Lenz et al. 2008): NH_4Cl 300, $CaCl_2 \cdot 2H_2O$ 10, $MgCl_2 \cdot 6H_2O$ 10, KCl 25, $NaHCO_3$ 40, $Na_2HPO_4 \cdot 2H_2O$ 53, KH_2PO_4 41 and trace elements. Trace acid and alkaline elements were added at 0.1 mL/L according to the composition described by Lenz et al. (2008). Phosphate buffer and alkalinity present in the mineral medium were used to maintain the pH at 7.5 ($\pm$ 0.2) for all experiments. Sodium lactate (60%; VMR Chemicals, France) was supplied as the sole electron donor and carbon source at 1780 mg/L (20 mM). SeO_4^{2-} and SO_4^{2-} were provided as sodium selenate (Sigma-Aldrich, Germany) and potassium sulfate (Merck, Germany), respectively.

8.2.4 Batch adsorption experiments

Batch experiments were conducted to evaluate the SeO_4^{2-} and SO_4^{2-} adsorption capacity of the IRA-900 resin under increasing complexity of solution composition mimicking mine wastewater (**Table 8.1**). Adsorption was conducted at 20°C, pH of 7.5, working volume of 100 mL, mixing at 150 rpm and a resin dosage of 5% (*w/v*).

Table 8.1 Parameters tested to study the effect of synthetic mine wastewater components on SeO_4^{2-} and SO_4^{2-} adsorption by IRA-900.

Batch bottles	Synthetic mine wastewater component
(1)	SeO_4^{2-} or SO_4^{2-} alone
(2)	SeO_4^{2-} + SO_4^{2-}
(3)	SeO_4^{2-} + SO_4^{2-} + mineral media
(4)	SeO_4^{2-} + SO_4^{2-} + minerals media + trace metal elements
(5)	SeO_4^{2-} + SO_4^{2-} + minerals media + trace metal elements + alkalinity (HCO_3^-)
(6)	SeO_4^{2-} + SO_4^{2-} + minerals media + trace metal elements + alkalinity (HCO_3^-) + COD (1335 mg/L lactate)

All batch experiments were performed in duplicate. To simplify the evaluation of ion competition in the synthetic mine wastewater solution, an interaction index was formulated based on the removal efficiency of the individual targeted components, i.e. SeO_4^{2-} or SO_4^{2-}. The interaction index was calculated according to Eq. (8.1):

$$Interaction\ index = \frac{Adsorption_{multi}\ \%\ -Adsorption_{single}\ \%}{Adsorption_{single}\ \%} \qquad\qquad \text{Eq. (8.1)}$$

This was used to determine the degree of negative interactions caused by the addition of the synthetic mine wastewater components on the SeO_4^{2-} or SO_4^{2-} adsorption (Strauss et al. 2004).

8.2.5 IX column experiments for process optimization

A glass column with an internal diameter of 2.9 cm and a height of 24.5 cm was used for the IX continuous operation experiments. The column working volume was 50 mL packed with 30 g (wet weight) resin and glass wool was used as a support at the bottom to hold the resin loss within the column. The solution was continuously fed to the column for a fixed operational period of 4 hours in a down-flow mode using a peristaltic pump (7528-30 model Masterflex, Cole-Parmer) to avoid bed fluidization. The IX column was optimized by testing 3 different flow rates: 2.5, 3.0 and 5.0 mL/min, corresponding to an empty bed contact time (EBCT) of 20, 17 and 10 min, respectively. Samples from the column outlet were collected once every 30 min. After resin saturation, the resin was regenerated according to the procedure described in **chapter 7** by incubating the resin in 100 mL of 0.5 M HCl for 20 min. For every adsorbed SeO_4^{2-} and SO_4^{2-} ion, four ions of Cl⁻ is required to fully regenerate the resin. Assuming 100% adsorption of SO_4^{2-} and SeO_4^{2-} at the highest flow rate (5.0 mL/min) for 4 h, that would equate to 18.12 mmol ions. Therefore, using 0.5 M HCl in 100 mL solution or 50 mmol Cl⁻ was used to regenerate the resin. The adsorption-desorption column tests were carried out for 6 cycles. For the integrated treatment system, the operational time was extended to 9 h and 60 g resin weight. The breakthrough point for this study was set at 0.5 C/C₀ (50% adsorption of oxyanions), where C is the effluent concentration and C₀ is the initial concentration.

8.2.6 Integrated treatment system (IX/UASB)

Process flow configuration Comb1: UASB1 → IX1

System Comb1 consisted of a UASB as a biological treatment process followed by a physical process (i.e. the IX column) as post-treatment (**Fig. 8.1a**). In order to study the influence of UASB1 effluent on the resin, IX1 was operated initially for 4 h in the first cycle before it was increased to 9 h for the remaining cycles. The UASB1 reactor was fed with 7.8 mg/L (0.1 mM) Se, 1441 mg/L (15 mM) SO_4^{2-}, mineral medium, trace elements and 1780 mg/L (20 mM) lactate. The UASB1 reactor was operated at 20°C, a HRT of 16 h, flow rate of 2.2 mL/min, recirculation ratio of 6.8, organic loading rate (OLR) of 7.7 $g_{COD}/L \cdot day$ and an upflow velocity of 0.35 m/h. UASB1 effluent was used as the influent feed for the IX1 column. This latter was operated for 6 h with a 3.0 mL/min flow rate and 60 g of IRA-900 resin. Resin regeneration was done using 100 mL of 0.5 M HCl for 20 min in batch mode. A maximum of 6 adsorption-desorption cycles was tested before the resin was replaced with fresh resin.

Process flow configuration Comb2: IX2 → UASB2

The Comb2 system used the IX column as a pre-treatment step prior to treating the discharged effluent with a UASB reactor (**Fig. 8.1b**). The feed solution used for the IX2 column was composed of mineral medium containing 7.8 mg/L (0.1 mM) Se and 1441 mg/L (15 mM) SO_4^{2-}. The IX2 column was operated similarly to the IX1 column with an EBCT of 9 h. The IX2 effluent was collected separately and supplemented with 1780 mg/L (20 mM) lactate before feeding the solution to the UASB2. The UASB2 reactor was operated similar to the UASB1 configuration under the same conditions.

8.2.7   Bioregeneration of IX resin in the UASB reactor

The possibility of bioregenerating the resin using anaerobic granular sludge (**Appendix 6, Fig. S8.1a**) as an alternative to chemical regeneration was evaluated. UASB1 was used as the bioregeneration unit, after the completion of experiments of the integrated treatment system. In this phase of the experiment, the UASB1 reactor was continuously fed with lactate at an OLR of 7.7 g_{COD}/L·day for 3 days to ensure no SeO_4^{2-}, SO_4^{2-} or other bioreduced products remained in the mixed liquor if the reactor prior to use as a regeneration unit. To facilitate the ease of repeated resin transfer from the IX column and UASB1 reactor, 30 g IRA-900 was packed into a 45 μm cloth mesh.

A resin saturation experiment was conducted by operating the adsorption phase (9 h, pH 7.5, 20°C and 3.0 mL/min) until the SeO_4^{2-} and SO_4^{2-} effluent concentration equaled the influent concentration, after which bioregeneration was applied. Bioregeneration was conducted in batch mode for 1 to 3 days in order to determine the most effective bioregeneration period. A 3 L feed solution containing 1780 mg/L lactate and mineral medium was recirculated in the UASB1 reactor at a flow rate of 4.4 mL/min where the saturated IRA-900 resin was placed. After each bioregeneration trial, the resin-packed cloth mesh was taken out from the UASB reactor, washed with pure water to remove the suspended solids and placed inside the glass column for the next adsorption phase. This was considered as one cycle of operation and a total of 6 cycles were performed.

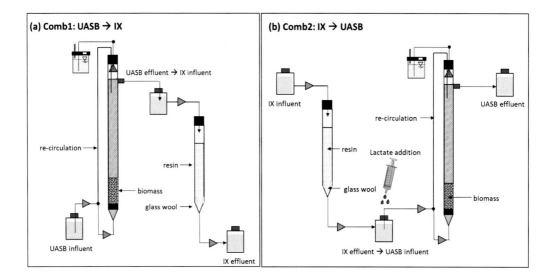

Fig. 8.1 Schematic of the integrated treatment system comprising of an IX column and a UASB reactor. Process flow configuration was evaluated as either (a) Comb1 system UASB1 → IX1 or (b) Comb2 system IX2 → UASB2.

8.2.8 Analytical methods

An ion chromatograph (IC, ICS-1000 Dionex) equipped with an AS4A 2 mm column was used to determine the SeO_4^{2-}, SeO_4^{2-} and lactate concentrations at retention times of 9.0, 7.2 and 1.3 min, respectively (Dessì et al. 2016). The mobile phase used was 1.8 mmol Na_2CO_3 and 1.7 mmol $NaHCO_3$ at a flow rate of 0.5 mL/min. The IC had a minimum detection limit of 2 mg/L SO_4^2 and 20 µg/L SeO_4^{2-}. Volatile fatty acids (VFA) were determined using gas chromatography (GC, Bruker 430, The Varian® 430-GC) according to the procedure described by Cassarini et al. (2017).

The total Se (Se_{tot}) concentration was determined by acidifying the sample with concentrated HNO_3 and analyzed using a graphite furnace atomic absorption spectrometer (GF-AAS, ThermoElemental Solaar MQZe GF95, Se lamp at 196.0 nm) with a detection limit ranging from 0.4-50.0 µg Se/L. Total dissolved Se was analyzed using the GF-AAS after the effluent samples were centrifuged at 21,000 g for 20 min and filtered with a 0.45 µm membrane filter (Dessì et al. 2016). The total dissolved sulfide (HS^-) concentration was measured using the procedure described in Standard Methods (APHA/AWWA/WEF 2005).

Changes in surface morphology of the resin after IX1, IX2 and the bioregeneration processes were analyzed using a scanning electron microscope (SEM, JEOL JSM6010LA) equipped with an energy dispersive X-ray spectroscopy (EDX). The resin samples were dried at room temperature prior to analysis.

8.3 Results

8.3.1 Effect of synthetic wastewater on oxyanion adsorption

The baseline adsorption efficiencies for SeO_4^{2-} and SO_4^{2-} in the single anion system were 97% and 73%, respectively, using an Amberlite® IRA-900 resin dosage of 5 g/L (wet weight) in pure water, pH 7.0 and 20°C. The interaction index value for SeO_4^{2-} (-0.3) showed that its adsorption was affected by the addition of other components (**Fig. 8.2a**). In contrast, SO_4^{2-} adsorption was only slightly affected by the components of the synthetic mine wastewater attaining an interaction index values ranging between -0.16 and -0.09 (**Fig. 8.2b**). Adsorption of lactate in the single anion system was low (< 10%), indicating that the IRA-900 resin has a low preference in adsorbing lactate.

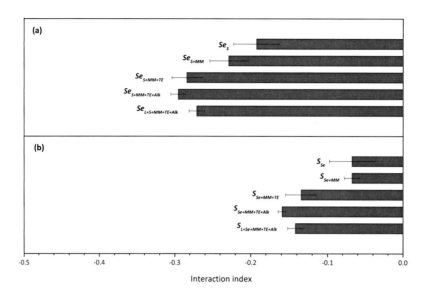

Fig. 8.2 Interaction indices showing the effects of synthetic mine wastewater components of mineral medium (MM), trace elements (TE), alkalinity (Alk) and lactate (L) as COD on (a) SeO_4^{2-} and (b) SO_4^{2-} adsorption efficiency. Se indicates the adsorption of SeO_4^{2-}, while S denotes the adsorption of SO_4^{2-}.

8.3.2 Optimization for continuously operated IX column

The three flow rates evaluated were 2.5, 3.0 and 5.0 mL/min for an operation period of 4 h and the breakthrough curve profiles are shown in **Fig. 8.3**.

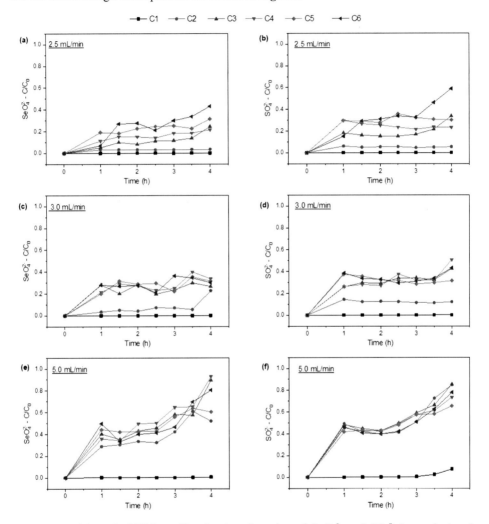

Fig. 8.3 Breakthrough (C/C₀) profiles for the adsorption of SeO₄²⁻ and SO₄²⁻ in synthetic mine wastewater onto IRA-900 resin for cycles 1–6 obtained at a flow rate of (a-b) 2.5 mL/min, (c-d) 3.0 mL/min and (e-f) 5.0 mL/min. Operating conditions: 4 h adsorption period and 30g adsorbent dose.

Fig. 8.3a-b show the breakthrough profile achieved at a flow rate of 2.5 mL/min. The high adsorption capacity due to the low flow rate was a consequence of the high EBCT, therefore the desired breakthrough profile was not reached. Similarly, the breakthrough profile at 3.0 mL/min (**Fig. 8.3c-d**) was close to reaching the 50% breakthrough point for both SeO₄²⁻ and

SO_4^{2-}. At a flow rate of 5.0 mL/min (**Fig. 8.3e-f**), breakthrough profiles showed that both oxyanions exceeded the breakthrough point already after 1 h adsorption. Therefore, the 3.0 mL/min flow rate, corresponding to an EBCT of 17 min, was selected for the integrated treatment system.

Since the 50% breakthrough was not reached for the 3.0 mL/min flow rate within 4 h, further adsorption experiments to determine the breakthrough profile and optimize the IX column for the integrated treatment were conducted at an adsorption time of 9 h for 6 cycles. The breakthrough point was reached between the 5th and 6th hour (**Fig. 8.4a**). Therefore, a 9 h adsorption period was chosen and the resin weight was doubled to 60 g resin, corresponding to a working volume of 100 mL, for the integrated treatment.

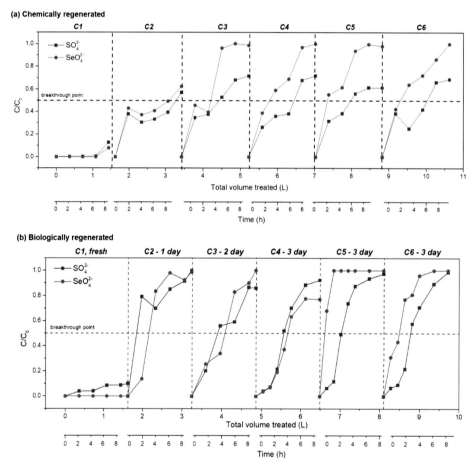

Fig. 8.4 Breakthrough (C/C_0) profiles of SO_4^{2-} and SeO_4^{2-} for cycles 1–6 obtained at a flow rate of 3.0 mL/min fed with synthetic wastewater containing SO_4^{2-} (1358 mg/L) + SeO_4^{2-} (7.8 mg Se/L), 30 g of IRA-900 resin and an operational period of 9 h for (a) chemically regenerated resin and (b) biologically

regenerated resin. Chemical resin regeneration was carried out using 100 mL of 0.5 M HCl for 1 h, while the biologically resin regeneration was conducted from 1 to 3 days. Note: "C" stands for cycle.

8.3.3   IX column performance of the Comb1 and Comb2 systems

Breakthrough profiles for IX2 and IX1 are shown in **Fig. 8.5a** and **8.5b**, respectively. **Table 8.2** summarizes the adsorption efficiencies of IX1 and IX2 for the entire 3 sets of 6 cycle operations, where the IX2 column had a significantly higher adsorption for both SO_4^{2-} and Se_{tot} than the IX1 column. The influent pH of the synthetic mine wastewater entering the IX1 column (Comb1: UASB1→IX1) from the UASB1 was 7.8 ($\pm$ 0.3). IX1 was operated initially for 4 h in the first cycle set before it was increased to 9 h for the second and third cycle set, while IX2 was operated for 9 h. The shorter adsorption period for IX1 was tested to determine the breakthrough behaviour of an influent that contains several compounds that could saturate the resin faster, such as suspended solids, VFA, total dissolved HS^- and reduced forms of Se.

Table 8.2 Adsorption efficiencies achieved by IX1 and IX2 operated at 3 mL/min flow rate, 60 g IRA-900 resin, 20°C and 9 h adsorption period.

Components	Adsorption efficiency (%)	
	IX1	IX2
SO_4^{2-}	57 ($\pm$ 35)	74 ($\pm$ 21)
SeO_4^{2-}	n/d [a]	72 ($\pm$ 30) [b]
Se_{tot}	44 ($\pm$ 22)	72 ($\pm$ 30) [b]
Lactate	12 ($\pm$ 10)	n/p [c]
Acetate	97 ($\pm$ 1)	n/p [c]
Propionate	95 ($\pm$ 1)	n/p [c]

[a] n/d - no SeO_4^{2-} was detected entering the IX1 after UASB1
[b] Only SeO_4^{2-} form was being adsorbed in IX2
[c] n/p - not present in the influent

Due to the varying effluent concentrations produced by the UASB1 (**Appendix 6, Table S8.1**), variations in the Se_{tot} and SO_4^{2-} adsorption performance in IX1 were observed (**Fig. 8.5a**). The adsorption capacity for IX1 ranged from 30-70% for SO_4^{2-} and Se_{tot}, respectively (**Table 8.2**). In contrast, IX2 showed a stable adsorption performance (**Fig. 8.5b**). The IX2 effluent concentrations were 367 ($\pm$ 91) mg/L SO_4^{2-} and 3.3 ($\pm$ 1.1) mg/L Se. The pH of the IX2 effluent was adjusted to 7.5 ($\pm$ 0.2) using 0.1 M NaOH and lactate (20 mM) was added before feeding it to UASB2. PO_4^{3-} was also adsorbed at ~85% by both the IX columns at the beginning of the cycles (C1-4), but this was lowered to 10-30% in the last two cycles (**Appendix 6, Table S8.2**).

Fig. 8.6 shows the surface morphology of the resin after the pre- (IX2) or post- (IX1) treatment process. **Table S8.3 (Appendix 6)** shows the EDX spectra of the resin surface. The surface of the IX2 resin (**Fig. 8.6b**) showed slight surface modifications as compared to the fresh resin (**Fig. 8.6a**), but had a cleaner and smoother surface compared to the IX1 resin (**Fig. 8.6c**). The surface of the IX1 resin exhibited more modifications and revealed the possible presence of bacterial growth on the resin surface (**Fig. 8.6c**). As shown in **Table S8.3 (Appendix 6)**, the elemental composition on the resin surface revealed traces of S and Se, particularly for IX1.

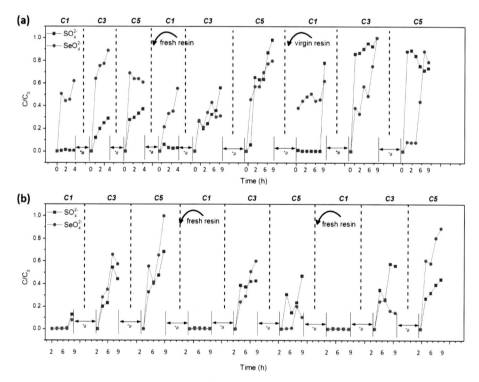

Fig. 8.5 Breakthrough (C/C_0) profiles of SO_4^{2-} and SeO_4^{2-} adsorption onto IRA-900 resin for cycles 1– 6 of the integrated treatment system obtained at a flow rate of 3.0 mL/min for (a) IX1 was fed with UASB1 effluent for adsorption period of 6 h. IX operation was maintained at 20°C, 3.0 mL/min flow rate and 60 g resin for a total of 6 cycles before changing to a fresh resin and (b) IX2 fed with a synthetic mine wastewater containing SO_4^{2-} (1358 mg/L) + SeO_4^{2-} (7.8 mg Se/L) solution for an adsorption period of 9 h. Resin regeneration was carried out using 100 mL of 0.5 M HCl for 20 min. Note: "*a" indicates that the cycles were operated overnight. Hence, samples were not collected during that time.

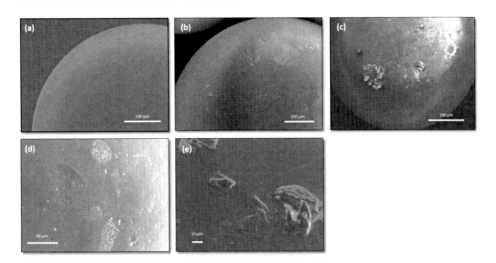

Fig. 8.6 Surface morphology of the IX resin IRA-900 (a) fresh resin, (b) resin exposed to the pre-treatment process for 9 h (IX2), (c) resin exposed to the post-treatment process for 9 h (IX1), (d) resin bioregenerated for 3 days and (e) a close up view of the resin surface bioregenerated for 3 days.

8.3.4 UASB reactor performance in the Comb1 and Comb2 systems

For UASB2, the average SO_4^{2-} and Se_{tot} concentrations were 367 ($\pm$ 91) mg/L and 3.3 ($\pm$1.1) mg Se/L, respectively, and an average pH of 7.5. **Fig. 8.7** shows the removal efficiencies for SO_4^{2-} and total Se of both UASB reactors. UASB1 shows a gradual decrease in the SO_4^{2-} removal efficiency at > 50% (**Fig. 8.7a**). In contrast, high SO_4^{2-} removal efficiencies at > 90% were reached for UASB2. The lactate concentration was increased on day 31, from 1780 to 2500 mg/L (OLR of 10.8 g_{COD}/L·day) to improve the SO_4^{2-} removal of UASB1. However, no change in removal efficiency for SO_4^{2-} and Se_{tot} was observed. The effluent HS^- concentration for both reactors was at < 60 mg/L HS^- (**Appendix 6, Fig. S8.2b**). The average Se_{tot} and SO_4^{2-} removal efficiency achieved in UASB2 was 85 ($\pm$ 10) and 66 ($\pm$ 21)%, while UASB1 achieved only 69 ($\pm$ 17) and 41 ($\pm$ 15)%, respectively, after 42 days of continuous reactor operation.

SeO_4^{2-} was not detected in the effluent since day 6 of operation for both UASB1 and UASB2. Colloidal biogenic Se (Se^0) concentrations in the effluent for UASB1 fluctuated from ~1.0 mg/L to 3.0 mg/L, while those of UASB2 decreased from ~2.0 mg/L to < 0.5 mg/L (**Appendix 6, Fig. S8.2a**). The total dissolved Se concentration in both reactor effluents were < 10% or < 0.2 mg/L of the Se_{tot} concentration. The acetate and propionate concentrations of both UASB reactors were similar, respectively, 3-30 mg/L and 150-600 mg/L (**Appendix 6, Fig. S8.2c-d**).

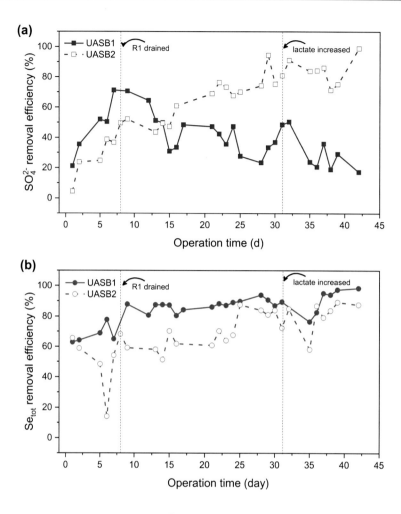

Fig. 8.7 Removal efficiencies (%) of (a) SO_4^{2-} and (b) Se_{tot} comparing UASB1 (1st treatment stage) and UASB2 (2nd treatment stage). Note that SeO_4^{2-} was no longer detected after day 6 and it was completely reduced during the entire operational time for both.

8.3.5 Bioregeneration of saturated resin

Fig. 8.4b shows the breakthrough profiles of the bioregenerated resin in the IX column during 6 cycles of operation. The bioregeneration period of 1 day for cycle 2 reached 50% breakthrough within 2 h operation. Increasing the bioregeneration period from 1 to 2 days for cycle 3 showed a slight improvement, reaching 50% breakthrough during 4 h of operation. However, resin saturation was reached by the end of the 9 h operation period. For cycle 4, the bioregeneration period was extended to 3 h and although 50% breakthrough was reached similarly within 4 h, resin saturation was not observed until 9th hour. Therefore, a 3 day bioregeneration period was used for cycles 5 and 6. The profile of the breakthrough curve was nearly similar from cycle 4 to 6, particularly for SO_4^{2-}. However, there was a ~5-10% decrease in the adsorption capacity after repeated bioregeneration from cycle 4 until 6, especially for SeO_4^{2-}. Compared to chemical regeneration (**Fig. 8.4a**), the reusability of the bioregenerated resin was lower, reaching the 50% breakthrough limit within < 4 h of IX operation (**Fig. 8.4b**).

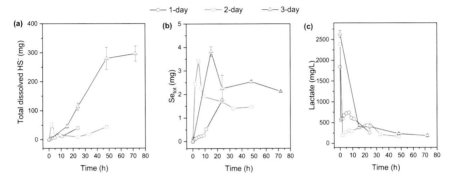

Fig. 8.9 Concentration of bio-reduced products: (a) total dissolved HS⁻, (b) Se$_{tot}$ and (c) lactate during bioregeneration of the resin in batch mode, at 20°C fed, with a 3 L lactate (1800 mg/L) solution at a recirculation flow rate of 4.4 mL/min. Total SO_4^{2-} and SeO_4^{2-} loaded during IX operation at cycle 1 was 115 mg Se and 7460 mg S, respectively.

The initial SeO_4^{2-} and SO_4^{2-} loaded onto the resin during cycle 1 was 115 mg Se and 7460 mg S, respectively (**Fig. 8.8**). The maximum Se$_{tot}$ and total dissolved HS⁻ concentration was only 4 mg Se and 300 mg HS⁻-S, respectively, after a bioregeneration period of 3 days, i.e. cycle 4 to 6 (**Figs. 8.8a-b**). The lactate concentration (**Fig. 8.8c**) decreased with increasing contact time during bioregeneration from 2000 to 250 mg/L.

The bioregenerated resins showed progressively a darker reddish coloration after each bioregeneration step (**Appendix 6, Fig. S8.1b**). SEM images of the bioregenerated resin showed surface changes from smooth (**Fig. 8.6a**) to rough (**Fig. 8.6d**) surface. A close-up SEM image (**Fig. 8.6e**) of the bioregenerated resin showed the possibility of a mineral deposit or microbial growth on the surface. EDX analysis on the surface of the bioregenerated resin showed the absence of Se, while S was present at < 2% mass elemental composition (**Appendix 6, Table S8.3**).

8.4 Discussions

8.4.1 Integrated process for the simultaneous removal of Se_{tot} and SO_4^{2-}

This study showed that the integration of an IX and UASB reactor can increase the SO_4^{2-} and Se_{tot} removal efficiency from synthetic mine wastewater treatment. Biological (i.e. UASB) treatments are considered to be economical and environmentally friendly processes for treating industrial wastewaters (Buyukkamaci et al. 2010). However, biological treatment systems likely do not reach discharge regulatory limits, especially for high strength wastewaters (Tan et al. 2016). This study investigated the integration of two treatment system, i.e. an IX and a UASB reactor, for the removal of SO_4^{2-} and SeO_4^{2-} from synthetic mine wastewater. Assessing both configurations in terms of its removal performance, the Comb2 configuration allowed for higher SO_4^{2-} and SeO_4^{2-} removal by the IX2 and as well as reducing the concentration load entering the UASB2. A similar approach was tested where the authors investigated the use of anion IX membranes as a pre-treatment step before the coagulation unit for the removal of arsenic from drinking water (Oehmen et al. 2011). This integrated treatment system was able to reduce the arsenic concentration to levels below the regulatory limits (10 μg/L) and after the coagulation process the arsenic concentrations averaged 6.6 (± 2.8) μg/L.

8.4.2 IX performance as pre- and post-treatment to biological treatment

A similar breakthrough pattern was observed for both IX columns, i.e. SeO_4^{2-} or Se_{tot} adsorption reached their breakthrough points faster than SO_4^{2-} (**Fig. 8.5**). This was further supported by the interaction index values that showed the effect of addition of mine wastewater components on the adsorption of SeO_4^{2-} and SO_4^{2-}. According to **Fig. 8.2**, SeO_4^{2-} adsorption was more affected than SO_4^{2-}, most likely due to the difference in equilibrium concentration between the two oxyanions. In a previous study, **chapter 7** investigated the applicability of IRA-900 to adsorb SO_4^{2-} or SeO_4^{2-} in either single or binary anion solutions in a batch system using pure water. The results showed that the use of about ~ 20% decreased in the adsorption efficiency was observed in binary anion solution, although IRA-900 was still able to effectively adsorb both SeO_4^{2-} and SO_4^{2-} at ~ 70% from an initial concentration of 237 mg Se/L and 1441 mg/L, respectively, at 5% (*w/v*) resin dosage.

Since IX1 was receiving the UASB reactor effluent, it is possible to compare these results with previously reported adsorption studies carried out with real wastewaters. Staicu et al. (2017) used real flue gas desulfurized (FGD) water and showed that other species present in the wastewater, mainly SO_4^{2-}, can saturate the resin Purolite® FerrIX A33E and limit the SeO_4^{2-} removal. Therefore, the authors suggested that a desulfurization process must be conducted to remove SO_4^{2-} prior to applying IX processes for Se removal. Although IRA-900 was capable of removing both SO_4^{2-} and SeO_4^{2-} (Mac Namara et al. 2015; **chapter 7**), the variety of other compounds such as VFAs and suspended solids in a UASB effluent entering the IX1 likely saturated the resin faster due to competition for active sites or blockage of active sites by the suspended solids. Additionally, Se_{tot} contains other Se forms such as colloidal Se^0 which was possibly not preferably adsorbed by IRA-900, therefore resulting in a lower Se_{tot} adsorption by the IX1 column. Martins et al. (2012) evaluated the treatment of Se-laden coal mine effluent using a zero-valent iron (ZVI) and IX process and observed that the strong base anionic resins gave a better performance and reached a final concentrations of 4 µg Se/L because the Se was all in its anionic form. These results relate to the Comb2 integrated treatment system, where the overall removal efficiencies for Se_{tot} and SO_4^{2-} concentrations were higher when the IX2 was used as a pre-treatment step.

An alternative method for the improvement of the IX process, particularly when using IX as a post-treatment (IX1), is the use of different resin types and selectivity. The use of resins which are selective for specific compounds others than SO_4^{2-} and SeO_4^{2-} will increase the adsorption capacity and the overall removal performance. In a recent study, Lim and Kim (2015) demonstrated a high removal efficiency of ammonium (~ 3000 mg/L) from real swine wastewater by stacking a range of cationic and anionic resins in an IX column. The authors used a cationic resin to adsorb ammonia in the first step, followed by biodegradation using nitrifying bacteria, which oxidized the remaining ammonia to nitrite/nitrate and after that the water was treated again using an IX column with an anionic resin to remove the residual nitrate/nitrite.

8.4.3 UASB reactor performance at different influent Se and SO_4^{2-} concentration

The possible reason for the poor SO_4^{2-} removal efficiency of UASB1 (**Fig. 8.7a**) could be due to inhibition by SeO_4^{2-}. The presence of SeO_4^{2-} (1-8 mg/L) can indeed inhibit biological SO_4^{2-} (~2500 mg/L) reduction, causing an incomplete SO_4^{2-} reduction (Lenz et al. 2008; Hockin and Gadd 2006). Although SO_4^{2-} reducers are able to reduce Se oxyanions, SO_4^{2-} reducers are susceptible to Se toxicity (> 1 mM) (Lenz et al. 2008). Selenium is a sulphur analogue and, therefore, it disrupts both the assimilatory and dissimilatory SO_4^{2-} reduction processes (Hockin and Gadd 2006).

Although UASB2 was receiving low SO_4^{2-} and Se_{tot} concentrations, the chloride (Cl⁻) concentration entering the UASB2 from the IX was high (~1 g/L) compared to concentration fed to UASB1 (~0.25 g/L). Though UASB2 showed no observable decrease in the removal efficiency due to high Cl⁻ concentrations, this does not rule out the possibility of a negative impact by high Cl⁻ concentration exposure over a prolonged period of operation. Cell plasmolysis and mortality of the microbial community may occur when high Cl⁻ containing influents are fed to UASB reactors (Jeison et al. 2008). This aspect should be taken into consideration for reactor scale-up where higher Cl⁻ concentrations are expected to be released from the IX process due to an increase in resin dosage for larger IX column size or resin swelling.

Despite high Se_{tot} removal efficiencies (up to 97%) in both reactor configurations, the effluent quality did not meet the discharge limit established by the USEPA (5 µg/L). SeO_4^{2-} reduction can have by-products such as other dissolved Se and colloidal Se compounds, which are still released into the effluent, thereby deteriorating the Se_{tot} removal process. Better retention of colloidal Se by the UASB reactor could possibly be attained by increasing the operation temperature from 20°C to > 30°C. Dessì et al. (2016) investigated the effect of temperature (35 and 55°C) on the UASB performance, where the reactor operation at 55°C showed lower Se_{tot} release into the effluent, possibly due to an increase in size and decrease in zeta-potential of the biogenic Se allowing for better retention in the biomass. However, increasing the temperature for any bioreactor operation would also increase the operation cost.

8.4.4 Bioregeneration of resin as alternative approach for chemical based resin regeneration

This study gave a "proof of concept" for the bioregeneration process as an alternative method for chemical resin regeneration. To the best our knowledge, no studies have so far used the bioregeneration technique in SeO_4^{2-} and SO_4^{2-} adsorption studies. Bioregeneration of resins is a new concept that uses biological processes to regenerate the resin through the use of microorganisms (Venkatesan et al. 2010) and have been mainly used in nitrate and perchlorate resin adsorption studies (Venkatesan et al. 2010; Meng et al. 2014; Ebrahimi et al. 2013, 2015). Biomass mixed with the resin allows for microbe-resin contact, permitting the bioreduction of the adsorbed oxyanion, thus eliminating the need for treatment of the brine solution after chemical regeneration and allowing for a more sustainable and cheaper solution. Bioregeneration can be improved by utilizing optimal biological conditions for the targeted ions. Meng et al. (2014) investigated multiple parameters, i.e. biomass content, pH, salinity and

238

molar ratio of organic carbon to nitrate, for the bioregeneration of a nitrate-loaded resin and was able to effectively regenerate the resin within 5 h for 5 cycles.

One of the key factors to be considered for the bioregeneration is the resin selection. IRA-900 has a spherical shape and therefore allows the adsorbed oxyanion to penetrate deep into the resin. This can render the bioregeneration process to be ineffective or slow, since the microbes will have less contact with the adsorbed oxyanion. Venkatesan et al. (2010) used suspended bacterial cultures for the bioregeneration of perchlorate-loaded resin (pore size < 1 µm) and reported that perchlorate ions present deep inside the resin core are not easily removed, while those present in the surface of the resin can be utilized by the bacterial culture and removed from the resin. Mac Namara et al. (2015) fabricated a new synthetic strong base anionic resin called Smopex® that is more selective for SeO_4^{2-} than SO_4^{2-}. Smopex® resin has a unique rod shaped feature, allowing more active sites to be placed on the outer surface. Using this type of resin morphology in an IX processes can increases the efficiency of the adsorption process and allows for easier resin bioregeneration.

Another aspect that should be considered is the material housing the resin to prevent biofouling and resin loss. Ebrahimi and Roberts (2013, 2015) applied bioregeneration for nitrate-loaded resin wrapped within a membrane using suspended denitrifiers. The authors observed high resin reusability after bioregeneration, even after 6 cycles with only < 6% decrease in the adsorption capacity. The authors suggested the use of membrane cartridges as a resin holder to pack the synthetic resin and can therefore be easily plugged in and out of the IX column and into a bioreactor for better bioregeneration process.

8.5 Conclusions

The integration of a chemical and biological process, i.e. an IX column and a UASB reactor, for the simultaneous removal of SO_4^{2-} and SeO_4^{2-} from synthetic mine wastewater was investigated. The interaction index values indicated that the adsorption of SeO_4^{2-} was more affected by the presence of other components in the synthetic wastewater compared to SO_4^{2-}. Optimal IX column continuous operation was set at 3 mL/min, 17 min EBCT, 6 h adsorption period and a working volume of 100 mL. The IX process as a pre-treatment step was found to be the best option because the targeted species (SeO_4^{2-} and SO_4^{2-}) were in their anionic form, allowing for an easier adsorption process, while also providing a lower concentrated influent to the UASB2. Overall, the integrated Comb2 treatment showed a higher performance for SO_4^{2-} and Se_{tot} removal achieving an overall removal efficiency of 99% and 97%, respectively, with complete removal/reduction of SeO$_4^{2-}$. Bioregeneration of the IRA-900 resin using previously acclimated biomass was capable of reducing the SO$_4^{2-}$ and SeO$_4^{2-}$ adsorbed onto the resin and it should be further optimized in order to be used as an alternative to chemical regeneration.

8.6 References

Association, A.P.H., Association A.W.W., Federation, W.E., 2005. Standard methods for examination of water and wastewater, 5th ed. American Public Health Association, Washington, DC, USA.

Butterman, W., Hilliard, H. 2004. Mineral commodity profiles: selenium report. US Department of the Interior US Geological Survey, pp. 1-20.

Buyukkamaci, N., Koken, E., 2010. Economic evaluation of alternative wastewater treatment plant options for pulp and paper industry. *Sci. Total Environ.* 408, 6070-6078.

Cassarini, C., Rene, E.R., Bhattarai, S., Esposito, G., Lens, P.N.L., 2017. Anaerobic oxidation of methane coupled to thiosulfate reduction in a biotrickling filter. *Bioresour. Technol.* 240, 214-222.

Dessì, P., Jain, R., Singh, S., Seder-Colomina, M., van Hullebusch, E.D., Rene, E.R., Ahammad, S.Z., Carucci, A., Lens, P.N.L., 2016. Effect of temperature on selenium removal from wastewater by UASB reactors. *Wat. Res.* 94, 146-154.

Ebrahimi, S., Roberts, D.J., 2013. Sustainable nitrate-contaminated water treatment using multi cycle ion-exchange/bioregeneration of nitrate selective resin. *J. Hazard. Mater.* 262, 539-544.

Ebrahimi, S., Roberts, D.J., 2015. Bioregeneration of single use nitrate selective ion-exchange resin enclosed in a membrane: kinetics of desorption. *Sep. Purif. Technol.* 146, 268-275.

Hockin, S., Gadd, G.M., 2006. Removal of selenate from sulfate-containing media by sulfate-reducing bacterial biofilms. *Environ. Microbiol.* 8, 816-826.

Jeison, D., Del Rio, A., van Lier, J.B., 2008 Impact of high saline wastewaters on anaerobic granular sludge functionalities. *Wat. Sci. Technol.* 57, 815-819.

Lemly, A.D., 2004. Aquatic selenium pollution is a global environmental safety issue. *Ecotoxicol. Environ. Saf.* 59, 44-56.

Lemly, A.D., 2002. Symptoms and implications of selenium toxicity in fish: the Belews Lake case example. *Aquat. Toxicol.* 57, 39-49.

Lenz, M., van Hullebusch, E.D., Hommes, G., Corvini, P.F., Lens, P.N.L., 2008. Selenate removal in methanogenic and sulfate-reducing upflow anaerobic sludge bed reactors. *Water Res.* 42, 2184-2194.

Lim, S.J., Kim, T-H., 2015. Removal of organic matter and nitrogen in swine wastewater using an integrated ion exchange and bioelectrochemical system. *Bioresour. Technol.* 189, 107-112.

Mac Namara, C., Torroba, J., Deacon, A., 2015. New Smopex® ion exchange materials for the removal of selenium from industrial effluent streams: Material characterization, modeling and process implementation. *Johnson Matthey Technol. Rev.* 59, 334-352.

Martins, K., Johnson, J., Leber, K., Srinivasan, R., Heller, B. 2012. Bench- and pilot-scale testing of ion exchange and zero valent iron technologies for selenium removal from a surface coal mine run-off water. *Proc. of the Wat. Environ. Federation.* 17, 318-338.

Meng, X.Y., Vaccari, D.A., Zhang, J.F., Fiume, A., Meng, X.G., 2014. Bioregeneration of spent anion exchange resin for treatment of nitrate in water. *Environ. Sci. Technol.* 48, 1541-1548.

Oehmen, A., Valerio, R., Llanos, J., Fradinho, J., Serra, S., Reis, M.A.M., Crespo, J.G., Velizarov, S., 2011. Arsenic removal from drinking water through a hybrid ion exchange membrane - coagulation process. *Sep. Purif. Technol.* 83, 137-143.

Roest, K., Heilig, H., Smidt, H., de Vos, W.M., Stams, A.J.M., Akkermans, A.D.L., 2005. Community analysis of a full-scale anaerobic bioreactor treating paper mill wastewater. *Sys. App. Microbiol.* 28, 175-185.

Spallholz, J.E., Hoffman, D.J., 2002. Selenium toxicity: cause and effects in aquatic birds. *Aqua. Toxico.* 57, 27-37.

Staicu, L.C., Morin-Crini, N., Crini, G., 2017. Desulfurization: critical step towards enhanced selenium removal from industrial effluents. *Chemosphere.* 172, 111-119.

Strauss, J.M., Riedel, K.J., du Plessis, C.A., 2004. Mesophilic and thermophilic BTEX substrate interactions for a toluene-acclimatized biofilter. *Appl. Microbiol. Biotechnol.* 64, 855-861.

Tan, L.C., Nancharaiah, Y.V., van Hullebusch, E.D., Lens, P.N.L., 2016. Selenium: Environmental significance, pollution, and biological treatment technologies. *Biotechnol. Adv.* 34, 886-907.

USEPA, 2001. Parameters of water quality: Interpretation and standards. Wexford, Ireland 1-133. (Available at) https://www.epa.ie/pubs/advice/water/quality/Water_Quality.

Venkatesan, A.K., Sharbatmaleki, M., Batista, J.R., 2010. Bioregeneration of perchlorate-laden gel-type anion-exchange resin in a fluidized bed reactor. *J. of Hazard. Mater.* 177, 730-737.

CHAPTER 9

General discussion and future perspective

9.1 General discussion

Se pollution has led to several cases of severe aquatic ecosystem deterioration due to Se poisoning. In locations in the US, there are still evident Se toxic effects to the biotic organisms even after Se inputs were stopped for 10 years (Lemly 1997, 2014). Se pollution and bioaccumulation in the aquatic environment can cause reproductive failures and teratogenic effects on fishes leading to both environmental degradation and monetary losses to fishery. Therefore, Se must be removed from wastewater before discharging to prevent environmental impacts. Additionally, fines and legislative problems for the industry can also occur due to the strict regulation on discharge limit imposed by regulatory on bodies, i.e. the US Environmental Protection Agency (USEPA) freshwater aquatic discharge limit of 5 µg Se/L. Biological Se treatment processes are attractive because of greener approach and lower capital/operation cost compared to physico-chemical methods (Nancharaiah and Lens 2015b). Despite the sustainable advantages of biological techniques, there is still a lack of application for large-scale processes, particularly in treating real Se-laden wastewater due to the complex wastewater characteristics and presence of co-contaminants that can inhibit Se oxyanion removal. Se importance, current available Se treatment removal strategies (bench- and large-scale), and challenges in biological Se removal are critically discussed in **chapter 2**.

In order to bridge the gap in the literature and bring to light further information on the Se removal in complex mine wastewater streams, the main objective of this research was to investigate the effect of co-contaminants on SeO_4^{2-} removal, biofilm-Se interactions and to evaluate different treatment processes for improving Se wastewater treatment. This research focused on Se removal under varying operational conditions such as co-contaminant concentrations, pH, and exposure of different mixtures of co-contaminants in order to elucidate their effects on the Se removal and subsequent biomass formation, growth, and composition (**Fig. 9.1**). In addition, various reactor configurations and unit process flow integration were explored in order to pinpoint an appropriate and optimized Se treatment process. A treatment scheme, based on the findings from this research, is provided for adopting to Se-wastewater treatment with co-contaminants (**Fig. 9.2**). Finally, recommendations for further research such as the use of innovative reactor configurations (**Fig. 9.3**) and possible recovery schemes for Se complexes in biomass (**Fig. 9.2**). This work provides valuable insights into the treatment process of SeO_4^{2-} with co-contaminants that can be exploited to better treat complex Se-laden mine wastewater as well as implicitly develop recovery systems for valuable materials.

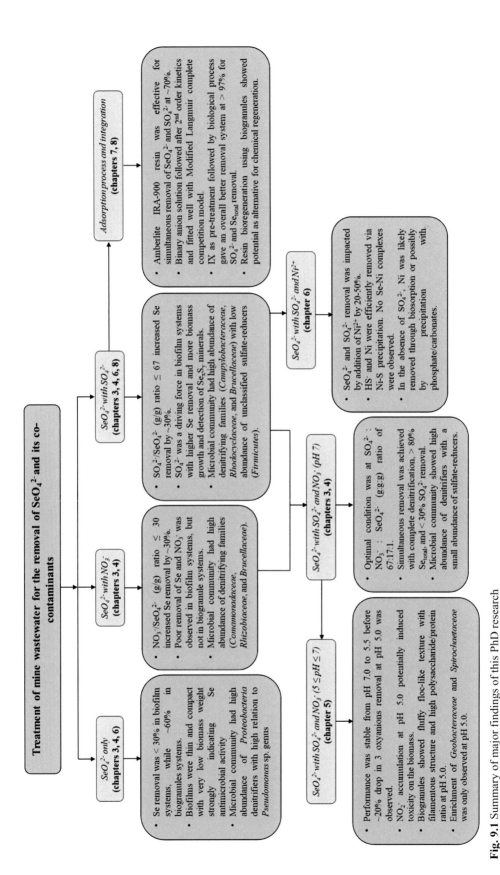

Fig. 9.1 Summary of major findings of this PhD research

9.1.1 Effect of co-contaminants and linking to bioreactor configuration

The practical application of biological treatment for Se-laden wastewater may have important limitations because of the presence of co-contaminants such as NO_3^- and SO_4^{2-} and heavy metal ions (e.g. Ni^{2+}). **Chapters 3, 4** and **6** investigated the performance of reactors (batch fed system, drip flow, biotrickling filter, and upflow anaerobic sludge blanket) receiving SeO_4^{2-} synthetic mine wastewater with a combination of different co-contaminants. It should be emphasized that the inoculum used in this study was not pre-adapted or acclimated to the presence of these targeted contaminants (i.e. SeO_4^{2-}, NO_3^-, SO_4^{2-}, and Ni^{2+}) before the start of each experiments. This was done to avoid the development of specific reducing microorganisms at the beginning of the operation and to clearly determine the changes in SeO_4^{2-} removal and the effect of co-contaminants on Se removal. Furthermore, this research not only aimed to achieve high SeO_4^{2-} removal efficiencies, but also to simultaneously remove the co-contaminants for developing a more suitable biological treatment process for Se-laden wastewaters.

Chapter 3 observed an optimal NO_3^- and SO_4^{2-} to SeO_4^{2-} ratio that allowed for a higher removal of SeO_4^{2-} of ~30% when compared it to SeO_4^{2-} alone in batch fed systems. Furthermore, the Se mass balance indicated that more Se was associated with the biomass in the presence of NO_3^- while more colloidal Se was found in the liquid phase in the presence of SO_4^{2-}. The optimal SO_4^{2-}:NO_3^-:SeO_4^{2-} (g/g) ratio was found to be at 67:17:1 and was used for all the reactor conditions conducted in the experiments of this study presented in the succeeding chapters. **Chapter 3** further demonstrated that an upflow anaerobic granular sludge blanket (UASB) reactor was capable of removing SeO_4^{2-} in the presence of NO_3^- and SO_4^{2-} without prior inoculum adaptation, achieving a 100% NO_3^-, 30% SO_4^{2-} and 80% Se_{tot} removal efficiency for 90 days.

Comparing the effectiveness of SO_4^{2-} and NO_3^- for Se removal in continuous operation, there seems to be a correlation with different reactor configurations, microbial growth system, and co-oxyanions. SO_4^{2-} showed to be a controlling factor in biofilm systems (drip flow reactor (DFR) and biotrickling filter (BTF)) achieving both higher Se removal and more biofilm growth/formation when SO_4^{2-} was present as observed in **chapters 4** and **6**. In contrast, the UASB reactor did not reveal any changes, whether increase or decrease, in Se removal efficiencies, when SO_4^{2-} was removed from (**chapter 3**) or included (**chapter 6**) in the feed solution. In contrast to SO_4^{2-}, there were indications that NO_3^- had a positive influence on the

UASB reactor performance. It was observed the removal of NO_3^- from the feed solution possibly caused the increase in Se_{tot} concentration in the effluent, thus negatively impacting Se removal efficiencies by the end of the reactor run (**chapter 3**). Interestingly, in a 10 day operated DFR, not only did NO_3^- show no positive influence on Se removal, NO_3^- removal was also very low in the biofilm system when fed with SeO_4^{2-} (**chapter 4**). However, complete denitrification was observed when SO_4^{2-} was present.

The influence of NO_3^- and SO_4^{2-} on SeO_4^{2-} removal is most likely interlinked with (i) the microbial growth type used, (ii) possible interaction among the bioreduced products of the oxyanions, and (iii) microbial community changes that occurred during the operation. SeO_4^{2-} reduction can be carried out by various microorganisms including denitrifiers and sulfate-reducers. The microbial community consisted of high relative abundances of denitrifiers along with small proportion of sulfate reducers irrespective of whether NO_3^- was present or not (**chapters 3** and **4**). SeO_4^{2-} reduction to Se^0 can be achieved by the action of either SeO_4^{2-} reductase, NO_3^- reductase or periplasmic NO_2^- reductase (DeMoll-decker and Macy 1993; Nancharaiah and Lens 2015a). It was hypothesized that the presence of NO_3^- can promote a higher state of metabolic activity in microorganisms (Oremland et al. 1999). This implies that NO_3^- can play a role in shaping up the microbial community in the bioreactor and specific metabolic pathways/activities of denitrifiers could be linked to the increase of SeO_4^{2-} removal in the presence of NO_3^- and its by-products (Lai et al. 2014).

In contrast to biogranules, biofilm systems (i.e. DFR and BTF) rely heavily on the biofilm formation and growth. Despite that denitrifiers were abundant in conditions without SO_4^{2-} (**chapter 4**), poor NO_3^- and SeO_4^{2-} removal efficiencies were observed. BTF reactor achieved Se removal faster and higher in SO_4^{2-} fed solutions compared with SeO_4^{2-} alone (**chapter 6**). These results are linked to the higher biofilm growth achieved. In the SeO_4^{2-} only system, biofilms were thin and compact (**chapter 4**) and had very low biomass weight with high concentrations of unaccounted $Se_{dissolved}$, possibly associated with organo-Se species (**chapter 6**). Other studies have confirmed that biologically synthesized Se^0 and organo-Se have the ability to both completely inhibit biofilm formation and eradicate established biofilms (Vercellino et al. 2013a; Zonaro et al. 2015). In reactors operating with SO_4^{2-}, Se-S mineral complexes were found (**chapters 4** and **6**), with a lower concentration of unaccounted $Se_{dissolved}$ detected in the effluent (**chapter 6**). It can, therefore, be hypothesized that the presence of SO_4^{2-} can mitigate the antimicrobial activity caused by biogenic Se species. Although it is still possible that Se^0 is still being formed under $SO_4^{2-} + SeO_4^{2-}$ conditions, it might be to a lesser degree allowing for higher biofilm growth. Overall, the presence of co-oxyanions showed surprisingly to be beneficial for SeO_4^{2-} removal. However, careful control of co-oxyanion concentrations is desired and an appropriate reactor configuration must be chosen.

Chapter 6 also demonstrated the importance of the COD/SO_4^{2-} ratio and the impact of Ni^{2+} on the BTF and UASB reactor. Poor SO_4^{2-} removal ($< 30\%$) was observed at ~ 1.8 COD/SO_4^{2-} while increasing the COD/SO_4^{2-} to 2.8 increased the SO_4^{2-} removal to $> 80\%$ with generation of high sulfide (HS^-) concentration of up to 250 mg S/L. Addition of Ni^{2+} to sulfidogenic conditions allowed simultaneous removal of Ni and sulfide via co-precipitation. Formation of Ni_3S_2 precipitates was confirmed by XRD. However, Ni^{2+} addition negatively impacted both SeO_4^{2-} and SO_4^{2-} removal by 20-50%, possibly due to the sudden metal toxicity to the biofilms/biogranules. Compared to BTF, the UASB reactor performance was able to recover quicker. This is perhaps due to the properties of biogranules that contain multi-layer microbial groups present allowing for better response to operational changes and stress. The microbial population and millimeter sized granular matrix was responsible for more resilient nature of biogranules possibly accounted for quick recovery of reactor performance.

9.1.2 Treatment operation at slightly acidic pH

One of the characteristics of mine wastewater is the low pH at which it is typically produced after mine processing. The effect of slightly acidic pH on the treatability of Se-laden wastewaters with NO_3^- and SO_4^{2-} was presented in **Chapter 5**. No changes in removal efficiency were observed when the pH was gradually lowered from 7.0 to 5.5 (30-129 days), before a 20-30% drop in performance was observed across all constituents during operation at pH 5.0 (130-168 days). Further operation of the UASB reactor at pH 5.0 for 98 days showed stable performance attaining 79% NO_3^-, 15%, SO_4^{2-} 43%, Se_{tot}, and 61% $Se_{dissolved}$ removal efficiencies.

Changes in granular sludge structure and microbial community were observed at low pH, showing the formation of filamentous floc biomass and enrichment of the family *Geobacteraceae* and *Spirochaetaceae*, both associated with acidic systems, was only observed at pH 5.0. Additional studies for reactor bioaugmentation with immobilized *Geobacter* and *Spirochaete* can be conducted to jumpstart the reactor operation at low pH without possible major reactor failure due to sudden exposure of the biomass to a low pH. Ecology profiling and identification of key microorganisms playing a key role in Se removal in low pH are useful in defining reactor strategies. By introducing isolated key microorganisms through bioaugmentation could accelerate the reactor start-up and increase the tolerance for the negative impact of the acidic solution.

9.1.3 Adsorption coupled with biological process

An ion exchange (IX) process was evaluated for removing both SeO_4^{2-} and SO_4^{2-} from synthetic mine wastewaters. **Chapter 7** validated the applicability of Amberlite® IRA-900, a strong anionic ion exchange (IX) resin, for the simultaneous removal of SeO_4^{2-} and SO_4^{2-} achieving > 70% adsorption capacities. Modified Langmuir multi-component isotherms using a complete competition model indicated that IRA-900 was not selective towards SeO_4^{2-} and that both oxyanions competed for the active sites. The non-selectivity of IRA-900 gives advantage of adsorbing both oxyanions saving the need for pre-treatment of SO_4^{2-} to avoid inhibition of SeO_4^{2-} adsorption. Chemical regeneration was optimal at 20 mins using 0.5 M HCl and resin was reusable up to 6 adsorption-desorption cycles.

Despite the advantages of biological techniques, there are still issues with reaching the regulatory discharge limit for Se_{tot}. **Chapter 8** evaluated the feasibility of combining two unit processes, an IX column and a UASB reactor, for the overall improvement of SO_4^{2-} and Se_{tot} removal. IX process was evaluated as a pre- or post-treatment process for the biological process. IX as a post-treatment demonstrated to be less effective due to competition with other bioreduced products and presence of suspended solids. On the other hand, IX as pre-treatment system allowed the UASB reactor to receive lower concentrations of SO_4^{2-} and SeO_4^{2-} facilitating for a better biological removal process and achieving an overall higher removal efficiency of 99% SO_4^{2-} and 97% Se_{tot}. The final treated effluent contained < 100 mg/L SO_4^{2-} and < 0.3 mg Se/L Se_{tot}. Though the Se_{tot} concentration still does not reach the regulatory discharge limit, it should be noted that all dissolved Se was removed and only colloidal Se was remaining. Therefore, the addition of coagulation process as finishing step could further lower the Se_{tot} concentration and achieve the regulatory limit (Nancharaiah and Lens 2015b). Additionally, with the lowering of the SO_4^{2-} concentration supplied to the biological process, less COD requirement would be needed, decreasing the operational costs. Finally, bioregeneration of resin by reduction of adsorbed SO_4^{2-} and SeO_4^{2-} by biogranules was feasible and could potentially be used as an alternative to chemical regeneration.

This study demonstrated that integration of unit processes is required to fully remove Se and SO_4^{2-} from effluent wastewaters. Based on the major findings obtained in this research, a

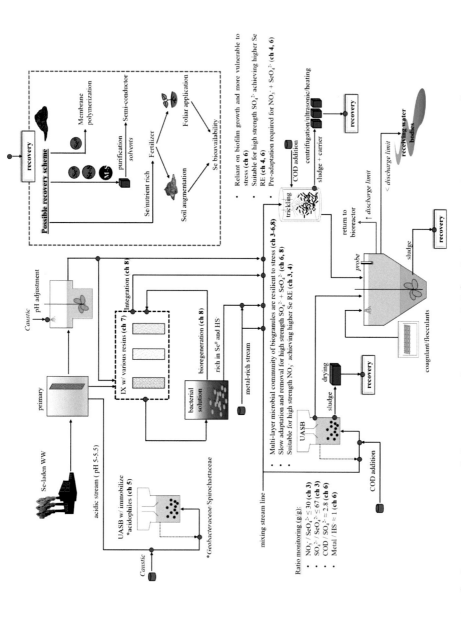

Fig. 9.2 Integrated strategy for the treatment process of mine wastewater for removing SeO$_4^{2-}$ and other co-contaminants (NO$_3^-$, SO$_4^{2-}$, and metal ions) at mesophilic conditions. The schematic treatment flow process was constructed based on the major findings taken from each chapter. A possible future recovery scheme and application are suggested.

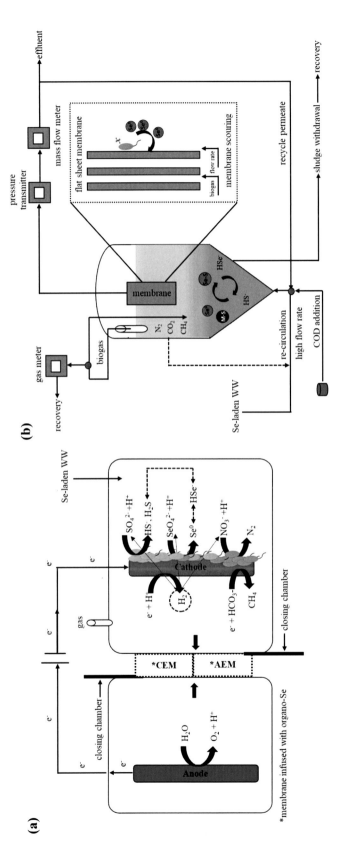

Fig. 9.3 Innovative reactor configurations proposed for Se-laden wastewaters with co-contaminants: (a) two-chamber bioelectrochemical system (BES) for autotrophic reduction of oxyanion utilizing hydrogen as electron donor generated electrochemically with an anionic/cationic exchange membrane (AEM/CEM) for ion transfer from one chamber to another; (b) modified upflow anaerobic sludge blanket reactor with submerged membrane module (e.g. flat sheet or hollow fiber membrane) utilizing biogas production and high flow rate for membrane scouring.

strategic treatment process flow is recommended for Se-laden wastewaters with co-contaminants (**Fig. 9.2**).

9.2 Future perspectives

9.2.1 Recovery of biologically produced materials for reuse applications

Resource recovery is an important aspect to consider when applying biological treatment techniques. Biomass from the bioreactors contains various valuable materials such as metals, sulfur or sulfur complexes, and Se components that have the potential to be recovered. However, the recovery process of Se and other complexes from biomass have not been explored. So far, a method for recovery and purification of Se^0 nanoparticles from biomass was suggested as a series of solid-liquid steps: high-speed centrifugation, sonication followed by hexane separation (Jain et al. 2015). As such, an eco-friendlier and simple recovery approach should be developed. Recovery and purification of Se^0 nanoparticles, along with metal and sulfur complexes, can be reused for various semiconductor application (i.e. batteries and catalyst, an adsorbent for toxic metals (i.e. mercury), and antimicrobial agents for medical devices and membranes (Staicu et al. 2015; Zonaro et al. 2015). In the case of biogranules, it is possible to simply dry the biomass and use it as a fertilizer for soil augmentation and foliar application for those regions with Se deficiency to increase the Se bioavailability (Alfthan et al. 2015; Wang et al. 2017). In the case of biofilms with carriers, detachment of biofilms through centrifugation/sonication or heating can be considered. However, it should be noted that the biomass produced while treating industrial mine wastewater contains also various other contaminants that can be harmful to the environment. Therefore, detailed studies on the toxic effect and health ramification of using Se-rich biomass as fertilizer should be conducted.

9.2.2 Innovative reactor configurations for SeO_4^{2-} removal with co-contaminants

Based on this PhD study, alternative reactor configurations or process system are suggested in (**Fig. 9.3**): (a) a bioelectrochemical (BES) process and (b) an anaerobic reactor with submerged membrane module. A BES system is an innovative process that employs microorganisms and electrodes to catalyze redox reactions and this system can be utilized in a two-chamber microbial fuel cell using either the anode, cathode or both as bioelectrodes (Hamelers et al. 2010). Electroactive microorganisms can extract electrons that can be utilized to generate electricity, treat wastewater by redox reactions and also recover nutrients (i.e. P, N) and metals (Nancharaiah et al. 2016). **Fig. 9.3a** suggests a possible reactor configuration that utilizes the ability of BES to generate hydrogen gas (H_2) through an electrochemical process.

Although lactate showed to be efficient and useful as the sole electron donor/carbon source in this research, alternative electron donors/carbon sources are recommended since using lactate adds operational cost considering that most industrial Se containing wastewaters have low a COD content. Autotrophic reduction of NO_3^-, SO_4^{2-} and SeO_4^{2-} using H_2 as the electron donor was found to be successful in membrane biofilm reactors (van Ginkel et al. 2011; Ontiveros-Valencia et al. 2016). It is, therefore, feasible to carry out autotrophic reduction of NO_3^-, SO_4^{2-} and SeO_4^{2-} at the biocathode side by autotrophs utilizing the generated H_2. An IX membrane (anionic or cationic) can also be used to separate the two-chambers and allow for charge balancing by moving H^+ and OH^- as well as other possible value-added chemicals produced in each chamber. An IX membrane can be infused with organo-Se to reduce the biofouling that can occur on the side of the bioelectrode. Removal of individual selenite, NO_3^- and SO_4^{2-} using BES systems has been reported in some studies (Catal et al. 2009; Hamelers et al. 2010; Luo et al. 2017). However, application of BES to complex Se-laden wastewater with co-contaminants stream has yet to be reported.

Membrane filtration technology using reverse osmosis (RO) and nanofiltration (NF) has been studied for treating Se-laden wastewaters (Staicu et al. 2017). However, despite the effectiveness of these technologies, the operational cost is very high due to the amount of the required hydraulic pressure. On the other hand, a membrane bioreactor is a cheaper option compared to NF or RO and combines the biological process and particle separation (Hu et al. 2017). Although membrane technology still has the major drawback of higher operating cost than conventional processes, this can be overcome by decreasing the downstream unit processes and overall reduction of reactor size (Hu and Stuckey 2006). **Fig. 9.3b** suggests the possible reactor configuration for the anaerobic reactor with a submerged membrane. One of the main disadvantages of using membrane technology, particularly in an anaerobic system, is the higher potential for biofouling. However, this could be mitigated by using a high flow rate and biogas recirculation as a means for membrane scouring. Additionally, it is possible to either fabricate membranes with covalent Se attached to the membrane or utilize the colloidal Se^0 as a means to reduce biofouling. It was reported in some studies that a reduction of the biofilm formation and thickness by > 5 logs, as well as a flux loss decreased from 55 to 15% due to the biofouling attenuation when organo-Se was covalently attached to the membrane surface (Vercellino et al.

2013a, 2013b). To the best of our knowledge, no studies have been reported utilizing a submerged membrane module in anaerobic reactor for treatment of Se-laden wastewater.

9.2.3 Understanding the biofilm ecology in the bioreactors

Aside from exploring innovative reactor configurations and process integration, additional studies are required to delve deeper into the Se-reducing biofilm/biogranules ecology for a better understanding of the metabolic processes and integration into wastewater process engineering. Considering the diverse presence of co-contaminants along with SeO_4^{2-}, development of a diverse microbial population is expected, as was shown in **chapters 3, 4,** and **5**. Ecology studies can relate the structure-function link of the different communities within an overall denitrifying and SO_4^{2-} reducing population, as well indicate the possible metabolic interplay within syntrophic associations. However, it is unknown in this research whether population abundance and microbial activity have a direct link with each other and if the syntrophic associations, if any, between different dominating microbial groups. For example, though SO_4^{2-} showed to be a driving force in Se removal, particularly in biofilm systems, the microbial community was composed of denitrifiers with only a smaller abundance of sulfate-reducers.

It would be interesting to identify the active population directly involved in the reduction of all three oxyanions or those that utilize/interact with their metabolic intermediates/products and how the individual populations (i.e. denitrifiers, sulfate-reducers, Se-reducers) interact with each other. It is possible to identify the active microbial fraction through isotope tracking based identification with qPCR quantification techniques in order to clarify the direct or indirect role of different microorganisms involved in the SeO_4^{2-}, NO_3^- and SO_4^{2-} metabolism (Hungate 2015). Distribution, abundance, and function of specific identified populations can be integrated with kinetic parameters obtain from reactor data experiments and can be input into design and assimilation software such as Biowin™ (EnvironSIM Associates Ltd.) for the modelling of Se reducing processes at different operational conditions (Moragaspitiya et al. 2017). To the best of our knowledge, no process modeling has been specifically defined for Se-laden wastewater treatment, unlike those with denitrification. Therefore, development of mathematical models for the Se-laden wastewaters is recommended for more accurate predictions of reactor performance (Boltz et al. 2011).

9.3 Conclusions

The removal of SeO_4^{2-} from wastewater streams with co-contaminants has been largely modeled as a black box in biological process systems using mixed consortia. This study demonstrated that the presence of co-contaminants can actually be beneficial for Se removal provided that concentrations are carefully monitored and appropriate treatment process, operating conditions, as well as configurations are used. The knowledge gain from this research can help in the advancement and application of biological processes such as predicting of reactor performance, solving specific design or practical problems and implement novel treatment techniques for Se-laden mine wastewater.

9.4 References

Alfthan, G., Eurola, M., Ekholm, P., Venäläinen, E.R., Root, T., Korkalainen, K., Hartikainen, H., Salminen, P., Hietaniemi, V., Aspila, P., Aro, A., 2015. Effects of nationwide addition of selenium to fertilizers on foods, and animal and human health in Finland: From deficiency to optimal selenium status of the population. *J. Trace Elem. Med. Biol.* 31, 142-147.

Boltz, J. P., Brockmann, D., Sandy, T., Johnson, B. R., Daigger, G. T., Jenkins, K., Munirathinam, K., 2011. Framework for a mixed-culture biofilm model to describe oxidized nitrogen, sulfur, and selenium removal in a biofilm reactor. In: WEFTEC 2011. WEF Proceedings. Presented at WEFTEC 2011 - the Water Environment Federation's Annual Technical Exhibition and Conference, Los Angeles, USA.

Catal, T., Bermek, H., Liu, H., 2009. Removal of selenite from wastewater using microbial fuel cells. *Biotechnol. Lett.* 31, 1211-1216.

DeMoll-decker, H., Macy, J.M., 1993. The periplasmic nitrite reductase of *Thauera selenatis* may catalyze the reduction of selenite to elemental selenium. *Arch. f* 160, 241-247.

Hamelers, H.V.M., Ter Heijne, A., Sleutels, T.H.J.A., Jeremiasse, A.W., Strik, D.P.B.T.B., Buisman, C.J.N., 2010. New applications and performance of bioelectrochemical systems. *Appl. Microbiol. Biotechnol.* 85, 1673-1685.

Hu, A.Y., Stuckey, D.C., 2006. Treatment of dilute wastewaters using a novel submerged anaerobic membrane bioreactor. *J. Environ. Eng.* 132, 190-198.

Hu, Y., Wang, X.C., Hao Ngo, H., Sun, Q., Yang, Y., 2017. Anaerobic dynamic membrane bioreactor (AnDMBR) for wastewater treatment: A review. *Bioresour. Technol.* (in press).

Hungate, B.A., 2015. Quantitative microbial ecology through stable isotope probing. *Soil Biol. Biochem.* 81, 7570-7581.

Jain, R., Jordan, N., Weiss, S., Foerstendorf, H., Heim, K., Kacker, R., Hübner, R., Kramer, H., Van Hullebusch, E.D., Farges, F., Lens, P.N.L., 2015. Extracellular polymeric substances govern the surface charge of biogenic elemental selenium nanoparticles. *Environ. Sci. Technol.* 49, 1713-1720.

Lai, C.-Y., Yang, X., Tang, Y., Rittmann, B.E., Zhao, H.-P., 2014. Nitrate shaped the selenate-reducing microbial community in a hydrogen-based biofilm reactor. *Environ. Sci. Technol.* 48, 3395-3402.

Lemly, A.D., 1997. Ecosystem recovery following selenium contamination in a freshwater reservoir. *Ecotoxicol. Environ. Saf.* 36, 275-281.

Lemly, A.D., 2014. An urgent need for an EPA standard for disposal of coal ash. *Environ. Pollut.* 191, 253-255.

Luo, H., Teng, W., Liu, G., Zhang, R., Lu, Y., 2017. Sulfate reduction and microbial community of autotrophic biocathode in response to acidity. *Process Biochem.* 54, 120-127.

Mal, J., Nancharaiah, Y.V., van Hullebusch, E.D., Lens, P.N.L., 2016. Effect of heavy metal co-contaminants on selenite bioreduction by anaerobic granular sludge. *Bioresour. Technol.* 206, 1-8.

Moragaspitiya, C., Rajapakse, J., Senadeera, W., Ali, I., 2017. Simulation of dynamic behaviour of a biological wastewater treatment plant in South East Queensland, Australia using Bio-Win software. *Eng. J.* 21, 1-22.

Nancharaiah, Y.V., Lens, P.N.L., 2015a. Ecology and biotechnology of selenium-respiring bacteria. *Microbiol. Mol. Biol. Rev.* 79, 61-80.

Nancharaiah, Y.V., Lens, P.N.L., 2015b. Selenium biomineralization for biotechnological applications. *Trends Biotechnol.* 33, 323-330.

Nancharaiah, Y.V., Mohan, S.V., Lens, P.N.L., 2016. Biological and bioelectrochemical recovery of critical and scarce metals. *Trends Biotechnol.* 34, 137-155.

Ontiveros-Valencia, A., Penton, C.R., Krajmalnik-Brown, R., Rittmann, B.E., 2016. Hydrogen-fed biofilm reactors reducing selenate and sulfate: Community structure and capture of elemental selenium within the biofilm. *Biotechnol. Bioeng.* 113, 1736-1744.

Oremland, R.S., Blum, J.S., Bindi, A.B., Dowdle, P.R., Herbel, M., Stolz, J.F., 1999. Simultaneous reduction of nitrate and selenate by cell suspensions of selenium-respiring bacteria. *Appl. Environ. Microbiol.* 65, 4385-4392.

Staicu, L.C., van Hullebusch, E.D., Lens, P.N.L., 2015. Production, recovery and reuse of biogenic elemental selenium. Environ. *Chem. Lett.* 13, 89-96.

Staicu, L.C., van Hullebusch, E.D., Lens, P.N.L., 2017. Industrial selenium pollution: Wastewaters and physical-chemical treatment technologies, in: van Hullebusch, E.D. (Ed.), Bioremediation of Selenium Contaminated Wastewater. Springer International Publishing, pp. 103-130.

van Ginkel, S.W., Yang, Z., Kim, B., Sholin, M., Rittmann, B.E., 2011. The removal of selenate to low ppb levels from flue gas desulfurization brine using the H2-based membrane biofilm reactor (MBfR). *Bioresour. Technol.* 102, 6360-6364.

Vercellino, T., Morse, A., Tran, P., Hamood, A., Reid, T., Song, L., Moseley, T., 2013a. The use of covalently attached organo-selenium to inhibit S. aureus and E. coli biofilms on RO membranes and feed spacers. *Desalination* 317, 142-151.

Vercellino, T., Morse, A., Tran, P., Song, L., Hamood, A., Reid, T., Moseley, T., 2013b. Attachment of organo-selenium to polyamide composite reverse osmosis membranes to inhibit biofilm formation of S. aureus and E. coli. *Desalination* 309, 291-295.

Wang, Q., Yu, Y., Li, J., Wan, Y., Huang, Q., Guo, Y., Li, H., 2017. Effects of different forms of selenium fertilizers on Se accumulation, distribution, and residual effect in winter wheat-summer maize rotation system. *J. Agric. Food Chem.* 65, 1116-1123.

Zonaro, E., Lampis, S., Turner, R.J., Junaid, S., Vallini, G., 2015. Biogenic selenium and tellurium nanoparticles synthesized by environmental microbial isolates efficaciously inhibit bacterial planktonic cultures and biofilms. *Front. Microbiol.* 6, 658.

Appendix 1

Supporting information for Chapter 3

Table S3.1 Batch test results with standard deviation after 5 days of incubation under different molar ratios of NO_3^- and SO_4^{2-} to SeO_4^{2-}.

Condition/molar ratio	Lactate Initial conc. (mM)	Lactate % Removal	NO_3^- * Initial conc. (mM)	NO_3^- * % Removal	SO_4^{2-} Initial conc. (mM)	SO_4^{2-} % Removal	TDS Produced (mM)	Se_{diss} Initial conc. (μM)	Se_{diss} % Removal
Killed biomass	21 (±1)	0 (±0)	2 (±1)	0 (±0)	3 (±1)	0 (±0)	0.0 (±0.0)	179 (±1)	0 (±0)
No lactate added	n/a***	100 (±0)	2 (±0)	0 (±0)	3 (±0)	0 (±0)	0.0 (±0.0)	163 (±1)	0 (±0)
NO_3^- only	12 (±1)	100 (±0)	4 (±0)	100 (±0)	n/a***	n/a***		n/a***	n/a***
SO_4^{2-} only	14 (±1)	100 (±0)	n/a***	n/a***	25 (±1)	15 (±1)	4.0 (±0.0)	n/a***	n/a***
NO_3^- + SO_4^{2-}	14 (±0)	100 (±0)	4 (±1)	100 (±0)	24 (±0)	6 (±2)	3.0 (±0.0)	n/a***	n/a***
SeO_4^{2-} only	25 (±4)	83 (±0)	n/a***	n/a***	n/a***	n/a***		108 (±0)	67 (±2)
**NO_3^- / SeO_4^{2-} — 1	13 (±0)	100 (±0)	0.5 (±0.0)	100 (±0)		n/a***		410 (±0)	62 (±12)
20	17 (±2)	85 (±3)	2 (±0)	100 (±0)		n/a***		96 (±0)	100 (±0)
40	17 (±2)	85 (±3)	4 (±0)	100 (±0)		n/a***		141 (±0)	91 (±12)
70	19 (±2)	90 (±1)	7 (±0)	100 (±0)		n/a***		117 (±0)	81 (±10)
100	18 (±1)	92 (±1)	10 (±0)	100 (±0)		n/a***		98 (±0)	59 (±4)
SO_4^{2-} / SeO_4^{2-} — 1	13 (±0)	91 (±2)	n/a***	n/a***	0.5 (±0.0)	63 (±9)	0.3 (±0.2)	410 (±0)	71 (±4)
50	24 (±4)	83 (±2)	n/a***	n/a***	4 (±0)	39 (±10)	0.5 (±0.2)	236 (±0)	85 (±0)
100	24 (±5)	92 (±0)	n/a***	n/a***	9 (±1)	28 (±1)	1.0 (±0.2)	237 (±0)	79 (±1)
150	24 (±5)	86 (±1)	n/a***	n/a***	14 (±1)	23 (±1)	1.0 (±0.2)	194 (±0)	40 (±0)
200	24 (±3)	92 (±1)	n/a***	n/a***	20 (±1)	25 (±7)	0.9 (±0.3)	235 (±0)	52 (±0)
300	22 (±4)	89 (±1)	n/a***	n/a***	28 (±1)	16 (±1)	1.7 (±0.9)	247 (±0)	39 (±1)
SO_4^{2-} / SeO_4^{2-} with co-exposure of 4 mM NO_3^- — 1	13 (±0)	92 (±1)	0.5 (±0.0)	100 (±0)	0.5 (±0.0)	27 (±4)	0.1 (±0.0)	500 (±0)	100 (±0)
50	25 (±3)	86 (±0)	3 (±1)	100 (±0)	4 (±0)	30 (±2)	0.3 (±0.1)	201 (±0)	92 (±1)
100	24 (±5)	84 (±1)	3 (±1)	100 (±0)	9 (±1)	25 (±5)	0.7 (±0.3)	197 (±0)	91 (±1)
150	24 (±4)	86 (±0)	3 (±1)	100 (±0)	14 (±1)	25 (±6)	0.6 (±0.2)	217 (±0)	61 (±2)
200	24 (±5)	92 (±1)	3 (±1)	100 (±0)	19 (±1)	20 (±4)	0.4 (±0.0)	240 (±0)	61 (±2)
300	24 (±5)	92 (±1)	3 (±1)	100 (±0)	31 (±1)	19 (±3)	0.5 (±0.1)	221 (±0)	32 (±0)

*Note: results are all reported as mean ± standard deviation, n = 4 replicates (total); *NO_3^- concentration reached 100% removal for all concentration at 48 h; ** Se_{diss} reported here was taken from IC measurement as SeO_4^{2-} peak; ***n/a – not applicable*

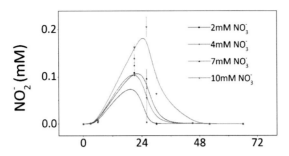

Fig. S3.1 NO$_2^-$ profile for batch experiment at NO$_3^-$ + SeO$_4^{2-}$ condition varying initial NO$_3^-$ concentration.

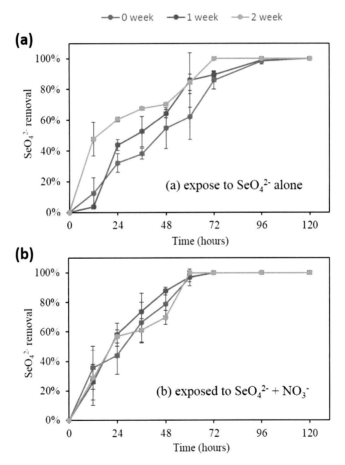

Fig. S3.2 SeO$_4^{2-}$ reduction using untreated sludge and treated sludge (reused from the previous week) (a) expose only to SeO$_4^{2-}$ and (b) expose to both NO$_3^-$ and SeO$_4^{2-}$. Batch experiments were conducted under excess lactate (10mM), 30°C, 200 rpm and pH 7.0.

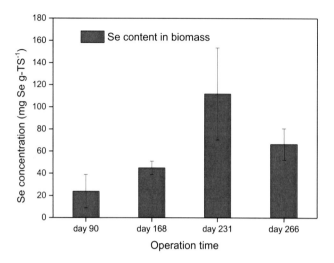

Fig. S3.3 Se concentration detected in the biomass at various operation times (n=3 biomass sample replicate at each time period).

Fig. S3.4 Red-color formation in the reactor walls and tubings due to Se^0 production.

Appendix 2

Supporting information for Chapter 4

Table S4.1 Overall Se mass balance for the biofilm system under different incubations.

	unit	SeO_4^{2-}	$SeO_4^{2-} + NO_3^-$	$SeO_4^{2-} + SO_4^{2-}$	$SeO_4^{2-} + NO_3^- + SO_4^{2-}$
Se fed into the system	mg Se	27.1 (± 1.7)	30.5 (± 0.7)	32.2 (± 1.0)	31.9 (± 1.3)
Effluent					
Se$_{tot}$ (discharged)	mg Se	23.0 (± 2.4)	26.3 (± 0.2)	13.8 (± 1.3)	13.1 (± 3.1)
Se in biomass	mg Se	0.5 (± 0.3)	0.3 (± 0.1)	5.4 (± 2.9)	2.0 (± 1.4)

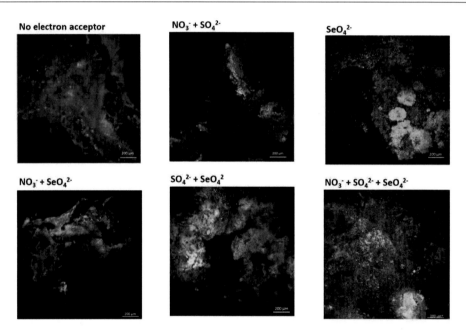

Fig. S4.1 CLSM (confocal laser scanning microscopy) images of bacterial biofilms taken at 100× magnification with SYTO 9 (green) and PI (red) staining. Scale bars represent 200 μm for all treatments. Both SYTO 9 and PI are DNA binding dyes. SYTO 9 can cross the cell membrane of normal (healthy) cells. PI, on the other hand, can only enter the cells with a compromised cell membrane (permeabilized) or stain extracellular DNA because PI is a large dye and cannot enter healthy cells with intact cell membrane.

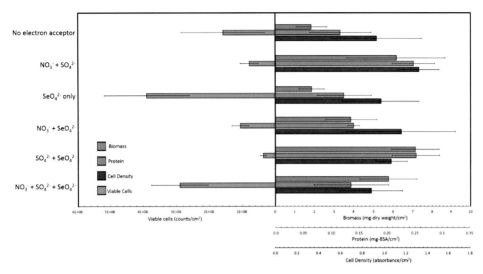

Figure S4.2 Physiochemical analysis of biofilms from each treatment for dry weight, protein content, cell density and viable cells reported per area.

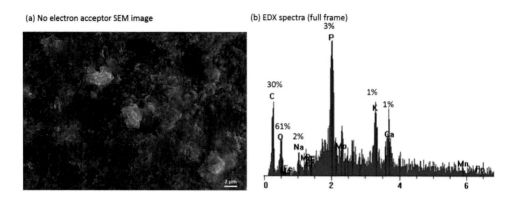

Fig. S4.3 SEM image (a) and full frame EDX spectra (b) of biofilm grown without external electron acceptor. Magnification is 10000×.

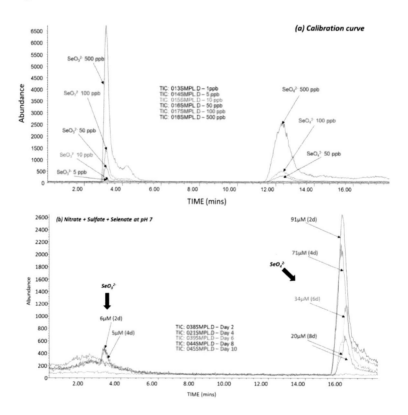

Fig. S4.4 HPLC-ICP-MS result for (a) calibration curve for SeO_3^{2-} and SeO_4^{2-} and (b) representative operational condition where all three oxyanions are present in the feed solution. Shift of SeO_4^{2-} peak in effluent samples was due to higher concentration of SeO_4^{2-} coinciding with the IC data. No other peaks were detected besides Se oxyanions.

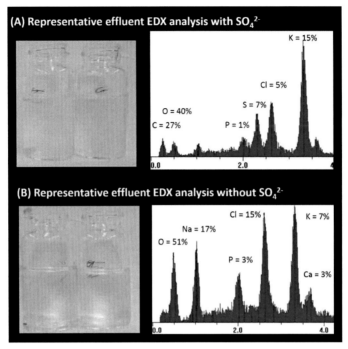

Fig. S4.4 EDX spectra of effluent samples (dried at room temperature for 5 days). Effluents with SO_4^{2-} in the system (A) showed yellow color in the liquid sample while those without SO_4^{2-} were whitish to clear (B). EDX spectra for treatments with sulfate showed sulfur peaks while those without showed no sulfur peaks.

Appendix 3

Supporting information for Chapter 5

Table S5.1 Summary of a statistical grouping for lactate, sulfate, sulfide, nitrate, Se_{tot} and Se_{diss} concentrations in the reactor effluent during the different phases of reactor operation. Statistical analysis was conducted using ANOVA with Tukey test. Different letters indicate statistically significantly different reactor performance.

Phases	Condition	Sample size	Lactate	Sulfate	Sulfide	Nitrate	Se_{tot}	Se_{diss}
Control reactor								
Phase I	pH 7.0, 12 mg Se/L	n = 69	A	B/C	B	A	A	A
Low pH reactor								
Phase Ia	Start-up, pH 7.0, 10 mg Se/L	n = 21	A	A	A	A	A	A
Phase Ib	pH 7.0, 12 mg Se/L	n = 24	A	B	B	A	A	A
Phase Ic	pH 6.0, 12 mg Se/L	n = 22	A	B	B	A	A	A
Phase Id	pH 5.5, 12 mg Se/L	n = 22	A	C	C	A	A	A
Phase Ie	pH 5.0, 12 mg Se/L	n = 33	B	A	A	A	B	B
Phase IIa	pH 5.0, < 5 mg Se/L	n = 31	C	A	A	B	B	B
Phase IIb	pH 5.0, 12 mg Se/L	n = 28	C	A	A	B	B	B
Phase III	pH 5.0, 50% SeO_3^{2-}	n = 26	D	C	A	B	B	B

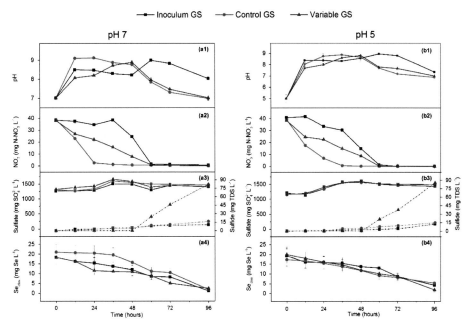

Fig. S5.1 Batch activity tests of sludge samples from the inoculum (black, square), control reactor (red, circle), and low pH reactor (blue, triangle) conducted at (a) pH 7.0 and (b) pH 5.0. The following measurements were conducted over time: (1) pH profiles and concentrations of (2) NO_3^-, (3) SO_4^{2-} (solid lines) and total dissolved sulfide (dashed lines), and (4) SeO_4^{2-}. Granular sludge samples were taken on day 164 of reactor operation.

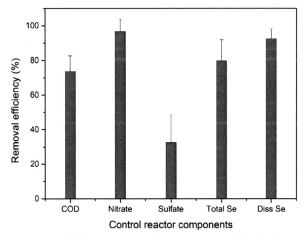

Fig. S5.2 Averaged removal efficiency of control reactor operated for 92 days at pH 7.0 fed with 2 g COD/L·d, 12 mg Se/L as SeO_4^{2-}, 42 mg NO_3^--N/L and 2 g SO_4^{2-}/L. Detailed removal performance is reported in **chapter 3**.

Fig. S5.3 Images of UASB reactors treating co-oxyanion synthetic acid mine drainage wastewater at 20 (± 2)°C (a) low pH reactor operated at pH 5.0 with floating sludge and (b) control reactor operated at pH 7.0 (without floating sludge).

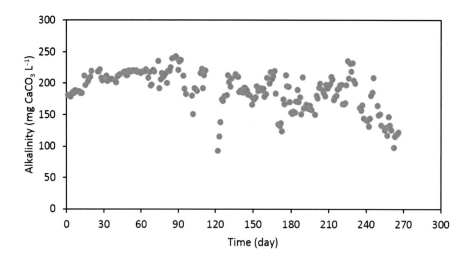

Fig. S5.4 Estimated alkalinity produced from only NO_3^- and SO_4^{2-} reduction

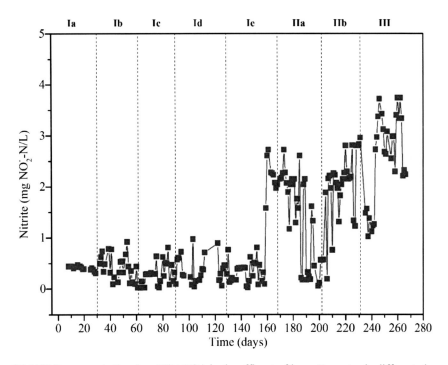

Fig. S5.5 Nitrite concentration (mg NO$_2^-$-N/L) in the effluent of low pH reactor in different phases of operation. Phases are the following: Ia - start-up, pH 7.0; Ib - pH 7.0; Ic - pH 6.0; Id - pH 5.5; Ie - pH 5.0; IIa - pH 5.0, < 5 mg Se/L; IIb - pH 5.0, 12 mg Se/L; III - pH 5.0, 50% SeO$_3^{2-}$ and 50% SeO$_4^{2-}$.

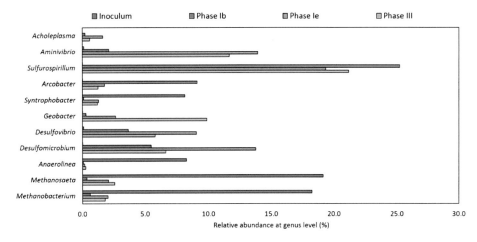

Fig. S5.6 Microbial community analysis of granular sludge reported as relative abundance at genus level. Phylotypes < 5% relative abundance (only if relative abundance are < 5% across all samples) are not displayed.

Appendix 4

Supporting information for Chapter 6

Table S6.1 Summary of a statistical grouping for SeO_4^{2-} and Se_{tot} for specific removal rates and removal efficiency among the reactors and between the reactors at different reactor conditions. Statistical analysis was conducted using ANOVA with the Tukey test. Same letters indicate statistically significantly similar values. Letters starting at A onwards indicate that values were the highest at A and decrease afterward.

	SeO_4^{2-}		Se_{tot}	
	BTF	UASB	BTF	UASB
Specific removal rates (SRR)				
RC1 (45 days)	C	A/B/C	D	B/C
RC2b (45 days)	B/C	C	B	B/C
RC2c	D	A/B	C/D	A/B
RC2d	D	A	C/D	A
Removal efficiency				
RC1 (45 days)	E	B/C	E	D
RC2b (45 days)	A	A/B	A	B/C
RC2c	A/B/C	C/D	A/B	A/B
RC2d	C/D	A/B/C	B/C/D	B/C

Table S6.2 Summary of a statistical grouping for SO_4^{2-}, Ni^{2+}, and lactate for specific removal rates and removal efficiency among the reactors and between the reactors at different reactor conditions. Statistical analysis was conducted using ANOVA with the Tukey test. Same letters indicate statistically significantly similar values. Letters starting at A onwards indicate that values were the highest at A and decrease afterward.

	SO_4^{2-}		Ni^{2+}	
	BTF	UASB	BTF	UASB
Specific removal rates (SRR)				
RC2a (COD/SO_4^{2-} = 1.1)	B/C	C	-	-
RC2b (COD/SO_4^{2-} = 2.8)	B	A	-	-
RC2c	C	A	C	A
RC2d	-	-	C	B
Removal efficiency				
RC2a (COD/SO_4^{2-} = 1.1)	B	B	-	-
RC2b (COD/SO_4^{2-} = 2.8)	A	A	-	-
RC2c	B	A	A	A
RC2d	-	-	B	B

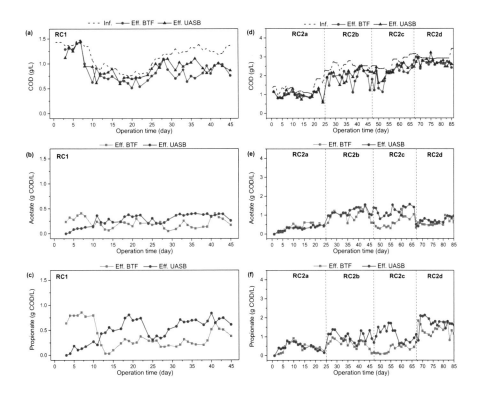

Fig. S6.1 COD and VFA profiles of the BTF and UASB reported in g COD/L for (a) RC1 COD, (b) RC1 acetate, (c) RC1 propionate, (d) RC2 COD, (e) RC2 acetate, and (f) RC2 propionate.

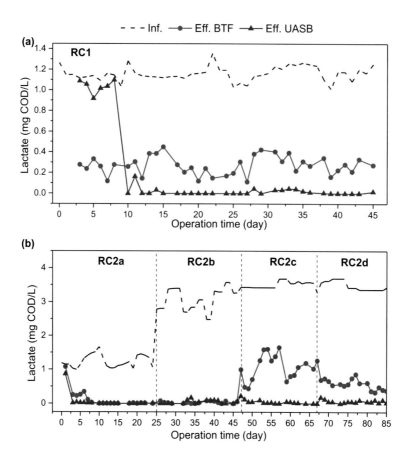

Fig. S6.2 Influent (black, dash line) and effluent lactate concentrations (mg COD/L) from the BTF (circle, red) and (b) the UASB (triangle, blue) reactors under different reactor conditions (a) RC1 and (b) RC2a to RC2e.

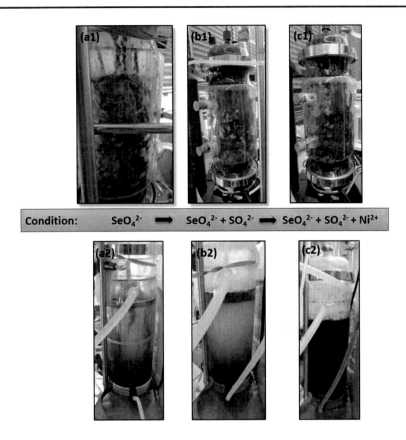

Fig. S6.3 Images of the (1) BTF and (2) UASB reactor at (a) SeO_4^{2-} alone, (b) $SeO_4^{2-} + SO_4^{2-}$ and (c) $SeO_4^{2-} + SO_4^{2-} + Ni^{2+}$.

Appendix 5

Supporting information for Chapter 7

Table S7.1 Pseudo-first order and Elovich kinetic model constants at an initial concentration of 428.9 mg/L SeO_4^{2-} and 1441.1 mg/L SO_4^{2-} at 2% (w/v) resin dosage, 20°C, and pH 7.5.

Anion solutions	Oxyanion	Experimental (24 h)	Pseudo-first order			Elovich		
		$q_{e, actual}$	k_1	$q_{e, cal}$	R^2	α_E	β_E	R^2
		mg/g	/h	g/g		mg/g	/h	
Single	SeO_4^{2-}	20.8	0.77	0.51	0.83	88	0.28	0.94
	SO_4^{2-}	67.1	0.81	12.7	0.99	174.4	0.087	0.92
Binary	SeO_4^{2-}	15.7	0.62	0.39	0.95	41.1	0.38	0.94
	SO_4^{2-}	42.4	0.79	2.7	0.96	244.9	0.15	0.96

Table S7.2 Average elemental mass composition from point EDS analysis of fresh resin and the resin regenerated with 0.25 M and 0.5 M HCl used in the SO_4^{2-} + SeO_4^{2-} solution after cycle 30.

Sample	Average percentage (%)					
	C	O	N	Cl	S	Se
Fresh resin	75 (± 19)	2 (± 1)	9 (± 2)	15 (± 23)	-	-
Resin treated w/ 0.25 M HCl	68 (± 13)	8 (± 9)	9 (± 3)	6 (± 2)	8 (± 1)	0.8 (± 0.1)
Resin treated w/ 0.50 M HCl	72 (± 7)	11 (± 1)	10 (± 2)	5 (± 1)	1 (± 1)	-

Table S7.3 Kinetic parameters calculated from pseudo-second order kinetic model for equimolar concentration experiment in binary anion solutions. Equimolar concentrations of SeO_4^{2-} and SO_4^{2-} were provided at 3.0 mM

Compound	$q_{e, actual}$ (mg/g)	k_2 (mg/g/h)	$q_{e, cal}$ (mg/g)	R^2
SeO_4^{2-}	18.5	0.045	19.6	0.99
SO_4^{2-}	13.5	0.039	14.4	0.99

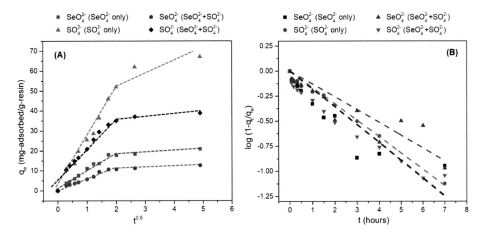

Fig. S7.1 Experimental and model fitted data: intra-particle diffusion (A) and liquid film diffusion (B) model as a function of time in single and binary anion solutions.

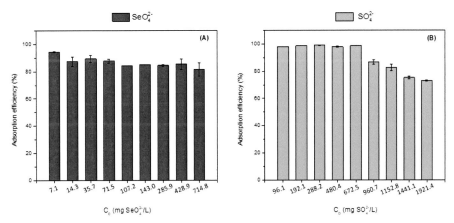

Fig. S7.2 IRA-900 removal efficiency with varying initial concentrations of SeO_4^{2-} and SO_4^{2-} in single anion solutions at pH 7.5, 20°C, and an adsorbent dose of 2% (*w/v*).

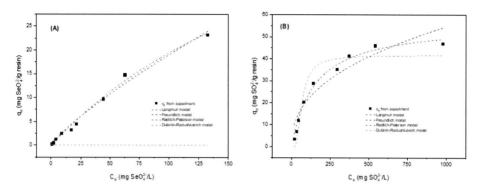

Fig. S7.3 Adsorption capacity of IRA-900 under the influence of varying initial concentrations of SeO_4^{2-} and SO_4^{2-} at pH 7.5, 20°C and an adsorbent dose of 2% (*w/v*). Dashed lines indicate the non-linear fit of Langmuir (red), Freundlich (blue), Redlich-Peterson (green) and Dubinin-Radushkevich (pink) isotherm models.

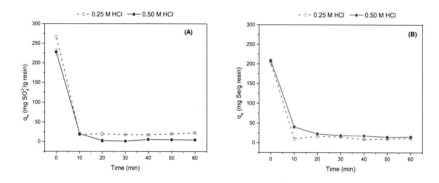

Fig. S7.4 Comparison of 0.25 M and 0.50 M HCl as regenerant solution for desorbing (a) SO_4^{2-} and (b) SeO_4^{2-}.

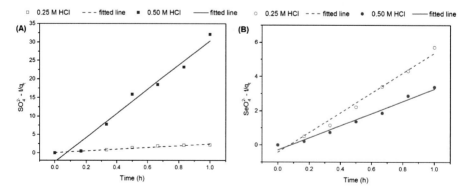

Fig. S7.5 Pseudo-second order kinetic fit of the desorption process done with 0.25 M and 0.5 M HCl regenerant solution for (a) SO_4^{2-} and (b) SeO_4^{2-}.

Appendix 6

Supporting information for Chapter 8

Table S8.1 Varying influent concentration entering the IX1 produced from the UASB1 effluent discharge

	Initial concentration coming from UASB (mg/L)	
Days	SO_4^{2-}	Se_{tot}
0-16	287 ($\pm$ 140)	0.78 ($\pm$ 0.23)
17-30	675 ($\pm$ 176)	0.81 ($\pm$ 0.10)
31-42	1030 ($\pm$ 377)	0.77 ($\pm$ 0.08)

Table S8.2 PO_4^{3-} adsorption onto IRA-900 resin.

	Cycles 1-4	Cycles 5-6
Pre-treatment IX1		
Influent PO_4^{3-} (mg/L)	19.0 ($\pm$ 8.7)	15.7 ($\pm$ 0.8)
Adsorption (%)	92 ($\pm$ 8)	10 ($\pm$ 1)
Post-treatment IX2		
Influent PO_4^{3-} (mg/L)	103.4 ($\pm$ 3.8)	103.4 ($\pm$ 3.8)
Adsorption (%)	83 ($\pm$ 5)	30 ($\pm$ 3)

Table S8.3 Average elemental mass composition from point EDX analysis of the fresh resin, the resin used in the IX1 and IX2 after regeneration with 0.5 M HCl after the 6 cycle and resin used in the bioregeneration process after 3 days.

Sample	Average elemental composition (mass %)					
	C	O	N	Cl	S	Se
Fresh resin	75 ($\pm$ 19)	2 ($\pm$ 1)	9 ($\pm$ 2)	23 ($\pm$ 15)	-	-
IX1 resin	70 ($\pm$ 12)	14 ($\pm$ 5)	17 ($\pm$ 7)	2 ($\pm$ 1)	4 ($\pm$ 2)	4 ($\pm$ 1)
IX2 resin	72 ($\pm$ 7)	11 ($\pm$ 1)	10 ($\pm$ 2)	5 ($\pm$ 1)	1 ($\pm$ 1)	-
Bioregenerated resin	62 ($\pm$ 7)	17 ($\pm$ 3)	19 ($\pm$ 5)	2 ($\pm$ 1)	1 ($\pm$ 1)	-

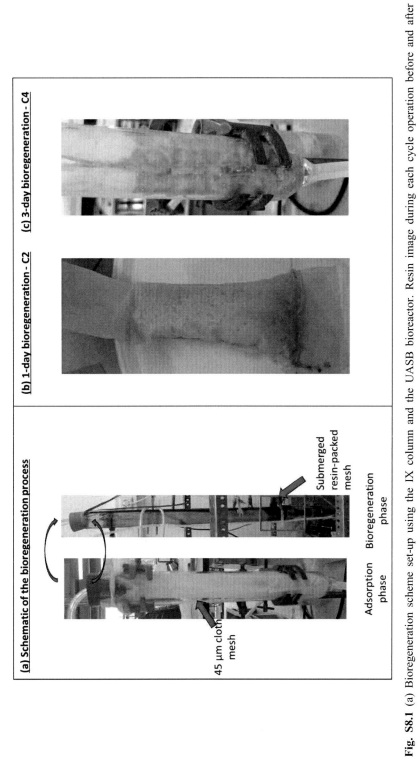

Fig. S8.1 (a) Bioregeneration scheme set-up using the IX column and the UASB bioreactor. Resin image during each cycle operation before and after bioregeneration using anaerobic granular sludge in the UASB bioreactor (b) after 1 day bioregeneration and (c) after 3 days bioregeneration.

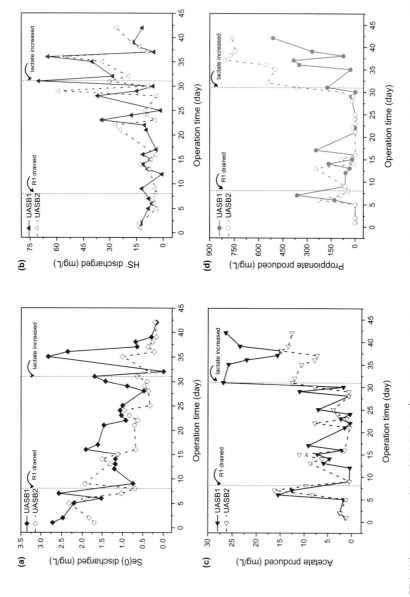

Fig. S8.2 UASB1 (1st treatment unit) and UASB2 (2nd treatment unit) effluent concentration for (a) colloidal Se, (b) HS⁻, (c) acetate and (d) propionate.

Appendix 7

Microbial community analysis

DNA extraction Protocol

- Add 0.5 g sample (wet weight) to a 2 ml screw cap tube.
- Add 0.7 g beads from tube.
- Add 750 μl 120 mM NaPO4 buffer (pH 8).
- Add 250 μl TNS (vial should be maximally filled to top of gripping ring with minimally 500 μl headspace).
- Bead beating: 45 s, 6.5 m/s, place immediately on ice.
- Centrifuge 20 min at maximum speed (20817 x g) and 4°C.
- Transfer the supernatant into a 2 ml DNase/RNase free SafeLock tube on ice.
- Add 1 volume (500 μl) Phenol/Chloroform/Isoamylalcohol (25:24:1; AppliChem A0837) pH 8, vortex, centrifuge 5 min at maximum speed (20817 x g) and 4°C.
- Transfer the supernatant into a 2 ml DNase/RNase free SafeLock tube on ice.
- Add 1 volume (500 μl) Chloroform/Isoamylalcohol (24:1; AppliChem A1935), vortex, centrifuge 5 min at maximum speed (20817 x g) and 4°C.
- Transfer the supernatant into a 2 ml DNase/RNase free SafeLock tube on ice, add 2 volumes PEG solution, mix thoroughly, precipitate by centrifugation for 90 min at
- maximum speed (20817 x g) and 4°C.
- Remove the liquid using a pipette and add 500 μl ice cold, DNase/RNase free 70 % EtOH to wash pellet.
- Centrifuge 30 min at maximum speed (20817 x g) and 4°C.
- Remove EtOH carefully using a pipette, dry pellet briefly at room temperature (max. 5 min).
- Resuspend DNA/RNA coextract with 50 μl EB buffer.
- Check coextract (2 μl) spectrophotometrically (using NanoDrop).
- Store coextract at -80°C.

PCR amplification protocol

- PCR with barcoded universal bacterial primers 8F and 529R.
- Run PCRs (3-5 reactions) for each sample.
- Each 20-μl PCR contains 1-5 ng of DNA, 1 μM of each primer, 10 uL of KAPA HiFi HotStart Ready Mix, 1x BSA, and an adjusted volume of sterilized water.

- PCR conditions are as follows: denaturation at 95C for 3 min, the number of optimal cycles of 98C for 20s, annealing at 58C for 15s, extension at 72C for 40s, and a final extension at 72C for 5 min.

- Negative PCR controls without DNA template are run concurrently for each sample.

- PCR products (5 µl) mixed with DNA loading buffer (2 µl) are visualized by agarose gel electrophoresis in a 1.0% agarose Tris-acetate EDTA gel stained with ethidium bromide.

- PCR products are pooled and cleaned using the QiaQuick PCR purification Kit.

- PCR products are quantified using the Qbit.

Illumina MiSeq protocol

Sequencing on Illumina MiSeq system for subsequent library preparation was carried out following the Illumina's protocol "16S Metagenomic Sequencing Library Preparation". PCR products were initially cleaned with Ampure XP beads followed by index PCR. Index PCR is the process of attaching Illumina-design dual indices to the respective amplicons allowing for the generation of a "library" for each sample. The samples were once again cleaned with Ampure XP beads after index PCR was done. The DNA from each library was then quantified using a fluorometric kit (QuantIT PicoGreen; Invitrogen) and was normalized to specifications according to Illumina's protocol. Finally, the DNA was mixed with Illumina-generated PhiX control library (10% PhiX) before being loaded for sequencing on the Illumina MiSeq v.3 platforms.

Sequence processing, bioinformatics, using MOTHUR platform

Sequence reads produced from Illumina MiSeq were processed using MOTHUR platform following MiSeq SOP pipeline available in the software package. The following steps were carried for the sequence processing: (1) assembly of read pairs into contigs; (2) contigs were screened for ambiguous base pairs, amplicon size, alignment positions, and chimeric sequences producing unique contig sequences; (3) the unique contig sequences were then classified using the MOTHUR software package training set using the Bayesian classifier at a 0.6 confidence score; (4) final screened and cleared sequences were binned into phylotypes following their taxonomic classification (*e.g.*, genus-level); and finally (5) resulting tax file was process in R software to produce a relative abundance plot for each library of each samples.

Biography

Lea Chua Tan was born on 19th March, 1987 in Manila, Philippines. Lea did her bachelor in science (BSc) in Chemical Engineering from De La Salle University - Manila and graduated with honors with a CGPA of 3.1 out of 4.0 on the 1st of September 2008. She earned her Chemical Engineering License on April 2009. Lea continued on to work in the industry from September 2009 until August 2012. She worked as a Wastewater Engineer Supervisor at Ibiden Philippines Inc., as a Field Service Coordinator (Specialist 2) at Emerson Electric Asia Ltd. Rosemount Analytical Division and as a Research Assistant at the Center for Sustainable Development (CeSDR) in De La Salle University - Manila. She then pursued her graduate studies from 2012 until 2017. Lea earned her master in science (MSc) degree in Environmental Engineering from Hokkaido University funded by the MEXT (*Monbukagakusho*) Japanese Government from September 2012 until 2014. During her MSc time, she worked on the biological-based response evaluation of different reclamation systems using a combination of conventional bioassay and transcriptome-based analysis as test battery. Lea was awarded a fellowship and admission into the Erasmus Mundus Joint Doctorate Program on Environment Technologies for Contaminated Solids, Soils and Sediments (ETeCoS^3) and started as a PhD Fellow at UNESCO-IHE, Delft, the Netherlands from October 2014. As part of this program, she has carried out two 6-month PhD research stays with partner universities at the Center for Biofilm Engineering in Montana State University, Bozeman, Montana, US as well as at the University of Naples Federico II, Naples, Italy. Her research was mainly focused on understanding the biological anaerobic treatment process for mine wastewater for the removal of selenate and its co-contaminants. She has successfully defended and earned her PhD degree on the 18th December 2017. She is currently working as a post-doctoral researcher at National University of Ireland, Galway focusing on anaerobic digestion.

Publications

- **Tan L.C.**, Calix E.M., Rene E.R., Nancharaiah Y.V., van Hullebusch E.D. and Lens PNL. 2018. Amberlite® IRA-900 ion exchange resin for the sorption of selenate and sulfate: Equilibrium, kinetic and regeneration studies. *Journal of Environmental Engineering.* doi: 10.1061/(ASCE)EE.1943-7870.0001453.

- **Tan L.C.**, Papirio S., Luongo V., Nancharaiah Y.V., Cennamo P., van Hullebusch E.D., Esposito G. and Lens P.N.L. 2018. Comparison of upflow anaerobic sludge blanket and biotrickling filter bioreactor for the simultaneous removal of selenate, sulfate, nickel in model mining wastewater. *Chemical Engineering Journal* 345; 545-555. doi: 10.1016/j.cej.2018.03.177.

- **Tan L.C.**, Espinosa-Ortiz E., Nancharaiah Y.V., van Hullebusch E.D., Gerlach R. and Lens P.N.L. 2018. Selenate removal in biofilm systems: effect of nitrate and sulfate on selenium removal efficiency, biofilm structure, and microbial community. *Journal of Chemical Technology and Biotechnology.* doi:10.1002/jctb.5586.

- **Tan L.C.**, Nancharaiah Y.V., van Hullebusch E.D. and Lens P.N.L. 2018. Effect of elevated nitrate and sulfate on selenate removal by mesophilic anaerobic granular sludge bed reactors. *Environmental Science: Water Research and Technology* 4; 303-314. doi: 10.1039/c7ew00307b.

- Fukushima T., Hara-Yamamura H., Nakashima K., **Tan L.C.** and Okabe S. 2017. Multiple-endpoints gene alternation-based (MEGA) assay: A toxicogenomics approach for water quality assessment of wastewater effluents. *Chemosphere* 188; 312-319. doi: 10.1016/j.chemosphere.2017.08.107.

- **Tan L.C.**, Nancharaiah Y.V., van Hullebusch E.D. and Lens P.N.L. 2016. Selenium: Environmental significance, pollution, and biological treatment technologies. *Biotechnology Advances* 34 (5); 886-907. doi: 10.1016/j.biotechadv.2016.05.005.

- **Tan L.C.**, Nancharaiah Y.V., Lu S., van Hullebusch E.D., Gerlach R. and Lens P.N.L. *under review 2018.* Biological treatment of selenium-laden wastewater containing nitrate and sulfate in an upflow anaerobic sludge bed reactor at pH 5.0. *Chemosphere.*

- Calix E.M., **Tan L.C.**, Rene E.R., Nancharaiah Y.V., van Hullebusch E.D. and Lens P.N.L. *under review 2018*. Integrated upflow sludge bed blanket bioreactor and ion exchange column process assessment for the removal of selenate and sulfate from model wastewater. *Separation Science and Technology.*

Conferences/Presentations

- **Tan L.C.**, Nancharaiah Y.V., van Hullebusch E.D. and Lens P.N.L. **Poster presentation** on "Selenate bioreduction in the presence of nitrate and sulfate" at 4th International Conference on Research Frontiers in Chalcogen Science and Technology (G16) at Delft, Netherlands, May 28-29, 2015.
- **Tan L.C. Oral presentation** on "Biological-based response assessment of membrane reclamation systems using bioanalytical tools" at 50th Annual Environmental Engineering Annual Environmental Engineering Research Forum, Japan Society of Civil Engineering, Sapporo, Japan, November 20, 2013.
- Lopez S., **Tan L.C.**, Tan H.C., Yu AK., Ubando A., Biona M., Culaba A., Chua A. and Tan R. **Oral presentation** on "Instrumentation of a solar dryer for microalgae-to-biofuel production: a green mechatronics approach" at Research@DLSU Congress 2012: Science and Technology Conference, De La Salle University, Manila, Philippines, February 15, 2012.
- **Tan L.C.**, Napasindayao T.G. and Bacani F.T. **Oral presentation** on "Sorption Behavior of Hibiscus Cannabinus L. Core in Simulated Bunker Oil C - Seawater Mixture" at 15th Regional Symposium on Chemical Engineering (RSCE), Kuala Lumpur, Malaysia, December 3-5, 2008.

*Netherlands Research School for the
Socio-Economic and Natural Sciences of the Environment*

DIPLOMA

For specialised PhD training

The Netherlands Research School for the
Socio-Economic and Natural Sciences of the Environment
(SENSE) declares that

Lea Chua Tan

born on 19 March 1987 in Manila, Philippines

has successfully fulfilled all requirements of the
Educational Programme of SENSE.

Delft, 18 December 2017

the Chairman of the SENSE board

Prof. dr. Huub Rijnaarts

the SENSE Director of Education

Dr. Ad van Dommelen

The SENSE Research School has been accredited by the Royal Netherlands Academy of Arts and Sciences (KNAW)

KONINKLIJKE NEDERLANDSE
AKADEMIE VAN WETENSCHAPPEN

The SENSE Research School declares that Ms Lea Chua Tan has successfully fulfilled all requirements of the Educational PhD Programme of SENSE with a work load 53.5 EC, including the following activities:

SENSE PhD Courses

- ○ Anaerobic wastewater treatment (2015)
- ○ Environmental research in context (2015)
- ○ Research in context activity: 'Co-organizing programme and abstract book of 4th International Conference on Research Frontiers in Chalcogen Cycle Science and Technology' (2015)

Other PhD and Advanced MSc Courses

- ○ Contaminated soil and remediation, Paris-Est University (2015)
- ○ Contaminated soil, Paris-Est University (2015)
- ○ Contaminated sediments – characterization and remediation, UNESCO-IHE (2016)
- ○ Online course (MOOC) on Municipal solid waste management in developing countries, Coursera (2016)
- ○ Mathematical models in environmental technologies, Entrepreneurship and innovation, and How to write successful proposal in Horizon 2020, University of Cassino (2017)
- ○ Biological treatment of solid waste, University of Cassino (2017)

External training at a foreign research institute

- ○ Microbial community analysis and bioinformatics Mothur usage and analysis, Center for Biofilm Engineering, Montana State University (2016)
- ○ Molecular analysis: Automated Ribosomal Intergenic Spacer Analysis (ARISA) and cloning technique, University of Suor Orsola Benincasa of Naples (2017)

Management and Didactic Skills Training

- ○ Assisting microbiology laboratory session of MSc programme 'Environmental science and urban water supply' (2014-2015)
- ○ Supervising MSc student with thesis entitled 'Integrated process assessment for the removal of selenium in the presence of sulphate from wastewater (2016-2017)

Oral Presentations

- ○ *Selenate bioreduction in wastewater containing competitive electron acceptors.* PhD Symposium: Integrating Research Water Sector, 28-29 September 2015, Delft, The Netherlands
- ○ *Comparison of UASB and BTF bioreactor for the removal of selenate with sulfate and nickel.* PhD Symposium: Climate Extremes and Water Management Challenge, 2-3 October 2017, Delft, The Netherlands

SENSE Coordinator PhD Education

Dr. Monique Gulickx

For Product Safety Concerns and Information please contact our EU representative GPSR@taylorandfrancis.com Taylor & Francis Verlag GmbH, Kaufingerstraße 24, 80331 München, Germany

Printed and bound by CPI Group (UK) Ltd, Croydon, CR0 4YY

01/05/2025

01858616-0003